*Tropical Agricultural
Hydrology*

The papers in this volume were initially presented at a Conference organized by the International Institute of Tropical Agriculture, Ibadan, Nigeria, in November 1979

Tropical Agricultural Hydrology

Watershed Management and Land use

Edited by

R. Lal
International Institute of Tropical Agriculture,
Ibadan, Nigeria

and

E. W. Russell
Professor Emeritus of the Department of Soil Science,
University of Reading

A Wiley-Interscience Publication

JOHN WILEY & SONS
Chichester · New York · Brisbane · Toronto

Copyright © 1981 by John Wiley & Sons Ltd.

British Library Cataloguing in Publication Data:

Tropical agricultural hydrology.
    1. Agriculture–Tropics
    I. Lal, R. II. Russell, Edward Walter
    III. International Institute of Tropical
    Agriculture
    630'.913   SB111   80-41590

    ISBN 0 471 27931 5

Typeset by Activity, Teffont, Salisbury, Wilts
and printed in Great Britain by The Pitman Press, Bath, Avon

Contributors

I. P. ABROL
: Central Soil Salinity Research Institute, Karnal 132001, Haryana, India.

J. B. BALL
: Federal Department of Forestry, PMB 5011 Ibadan, Nigeria

R. G. BARBER
: Department of Agricultural Engineering, University of Nairobi, P.O. Box 30197, Nairobi, Kenya.

J. R. BLACKIE
: Ministry of Water Development, P.O. Box 30521, Nairobi, Kenya.

F. W. BLAISDELL
: St. Anthony Falls Hydraulic Laboratory, Third Avenue SE at Mississippi River, Minneapolis, Minnesota 55414, USA.

S. L. CLAASSEN
: International Institute of Tropical Agriculture, Oyo Road, PMB 5320, Ibadan, Nigeria.

D. C. COUPER
: International Institute of Tropical Agriculture, Oyo Road, PMB 5320, Ibadan, Nigeria.

M. DE KESEL
: P.R.B. Avenue de Broqueville 12, Bte 1, 1150 Bruxelles, Belgium.

D. DE VLEESCHAUWER
: Department of Soil Physics, Coupure Links 533, University of Ghent, B. 9000 Ghent, Belgium.

V. V. DHRUVA NARAYANA
: Central Soil Salinity Reseach Institute, Karnal 132001, Haryana, India.

A. J. DYE
: USDA Forest Services, P.O. Box 2417, Washington, DC 20013, USA.

K. A. EDWARDS
: Ministry of Water Development, P.O. Box 30521, Nairobi, Kenya.

N. EGBUNIWE
: Department of Civil Engineering, University of Nigeria, Nsukka, Nigeria.

R. F. FISHER
: Centre for Tropical Agriculture, University of Florida, Gainesville, Florida 32611, USA.

G. R. FOSTER
: Department of Agronomy, Lilly Hall, Purdue University, West Lafayette, Indiana 47907, USA.

K. FURCH   *Max-Planck-Institut für Limnologie, D-2320 Plon, Postfach 165, Federal Republic of Germany.*

J. HARI KRISHNA   *International Crops Research Institute For The Semi-Arid Tropics, 1-11-256 Begumpet, Hyderabad 500-016, A. P., India.*

I. G. G. HOGG   *Department of Agricultural Engineering, University of Nairobi, P.O. Box 30197, Nairobi, Kenya.*

N. W. HUDSON   *National College of Agricultural Engineering, Silsoe, Bedford, MK U5 UDT, UK.*

U. IRMLER   *Max-Planck-Institut für Limnologie, D-2320 Plon, Postfach 165, Federal Republic of Germany.*

W. S. JUNK   *Max-Planck-Institut für Limnologie, D-2320 Plon, Postfach 165, Federal Republic of Germany.*

J. KAMPEN   *International Crops Research Institute For The Semi-Arid Tropics, 1-11-256 Begumpet, Hyderabad 500-016, A. P., India.*

B. T. KANG   *International Institute of Tropical Agriculture, Oyo Road, PMB 5320, Ibadan, Nigeria.*

H. KLINGE   *Max-Planck Institut für Limnology, D-2320 Plon, Postfach 165, Federal Republic of Germany.*

S. H. KUNKLE   *Timber and Watershed Laboratory, P.O. Box 445, Parsons, West Virginia 26287, USA.*

R. LAL   *International Institute of Tropical Agriculture, Oyo Road, PMB 5320, Ibadan, Nigeria.*

L. J. LANE   *Southwest Watershed Research Center, 442 East Seventh Street, Tucson, Arizona 85705, USA.*

C. L. LARSON   *Department of Agricultural Engineering, University of Minnesota, 1390 Eckles Avenue, St. Paul, Minnesota 55108, USA.*

T. L. LAWSON   *International Institute of Tropical Agriculture, Oyo Road, PMB 5320, Ibadan, Nigeria.*

L. LUNDGREN   *Department of Physical Geography, University of Stockholm, Box 6801, 113 86 Stockholm, Sweden.*

E. M. MORRIS   *Institute of Hydrology, Maclean Building, Crowmarsh Gifford, Wallingford, Oxfordshire, OX10 8BB, UK.*

S. NORTCLIFF   *Department of Soil Science, University of Reading, London Road, Reading, RG1 SA2, UK.*

K. ODURO-AFRIYIE   *International Institute of Tropical Agriculture, Oyo Road, PMB 5320, Ibadan, Nigeria.*

O. OYEBANDE   *Department of Geography, University of Lagos, Lagos, Nigeria.*

P. PATHAK	*International Crops Research Institute For The Semi-arid Tropics, 1-11-256 Begumpet, Hyderabad 500-016, A.P., India.*
C. PEREIRA	*Peartrees, Tesdon, Maidstone, Kent ME18 5AD, UK.*
M. A. ROCHE	*Office De La Recherche Scientifique et Technique Outre-Mer, 24 Rue Bayard, 75008 Paris, France.*
E. W. RUSSELL	*31 Brooklyn Drive, Emmer Green, Reading RG4 8SR, Reading, UK.*
N. SENGELE	*Institut National Pour L'Etude et la Recherche Agronomiques, B.P. 1513, Kisangani, Zaire.*
T. C. SHENG	*FAO, P.O. Box 1136, Kingston, Jamaica, W.I.*
D. C. SLACK	*Department of Agricultural Engineering, University of Minnesota, 1390 Eckles Avenue, St. Paul, Minnesota 55108, USA.*
L. STROMQUIST	*Department of Physical Geography, Box 554, S-751 22 Uppsala, Sweden.*
K. G. TEJWANI	*Central Soil and Water Conservation Research and Training Institute, Dehra Dun 248195, India.*
D. B. THOMAS	*Department of Agricultural Engineering, University of Nairobi, P.O. Box 30197, Nairobi, Kenya.*
J. B. THORNES	*Department of Soil Science, University of Reading, London Road, Reading RG1 5AQ, UK.*
F. ZADROGA	*Centro Agronomico Tropical de Investigacio y Ensenaza, Turrialba, Costa Rica.*

Contents

ix

PART 8
Research and Development Needs

Foreword

H. L. PENMAN, O.B.E., F.R.S.

Even for a primitive biological existence at subsistence level mankind needs air, water, food, and energy, adequate in quantity and in quality. There is here no special order of importance, but putting water first or second could be defended. What does matter is that getting access to any of the four may produce conflict with attempts to ensure access to the others, and in our progress towards civilization there has been occasional political recognition that controlled management of our environment is desirable to minimize the effects of conflict. Recently there have been important international attempts to see what might be done: the International Hydrological Decade and its successor the International Hydrological Programme are two of these sponsored by UNESCO with support from other UN agencies; there are non-governmental learned societies, mainly within the International Council of Scientific Unions, that organize symposia at intervals of a few years; and there are universities or institutes of kindred status that occasionally offer hospitality to world experts to hold discussion meetings or workshops at which some teaching content is an acceptable part of the programme. There will be general gratitude to the International Institute of Tropical Agriculture for acting as host—apparently splendidly so in both scientific and social aspects—for a five day Conference on the interactions of land use and catchment management. Much of the material submitted is likely to have value outside the place and time that produced it and this collection of selected papers will be welcomed both by participants and by others who could not take part.

The title *Tropical Agricultural Hydrology* might seem to be superficially restrictive, but this is not the case. The 40 per cent of the earth's surface between Cancer and Capricorn not only includes the world's largest river but also several very famous deserts, so there is no lack of variety in the range of inputs to the hydrological cycle. On the other hand, there is astonishingly little variation in the inefficiency of the farming systems of the area, as measured by the fraction of solar radiation that is fixed in harvested staple crops. Perhaps there are more taboos and fetishes in intertropical farming than there are in the water management, so it is to be expected that there are more occasions on which the impact of traditional land use on water resources provokes discussion, sometimes heated, than there are where a planned major change in hydrology is closely scrutinized for its possible impact on land use. Against the known poor achievement of the farming it might be fairly urged that good water husbandry of a catchment will help to improve the management of its soil and of the crops and stock the soil supports: at least we can assert that it is not inevitably harmful to land use.

What is 'inevitable'? What is 'harmful'? The first question is partly science and partly technology: the second is mainly value judgement, with technical components. They both arise whether the start is in a water management scheme that may affect the way of life of whose who live on or near the area to be managed; or a proposed change in land use that may affect surface and groundwater in quantity and in quality. Examples of the interactions occur many times in conference reports and in text books—and there are new ones in the papers that follow. There are also forward looks at research needs and it may be worth while making some personal comments here on a few general points about research. The answers to the questions may come from existing experience acquired locally or elsewhere and skilfully interpreted on the basis of sound ideas; or from completely fresh experience newly sought; or from a combination of the two. It should not be as rare as it now is for the research worker to be left to decide for himself what new work would be needed to answer a particular question—which should be posed to him as a technical problem and not as a desired solution. Further, some of the most important work may need continued support over many years to ensure that within the records made there is some representation of long period components in the hydrological cycle: equally desirable, there should be some long term continuity in the composition of the research team, and this limits the possible sources of effective sponsorship. But whatever the time scale, or the supporting agency, in addition to the self-evident first priority of getting a technical solution to the technical problem, there should be room within the planning of the work for some extra time, equipment, and effort to go into solving the intellectual problem that often lies beyond the technical puzzle. There are at least two reasons. First, this kind of study in depth is the most likely way to appeal to bright young scientists looking for a career—the prospect of a challenge in the basic principles of their own specialist subject. Second, the thrashing out of interpretative ideas is—or should be!—the stimulus at the heart of organized international activity. Shared problems usually have something in common in their solutions, and it is a counsel of despair to suggest that every site of a land management is unique: accepting necessary minor qualifications, the 'physics' and 'biology' of water, soil, plants, and animals are the same everywhere and, though the field ecology of their interactions may be unique in some places, the search for reasons at a few well-equipped and well-staffed centres of excellence may produce better general answers at less cost than many smaller units aiming at no more than local relations to describe local catchment behaviour.

PART 1

Watershed Management in the Tropics

Tropical Agricultural Hydrology
Edited by R. Lal and E. W. Russell
© 1981, John Wiley & Sons Ltd.

1.1

Land-use Management on Tropical Watersheds

CHARLES PEREIRA

1.1.1. SOIL AND WATER MANAGEMENT STUDIES IN THE CGIAR SYSTEM

The International Institute of Tropical Agriculture (IITA) and other international centres are now turning to the broader issues of watershed management. The summary published in 1977 of the Institute's first decade of progress demonstrated a pattern of programme development which has characterized the whole CGIAR system. The initial determined attack on the low yield potential of traditional crop varities brought successes which made the use of fertilizers profitable. The vigorous biological response of pest and disease organisms favoured by the enriched environment set the next priority as crop protection. Successful reduction of these three constraints has focused attention on the next major limiting factor, which is the management of soil and water resources.

Pressure of human population growth is forcing subsistence populations in many tropical and subtropical countries on to marginal lands which are at one extreme steeper, shallower, or more drought-prone, and at the other extreme subject to waterlogging and salinity. The need to develop better technology for improvement and conservation of the physical environment of crops and of livestock is becoming increasingly urgent.

In the first six years of the work of the Technical Advisory Committee (TAC), the policy has been maintained that the physical aspects of tropical agriculture are largely specific to the tasks of the International Centres and should be an integral part of their work. Indeed, in 1976 the Consultative Group issued a direct request that all IARC's should give increased attention to soil and water management. The results, over the whole system of Institutes, were somewhat meagre. This is in part due to the shortage of soil physicists and soil conservation engineers with experience of tropical conditions or readiness to acquire such experience. TAC therefore accepted offers both by FAO and by IDRC to undertake enquiries as to needs and

Table 1.1. International research in water management for broad climatic zones

Climatic characteristics	International agricultural research centre
(a) Humid regions where rice is the dominant crop. Rainfall of 2000 mm or more	IRRI
(b) Humid tropics where shifting cultivation is dominant under rainfall of 1000–2000 mm.	CIAT and IITA
(c) Semi-arid areas of rainfed agriculture with some supplementary irrigation under rainfall of 400–1000 mm.	ICRISAT
(d) Dry areas of rainfed agriculture, with some supplementary irrigation from groundwater, under rainfall of 200–400 mm.	ICARDA
(e) Dry areas where rainfed cultivations are not effective but major production is achieved from irrigation.	No international research support for water management of this important cropping area.

opportunities in this field. The FAO made a brief and rather generalized desk study, recommending the establishment of a series of regional institutes for research and training in soil and water management. The subsequent IDRC study sent a team of five experienced people to investigate in some sixteen countries. International centres being mandated with activities in specific ecological regions (Table 1.1), concentrated attention is needed on the major climatic zone in developing countries which no centre is as yet designed or sited to serve. This is the zone in which national economics are dependent on irrigation in climates too dry for rain-grown crops. They include the world's five great arid-land river-irrigation systems of the Upper Ganges, Indus, Nile, Euphrates, and Tigris.

Crop production statistics published by FAO report that some 92 m ha of these commanded areas produce one annual crop of 1½ tonnes p.a. In all of them the ability to produce two (sometimes three) crops in a year, each from 3 to 5 tonnes per hectare has already been demonstrated. It may therefore seem obvious that the first priority is to apply the knowledge already available, but in practice the reclamation problems are too complex for conventional extension services. On-site investigation, diagnosis and quantitative specification of treatments are essential. This requires the training of several thousands of field specialists, to give them a working knowledge of field engineering and water control, together with an understanding of crop rotations, rooting depths, soil moisture storage, and crop water requirements and salt and alkali control. We were able to discuss this training with authorities in many countries, and found their experiences rather similar. Although agricultural engineers trained in the USA succeed in spanning the wide syllabus covering both agriculture and mechanical engineering. The solution is not to set up mixed-subject undergraduate courses to combine civil engineering and agriculture. The syllabus is too broad. Two countries had set up such courses and both had been abandoned because their graduates were not recognized as fully qualified either in civil engineering or in agriculture. The agreed solution is post-graduate courses for a limited number of civil engineers and for a much larger of agronomists, studying together in an environment of subsistence irrigation villages.

The effect of providing such instruction would be to achieve a welcome increase in the world supply of trained soil and water management specialists in tropical agriculture, but this would in no way decrease the urgent need for each centre to press ahead with improvements in the physical technologies of the crops or the regions for which the consultative group has assigned them the responsibility. IITA, IRRI, and ICRISAT have some active work in progress in soil and water management technology, but the total resources as yet allocated throughout the CGIAR system are seriously inadequate. Our challenge is the vast scale on which mis-use of land and water is destroying the basic agricultural resources of developing countries.

1.1.2. CHANGING CLIMATES: MYTH OR REALITY?

The international institutes are seeking optimum solutions in the present climates of their regions or the areas of their crops. But, say the armchair critics, what happens if these climates are changing? They are joined by elderly subsistence farmers whose semi-arid lands are deteriorating under the usual combination of exhaustive cropping and relentless overgrazing. 'This was once good land but we have had poor rains for many years' expresses a genuine belief of those who do not understand the effects of changes in rainfall–runoff relationships. The drying up of streams and the wilting of crops are such convincing symptoms of drought that many an agricultural officer has spent long evenings in plotting the rainfall records in search of evidence of a change in climate.

Meteorologists have indeed worked on the problem for a century or more, but it is only in the last three or four years that they have finally achieved a scientific explanation of the Ice Ages. The hypothesis advanced 38 years ago by Milankovitch (1941) has been substantiated by evidence from the oxygen-isotope ratios of deep-sea sediment cores, from raised beaches and from deep drilling of the polar icecaps. These give a time-scale to the sea-level fluctuation which accompanied the Ice Ages, and confirm that they correspond to changes in the geometry of the earth's orbit which affect the intensity of solar radiation reaching the outer atmosphere. (Mason, 1976; Hays, Imbrie, and Shackleton, 1976). Fortunately the next Ice Age is a comfortable interval of many centuries away but this time-scale shows that we are well past the warmest part of the current interglacial period and are now in the slow

cooling stage. Evidence from past interglacial periods suggests that violent fluctuations, ranging from temperatures warmer than our present climate, to glacial
conditions can occur within a century (Dansgaard, Johnsen, Klaussen, and Langway,
1972).

Our meteorological data records only the brief history of the past two centuries,
but with the help of modern satellite photography we have an increased understanding of the behaviour of the atmosphere. Two of the world's major meteorological
computer centres at Bracknell in Britain and at Boulder City in the USA have made
very thorough studies of the world data. They have detected no evidence of overall
significant trends during the past 200 years, either in the means of climatic values or
in variability about these means.

Although the world climate shows no contemporary evidence of change, the
mis-use of land can produce severe changes in local environments. The processes of
deterioration of tropical soil and water resources under the mis-uses of subsistence
agriculture are now fairly well documented. It is now accepted that the decreased
infiltration and increased runoff which simulate the effects of a declining rainfall
are due to the destruction of protective vegetation, the trampling or over-cultivation
of the surface soil and its alternative exposure to radiation and to rainfall of high
intensities. Our saddest example in Africa, the recent disaster in the Sahel, has since
been shown to have been due to a drought no worse than those previously recorded.
The disaster was due to the impact of excessive populations of people and livestock
in a fragile ecosystem.

1.1.3. A WORLD-SCALE IN THE IMPACT OF UNSTABLE CLIMATES ON AGRICULTURE

While there is no evidence of change in the world's weather system there is very
serious evidence of a massive change in the exposure of the world's farmers to
climatic variability. The public of the western world are becoming accustomed to
satellite pictures of tropical storms and thus to the vast scale of global atmospheric
instabilities. The global characteristics of the atmosphere result in a persistent belt
of large-scale disturbances known as the Inter-Tropical Convergence Zone (ITCZ).
Within this zone, long hot dry seasons are interrupted by characteristically irregular
and violent conventional storms. It is into this zone that, according to UN statistic
some 80 per cent of an extra three billion people are expected to be born in the
final quarter of this century. Most of these people will live by subsistence agriculture,
or in cities dependent on supply from primitive farming technologies. Since the
population of the temperate zone is approaching stability the net effect will be a
vast increase in the proportion of humanity which is affected by drought and flood.
In the world literature on food supply I do not believe that enough weight has been
given to this prospect of an effective deterioration in the average agricultural climate.

1.1.4. WATERSHED MANAGEMENT AGAINST CLIMATIC HAZARD

As world opinion becomes more conscious of this ominous trend, what can be done to alleviate it? To many scientists the first task would appear obvious, i.e. to use all the resources of science and humanity to abate the human fecundity which will make such massive additions to the problems of poverty and the dimensions of emergencies. The second, and easier, task is to strive for a radical improvement in the management of the soils, vegetation, and water supplies in the developing countries of the tropics and subtropics and thus to arrest the current destruction of those resources on which the future additional multitudes must depend. To all who have striven with the problems of resource management under subsistence agriculture this second task is also full of uncertainty depending as it must do on the winning of hearts and minds as well as on competence in technology.

There exist many examples that emphasise the increasing gravity and urgency of watershed management problems in areas of subsistence agriculture. In 1978, some parts of northern India suffered from the worst floods ever experienced on the Yamuna River, a tributary to the Ganga. These disastrous floods, however, have not been deterrent to large scale deforestation and misuse of land in the headwaters of the Yamuna.

Similar misuse of land in the headwaters of the Indus Basin have also created serious problems of land and water management. In spectacular mountain slopes, crops and habitations grow like eyebrows on steep cliffs. Large flocks of sheep and goats climb long unprotected 45° slopes, stripping the vegetation and trampling the exposed soil. Already the important new reservoirs behind the Mangla and Tarbella dams are filling with sediment far more rapidly than the designers had expected. Pakistan is dependent on the great irrigation network of the Indus not only for prosperity but for survival. The need to reduce floods and to increase storage is inescapable. Government efforts through both the Forest Department and the Water and Power Authority are reforesting limited areas and introducing fruit trees as an alternative livelihood, but the effects are small in comparison with the vast scale of the active damage. Until grazing is stopped and cultivation is fully protected against soil erosion the damage will continue to grow. There is a positive remedy. It is to return the land-use in these headwaters to hydrological stability by exploiting their natural ecological role. Pakistan imports both fuel and timber. The planting, tending, and harvesting of trees with full precautions against erosion, can provide employment far closer to the national interest than the tending of livestock.

The planting of trees is not the panacea for all streamsource areas. The growth of cities, roads, and enterprises in many developing countries is impressive but the accompanying growth of population has increased the pressure on areas of subsistence agriculture, while the care of soil and water resources has declined. For example, the results of some long-term watershed experiments established some 25 years ago are being applied without adequate control. A scientific summary by Drs. Edwards and Blackie of the analysis of 16 years of data on these studies of land-use changes on complete watersheds is included in this volume (Chapter 3.7).

One of these studies had demonstrated that a complete commercial tea estate, with roads, houses, factory, offices, and workshops can be developed in a stream source area of tall forest without long-term damage either to soil stability or to the amount and regulation of streamflow. This result was achieved by meticulous planning for soil conservation and runoff control, followed by highly competent execution and subsequent management by the Brooke-Bond Tea Company. The Kenya Government has subsequently developed some twenty tea estates, with support from the Commonwealth Development Corporation, in a highly successful pattern of nucleus plantations with factories designed to accept and to process green leaf from large numbers of smallholders. This system by which more than 150,000 smallholders supply leaf to the Kenya Tea Growers Authority (KTGA) is of worldwide interest to development agencies.

However, it is startling to observe that in the headwaters of the Thika and Chania Rivers whose watersheds hold about one fifth of Kenya's population and a high proportion of the country's resources of hydro-power, the smallholders have been allowed to plant tea on astonishingly steep hillsides. Soil conservation is taken seriously by the KTGA and planting material is issued only after terraces have been built and inspected. But each farmer's food crops are then planted, and his goats are grazed, on the same steep slopes, alongside the terraced tea, but without any soil conversation measures whatever. As a result these important rivers are red with eroded soil far higher up in the catchments than they were 20 years ago. Lower down in the catchment the reservoir behind the well-built dam which supplied Kitui township and hospital is now completely filled with transported soil and is being grazed by cattle.

1.1.5. PROTECTION OF SEMI-ARID WATERSHEDS

Both experiment and experience give some confidence that watershed management policies can now be drafted for the high-rainfall zones where the ecological climax is a closed forest. Far more difficult both technically, socially, and economically, the watershed problems of the semi-arid climates now present the most urgent challenge. With their characteristically higher rainfall variability and greater stresses of temperature and dessication they are prone to very rapid deterioration under misuse. Here we have rather unique evidence from Kenya's eastern province that restructuring with bulldozer and grader can provide only temporary protection unless there is a sound level of both crops and livestock husbandry. The Machakos areas was surveyed, planned and a full network of cut-off drains, bench terraces, and protected waterways was constructed in the 1930s, with 230 m ha of terracing. There was a negligible improvement in the farming system, which remained that of shifting cultivation with severe overgrazing of the abandoned areas. In the 1950s a thorough reconstruction of the earthworks was completed under the skilful and experienced leadership of Colin Maher. Terracing was increased to 69 m ha. There was again no corresponding effort to improve the farming methods. Two sets of aerial photographs of the Machakos District were taken in 1948 and 1972. These were thoroughly analysed on stereo-projection by Donald B. Thomas at the National College of Agricultural Engineering in the UK. Sheet erosion from overgrazed hillsides had increased by one third in the 24 years interval and many terraces had failed through soil erosion from the overgrazed areas. The engineer alone cannot win. The agriculture must be improved by more vigorous crops at higher planting densities and by productive forages and grasses giving better soil protection and more root residues. Twenty years ago direct experiments in this area showed good responses to legumes, cattle manure, and small dressings of phosphates. (Pereira *et al.*, 1952; 1961). Presently about one third of the farmers are maintaining their terraces and growing good crops, while all around them erosion and overgrazing are severe, and steep slopes are being newly cultivated without any precautions. This situation is very similar to that of much of the USA half a century ago. The remedy is now well known and widely tested. It is the organization of watersheds into *Intensive Conservation Areas*. A fairly brief concentration of extension staff on the better farms demonstrates the advantages of countour planting and the familiar package of improved inputs. Meetings of all the farmers on the watershed (from 50 to 200 approx.) are held to offer technical help and a small degree of subsidy and the farmers are encouraged to elect a committee to supervise and adjudicate. When two thirds of the farmers have joined, the conservation plan becomes compulsory, so that drainage ways follow optimum courses, crossing farm boundaries as necessary.

Minor subsidies are needed to enable the ICA committes, guided by extension staff, to employ farmers from within the watershed to carry out simple gulley-stopping work and to plant trees for fuel, and building poles in eroded areas and in small pockets of soil between boulders. Very encouraging results are reported by Tejwani *et al.* (1975) from the Soil and Water Conservation Research Station at Dehra Dun in India (and in Chapter 4.1 of this volume) where severe gully erosion was checked and stabilized by the planting of hardy trees. Subsequent sales of firewood poles and fodder from Leucaena and Posopis species paid twice over for the costs of the operation. The most difficult problem is to limit the numbers of livestock on the areas which are too steep to cultivate, but the planting of fodder grasses in drainage lines and the feeding of crops residues, together with communal efforts to check sheet erosion from drainage areas, have proved effective under ICA's strong leadership. This system is also successful in African subsistence farming areas, and needs major emphasis in national policies throughout the tropical and subtropical world.

1.1.6. DEVELOPMENT WITH MEASUREMENTS

Where does the scientist come into the intensive conservation area scheme? It may at first appear to be a purely extension effort, but the technical support must be good enough to ensure success for the farmers. Crops that have failed (for any reason) will kill the enthusiasm of extension assistants, so frequent visits of specialists in soils, crops, and livestock are needed to help them. Secondly the ICA system provides the critical assessment of 'packages' of practices recommended from the research centres. Thirdly the ICA system provides the opportunity for monitoring quantitatively the extent to which deterioration of soil and water resources can be arrested and reversed. The streamflow integrates the success or failure of soil and water management over the watershed. This needs the cooperation of the engineer hydrologist, for wide flumes are needed for soil-laden stormflow. Modern electronically triggered suction-type sediment samplers programmed to sample the peak flow which carry the main soil load, can give quantitative evidence vital to the design of reservoirs for irrigation or power. Monitoring of both rates of deterioration under population pressure and rates of recovery under ICA development will provide information of critical importance for the allocation of national resources in countries facing a doubling of their populations in the next 25 years or less.

This difficult and challenging field should have the top priority in our current studies of watershed management.

REFERENCES

Dansgaard, W., Johnsen, S. J., Klaussen, H. B., and Langway, C. C., 1972, Speculations about the next glaciation. *Quaternary Research*, **2**, 396–398.

Hays, J. D., Imbrie, J., and Shackleton, N. J., 1976, Variations in the earth's orbit: Pacemaker of the Ice Ages, *Science*, **194**, 1121–1132.

Mason, B. J., 1976, Royal Meteorological Society Symons Memoraal Lecture, *Quart. J. Roy. Met. Soc.*, **102**, 473.

Milankovitch, M., 1941, *K. Serb. Akad* Beogr. Spec. Publ. 132. (Translated 1969 by Israel Program for Scientific Translations).

Pereira, H. C., and Beckley, V. R. S., 1952, Gross establishment on an eroded soil in a semi-arid African reserve. *Emp. J. Exp. Agric.*, **21**, 1–14.

Pereira, H. C., Hosegood, P. H., and Thomas, D. B., 1961, Productivity of tropical semi-arid thornscrub country under intensive management. *Emp. J. Exp. Agric.*, **29**, 269–286.

Tejwani, K. G., Gupta, S. K., and Mathur, H. N., 1975, Soil and Water Conservation Research. *Indian Council Agric. Res.*, New Delhi.

Thomas D. B., 1975, *Thesis: Nat. Coll. Agric. Eng. Cranfield Inst. of Technology*, Bedford, England.

Tropical Agricultural Hydrology
Edited by R. Lal and E. W. Russell
© 1981, John Wiley & Sons Ltd.

1.2

Role of Watershed Management for Arable Land Use in the Tropics

E. W. Russell

Watershed management is concerned with control of water, and particularly with the control of water transfer from the upper to the lower parts of a river's catchment area; thus it can directly affect all the people living in the whole region. The amount of water transferred or moving out of the upper part depends on the difference between the rainfall and evapotranspiration, and the evapotranspiration itself depends on the vegetation, the depth, and the water holding capacity of the soil, and the surface run-off. Thus both the amount of water leaving the area, and the seasonal flow of the river, are dependent on the land use management there. Thus the amount of water entering the lower reaches of the river is dependent on management practices higher up the river, and these can affect the liability of the river to flood, the magnitude of the flood and the sediment load it carries, and the dry season flow; so the effects of these practices can have very serious consequences for the management of arable land in the lower reaches of the rivers. Since the soils in these areas are often deep alluvial silts of high agricultural potential, very serious economic damage can arise from inappropriate systems of land use in the higher regions of the basin.

Most perennial rivers in the tropics arise in highlands with an excess of rainfall over transpiration, and in many parts of the tropics, they then run through regions where the rainfall is less than the potential transpiration for periods of the year, so that agricultural productivity there is limited by lack of water. In these circumstances it should be possible to raise the agricultural productivity of the lower areas by using some of the surplus water from the upper areas in suitable systems of irrigation; and if the potential productivity of the uplands is low compared with the lowlands, it may be profitable for the river basin as a whole to develop land use systems for the uplands that minimize the transpiration demands there, so increasing the amount of water available for agricultural use on the productive lowland soils.

The success of any scheme to increase the agricultural production in the lower reaches of the river depends on the water leaving the uplands in rivers with a fairly even flow of clear water, and this can only occur if the system of land use there maintains the infiltration rate of rainwater into the soil at least as high as the normal maximum intensity of rainfall as measured over a suitable time period. These conditions usually prevail where the natural vegetation has been little disturbed, for this normally produces a soil surface capable of absorbing the rainfall and of allowing it to percolate into the deeper subsoil and seep out into the river as springs. In addition there are often a series of swamp basins and peats that can hold appreciable quantities of water that also slowly drain out into the main river. Unfortunately only too often the current system of land use is one in which the natural vegetation is so seriously disturbed that both the amount of water that these upper catchments can temporarily hold during prolonged or heavy rain, and the proportion of the rain that percolates into the soil, are seriously reduced, resulting in much greater variations in the seasonal flow of the river, with much greater sediment loads during peak flows, and often with a greater proportion of the total rainfall leaving the region in the river.

The primary cause of land misuse in the upper reaches is that the current systems appear to give the greatest profit in the short term to the local population in their present economic and social conditions; so any new system which reduces flood hazards and increases dry-season flow out of the area will only be willingly accepted by these people if they can see it will increase their standard of living in the short term.

The solution to these problems of the misuse of land depends first of all on developing new or improved methods of land use that are both more profitable to the local community and also give appreciably better control of the river flow, and secondly on getting the new methods adopted by the local community after they have been tested and approved. It should not, in principle, be too difficult to get these adopted if the principal land use is for a product that is exported from the region, provided the local government has sufficient competent technical staff to help with the solution of the many problems likely to be involved in the change-over from the current to the new system, and provided they also have the necessary administrative staff in the field to ensure that new policies are competently carried out. There may also be difficulties for the government if the new system increases the cost of production, for this may reduce the royalties and taxes they receive from these producers.

These problems in the control of land use are much more difficult of solution when the local population is living at a subsistence or semi-subsistence level, for the introduction of new more conservative systems usually involve increased labour and financial inputs, and those will only be provided if the population is convinced of their profitability. Unfortunately it is very rarely possible to develop conservative methods of land use in these areas of semi-subsistence farming that are sufficiently profitable to be self-financing, so they can only be adopted if outside money can be brought in to help pay for the additional cost incurred. This raises the very difficult problem of who pays for these additional costs. In developing countries farmers in the upper reaches of a river system have never accepted responsibility for any damage their system of land use may cause to the farmers in the lower reaches, and conversely the lower farmers have rarely considered they have a duty to help finance the higher farmers to adopt better systems of land use for their benefit. There is need for more discussion on the basis on which the costs to the lower farmers of flood damage and reduction in the dry-season flow of the river can be compared with the costs to the higher-up farmers of adopting improved land use practices so agreement can be reached on the amount of money that should be transferred to the communities in the upper reaches for a given improvement in the river regime. However, this would be a purely academic exercise unless the local government has competent technical staff to make reliable probability estimates of the likely effects on river flow of a proposed system of land use when adopted by the local population. As I think will be evident from the discussion in this volume, the necessary relevant and reliable information for such estimates to be made, is only available for very restricted areas of the tropics, though a number of experimental programmes for obtaining the necessary data have been developed.

1.2.1. SYSTEMS OF LAND USE IN THE UPLANDS

The typical natural vegetation at high altitudes in the tropics is closed canopy evergreen forest, often interspersed with areas of grass and swamp. In some areas these forests are being exploited for any valuable timber they contain under conditions that minimize damage to the canopy or the forest floor, and here the forest is managed so as to protect the natural flow of the rivers leaving the area. But in this system, the land is used at a very low level of intensity, for many of the valuable trees are hundreds of years old and are very slow growers so once they have been cut out the annual increment of marketable timber will be very small.

Planted production forests are sometimes used to replace indigenous forest, particularly when it contains few marketable trees. The planted trees are chosen for their rapid growth and production of a marketable quality of wood. At one time there was a fear that these fast-growing trees might use more water than the slow-growing trees in the natural forest, but extensive experimentation in East Africa has shown that these fast growing planted forests need not cause any disturbance to the river flow compared with the original natural forest, so there is no conflict between their role for production and for watershed production. But it is only possible to get full financial return from these forests if they have been properly planned and costed from the very beginning, and this must include the alignment of the roads needed for access when planting and tending the forest, and for extraction of the trees to the mill. The two principal uses of these large production forests are not for pulp or timber, but for wood for fuel (as charcoal for domestic cooking and heating) and for building material for villages. Both are becoming of increasing importance. Only rarely can indigenous species of trees be used, as in most areas suitable exotics (usually softwoods) have a much greater productivity for the particular purpose for which the wood is required; and commonly only a few species, and often only one, are planted.

The use of land for forests comes under great strain if the local population in the area begins to increase too rapidly, for this will increase the demand for land on which they can grow their food crops, or for land to increase the number of livestock they want to keep. This in turn increases the grazing pressure in the forest, so reducing its inflitration capacity. One advantage of productive forests is that they have a production cycle of a few decades and so have an appreciable labour demand. However, this involves providing land in the forest domain for the population to grow their own food. It is not too difficult to provide land for food production when the forest is just planted, for the families working in the forest are given land on which to clear, plant, and tend the young trees, and to grow their crops between them until the trees get so tall that they overshadow the land. This may occur two to four years after planting. Short cycle production forestry will carry a sufficiently large population to justify the creation of proper villages, but this will almost certainly involve the setting aside of land near the village for growing some of the food crops needed by the village. The whole problem of the introduction of food crop production within production forests, managed for good river control, is now being actively studied under the name agro-forestry, and a conference on this was recently held in Nairobi. An International Council for Research in Agro-forestry ICRAF has been set up to encourage development of these practices. Most of the problems of introducing soundly based agro-forestry practices in high rainfall areas are administrative, that is ensuring the local population willingly accepts the restrictions that must be placed on their system of food production and on the areas of land they can use, to minimize the risk of seriously increased soil and water loss during periods of heavy storms. However much local work is often needed to translate general principles into detailed recommendations and instructions suitable for each particular locality.

It is not always necessary to use plantation forests for the protection of the water-gathering areas of a river basin, for in suitable areas perennial crops such as coffee, bananas, and tea can take the place of trees. A banana–coffee economy, where the coffee is grown under bananas and the soil surface is kept mulched with banana trash, is very suited to a peasant economy, as no particular care is needed to prevent soil erosion if the soil is fertile enough for the crops to grow strongly. Tea is another suitable crop. It is increasingly being grown unshaded in Africa, and can be grown in forest clearings provided the size of the clearing is not too large and the area under food crops is strictly controlled. Tea can also be used as a plantation crop but only if the estates practice very good conservation management from the very beginning of the clearing operation. As an example a 750-hectare tea estate was excised from a high evergreen forest near Keriche in Kenya and just over half was planted to tea. This was in a catchment whose river was gauged at various points and where many agro-meteorological records were kept, and the river flows were compared with those from a similar catchment kept in undisturbed forest. During the actual establishment the flow from the cleared area was noticeably larger than that from the uncleared area, but once the tea was established the planted area became very similar to that from the forest. But the steeper slopes were kept in natural or planted forest, and great care was taken to ensure that all the runoff water from the settled areas—the factory site and the village—ran into very well protected waterways. It was this water that caused the somewhat larger flow in the river at the base of the tea catchment during heavy storms.

It is possible to increase the outflow of water from a forested catchment by replacing the trees with short season arable crops, particularly in areas with good rains but a pronounced dry season. Natural forest and many perennial grasses growing on deep soils are very deep rooting—they will dry the soils to depths of four to five metres during dry spells without their transpiration rates being seriously reduced; whilst short season food crops, such as the cereals and root crops, take most of their water from the top (up to one-and-a-half metres deep and little if any from below two metres). In theory, land in arable crops under very good soil and water conservation management should be able to accept most of the rain and thus allow the water to percolate into the groundwater and to seep slowly into the river; since transpiration will only be taking place for a part of the year, and then only from the top one-and-a-half metres, more water will be available to feed into the river. In practice this procedure can only be used in very special circumstances. In general it is far too dangerous, for it is almost impossible to enforce the very strict conservation measures needed to prevent very seriously increased erosion and runoff occurring, with devastating results for river flooding and heavily increased sediment loads.

There is less experience with the effects of replacing forest by pasture. It is probable that well-managed pasture in deep-soil areas which do not have a severe dry season will give an annual transpiration similar to (but probably rather less than) that from the forest. However, it will probably give more runoff but not necessarily more soil loss. There is still little experience of the consequences of this replacement in practice, when the quality of management is indifferent. Any management practice that encourages bunch grasses (with bare soil between the bunches at the expense of a uniform cover) will naturally increase the liability to runoff and soil loss. In an experiment in Kenya, grazing itself appears even on a uniform sward to encourage runoff for a period after the grazing has ceased. This is due to the trampling of the grazing animals.

Arable agriculture in the uplands, particularly on sloping land, can lead to (and has led to) very serious erosion in many parts of Africa. In many areas the slopes being cultivated must be terraced and the terrace outlets must discharge into protected waterways, which may require simple structures for reducing the velocity of the water if erosion is to be controlled effectively. This must usually be coupled with contour cultivation and planting if only a single crop is being grown. Even then the steeper slopes must usually be taken out of cultivation and planted with a protection cover. There are examples in Africa of the successful application of such well-tested methods for soil conservation, particularly where a high-priced cash crop is being introduced, such as coffee or milk production, for here it is quite obvious to the farmers that conservation pays. There has been much less success when no new cash crop can be introduced at the same time; and unfortunately there have been examples where conservation measures have been forced on an unwilling local community by the Government edict, which resulted in increased erosion damage due to neglecting the necessary maintenance of terraces and waterways. It is particularly in these arable areas on rolling topography and with weakly-structured soils that it is so essential to set up carefully planned research projects to discover the best ways of interesting the farmers in adopting conservative systems of farming and of developing appropriate new systems to fit in with the customs and aspirations of the local communities. Mere terrace building and waterway construction, without any other change in the local system of farming, is unlikely to lead to a long-term reduction in soil loss or flood damage. However, the development of cropping systems based on maintaining a surface mulch of the previous crops' residues on the soil surface (perhaps coupled with zero tillage as is being developed by Lal in IITA) could make a great contribution to the stabilizing of arable agriculture in these areas, provided they do not cause difficult problems of pest or weed control. Nevertheless, difficulties arise when these residues are put to any important use by the local community, as this reduces the amount of residues available for mulching very considerably.

One point that has given rise to much discussion concerns the effect of drainage swamps in the upper reaches of rivers. These areas will hold considerable volumes of water; these are slowly released as inflows are reduced, so they can form very valuable buffers against sudden storms. The typical vegetation is various species of coarse grasses and rushes, and their contribution to the productivity of the region is small: it is normally confined to dry season grazing, or the grasses are cut for mulching when this is necessary. But many of these swamps can be drained and have soils suitable for food crops, and this may result in a deterioration in the river flow regime. I believe there is still little direct evidence of how much damage is done to river flows by the drainage of such vlei areas in Africa, when the soils in the vlei are suited to arable cropping.

In conclusion, the principal problems likely to arise in practice for managing the upper reaches of watersheds for the benefit of farmers lower down are now known, and usually there exist also a number of alternative systems available which will maintain reasonable flows in the river leaving the watershed. The major problems for the future lie more in the realm of rural sociology and rural economy, for a recommended land use will only be willingly and effectively adopted by the local community if it increases the reliability of the yield of their food crops and of their income. The greater the number of options the land use research workers and the economists can offer, the easier it will be for the technical staff to develop a system adapted to the particular customs and aspirations of the community, and the easier it will also be for the local administration to introduce whatever measures are necessary for its implementation.

PART 2

Ecological Conditions in a Forest Ecosystem

Tropical Agricultural Hydrology
Edited by R. Lal and E. W. Russell
© 1981, John Wiley & Sons Ltd.

2.1

Fundamental Ecological Parameters in Amazonia, in Relation to the Potential Development of the Region

H. KLINGE, K. FURCH, U. IRMLER, AND W. JUNK

2.1.1. INTRODUCTION

The Amazon region still contains the world's largest area of tropical rain-forest and still carries a sparse population who earn their living mainly from gathering forest products: hunting, fishing, and practising shifting cultivation (Figure 2.1). The Brazilian government in recent years has attempted to develop this territory, often incorrectly referred to as Transamazonia, by constructing a major road network (Kohlhepp, 1976; Smith, 1976a). But the policy of developing this region has aroused strong controversy both in Brazil and elsewhere (Eden, 1979; Goodland and Irwin, 1975; Kleinpenning, 1977; Lovejoy, 1973; Sioli, 1977a; 1977b; 1979), for it could result in the disappearance of the Amazon forests, which cover most of the river basin, with the loss of its rich animal life.

It is now recognized that the development projects to date have not been as successful as expected, (Dickson, 1978; Jahoda and O'Hearn, 1975; Smith, 1978a). Thus the Ford rubber plantation at the Tapajos River has had to be handed back to the national government because it was economically unsuccessful (Russell, 1942). Another example is a large scale Gmelina and pine afforestation project for paper pulp production in the Jari River catchment, with an associated rice scheme on periodically inundated land (Alvim de Tarso, 1976; Bousfield, 1979a; FAO, 1973). In addition two projects aimed at developing the river transportation facilities and producing hydro-electric energy have been studied. During World War II the Orinoco–Casiquiare–Negro Waterway study was undertaken (US Army Corps of Engineers, 1943), and in more recent years the spectacular plan of the Hudson Institute (Panero, 1967; 1969) proposal to convert the Amazon lowlands into a huge inland lake.

This state of affairs of the actual development situation in the Amazon region raises the question whether there is any real chance to develop it further. In attempting to answer the question it is useful to describe its natural ecological conditions.

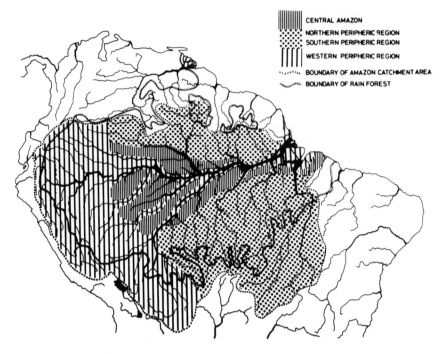

CENTRAL AMAZON
NORTHERN PERIPHERIC REGION
SOUTHERN PERIPHERIC REGION
WESTERN PERIPHERIC REGION
BOUNDARY OF AMAZON CATCHMENT AREA
BOUNDARY OF RAIN FOREST

Figure 2.1 The Amazon region

2.1.1.1. Amazon Ecology

Systematic research in the field of Amazon ecology began about 50 years ago. Our department has contributed to it, under its former director Herald Sioli, particularly in the field of freshwater ecology.

Reviewing the available literature with respect to the above question, it becomes quickly evident that two parameters are fundamental for both understanding the Amazon ecology and its potential development. These are:

(i) the water level fluctuation of the Amazon river system, for the flood plains cover very large areas, and:

(ii) how the composition of the waters, the soils, and the vegetation is related to the geological history of the basin and to the structure and functioning of the ecosystems composing the Amazon biosphere.

2.1.1.2. Hydrology and Climate

The water discharge of the Amazon system is in the order of $100\text{-}200 \times 10^3$ m^3 sec^{-1} (Davis, 1964; Oltman, without year; Oltman *et al.*, 1964; Rzóska, 1978), which reflects the humid tropical climate of this large watershed (Eidt, 1968; Schwerdtfeger, 1976). The average annual rainfall varies from about 2000 to almost 4000 mm (Walter *et al.*, 1977). There is a tendency of the rainfall to increase from the southeast toward the northwest (Salati *et al.*, 1978).

Salati *et al.* (1978) estimate that about 50 per cent of the water falling as rain in the basin is derived from water transpired by the forest, a figure recently confirmed by a study of the water budget of a tropical rainforest in Amazonian Venezuela (Jordan and Herrera, in press).

The distribution of the rainfall is uneven (Walter *et al.*, 1977). While a monthly rainfall up to 5000 mm is observed in most months, it is below 200 or even 100 mm in the remaining ones. In the drier months there can be periods of up to 20 days without any rain, even at localities with high annual rainfall, (Medina *et al.*, 1978; Herrera, 1979a), so that it is correct to distinguish a rainy season from a less rainy one, which in the lower rainfall area is almost a dry season. The dry season is never so severe that the evergreen rainforest is replaced by a deciduous forest. Even on moist sites however the rainforest is characterized by seasonal leaf fall and flowering and is therefore best classified as seasonal evergreen rainforest (Brünig and Klinge, 1975).

The seasonal distribution of the rainfall is reflected in the fluctuation of water levels of the Amazon river system (Oltman *et al.*, 1964; O'Reilly Sternberg, 1975). At Manaus in the central part of the Amazon region the difference between high and low water of the Rio Negro is 10.0 ± 2.1 m while at San Carlos de Rio Negro in South Venezuela it is 5-6 m only (Herrera, 1979b). A ratio of 1:3 of the seasonal water discharges is reported by Oltman *et al.* (1964) for the Amazon river itself.

The major southern tributaries of the Amazon have their highest water levels between March and April and their lowest between August and October, whilst major northern tributaries have their highest in June and July and their lowest between December and March. The Amazon itself is highest in June/July and lowest in October (Wilhelmy, 1970).

Rise and fall of the river level have a pronounced impact on the distribution of dry land and water surface in a lowland region like Amazonia where much of the land lies below 300 or even 100 m a.s.l. (Alvim de Tarso, 1973a; Fittkau, 1974; Irion, 1979). Thus, much land being dry at low water becomes flooded if the river level rises. The tidal influence from the Atlantic Ocean reaches 800 km upstream (Wilhelmy, 1970).

The area never exposed to flooding is 'terra firme' in the local language, the flooded areas are either 'varzea' or 'igapo' (Prance, 1978; 1979). Pires (1973) has estimated that the seasonally flooded areas constitute about 3 per cent of the entire watershed.

These seasonally flooded areas carry indigenous plants and animals communities well adapted to this environment, suggesting that these conditions have lasted for a long period of time. Figure 2.2 summarizes the present knowledge on the survival strategies of both terrestrial and acquatic animals (Irmler, 1979). The local human

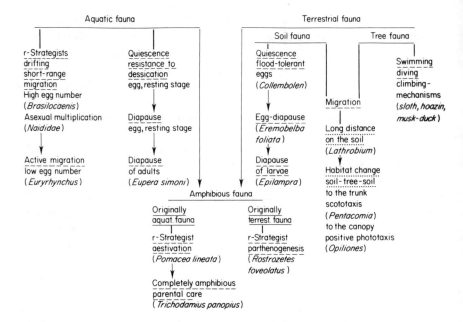

Figure 2.2 Synthesis of progressive survival strategies in relation to flooding, observed among several animal species of Amazon inundation forests (After Irmler, 1979)

populations, both aboriginal and more recent immigrants have also learned how to adapt themselves to these conditions in exploiting the regional resources (Meggers, 1971; O'Reilly Sternberg, 1975; Wilhelmy, 1970).

2.1.2. TROPHIC CONDITIONS

2.1.2.1. Hydrological and Chemical Observations

The term varzea is most often applied to the area flooded by rivers carrying turbid water rich in suspended inorganic matter, an example being the Amazon itself (Gibbs, 1967; Meade *et al.*, 1979; O'Reilly Sternberg, 1975; Sioli, 1957; 1964; 1967; 1976; Schmidt, 1972; Williams, 1968; and others). The typical vegetation on the area flooded by such 'white waters' is a tall closed forest; the varzea forest with its own characteristic flora (Hueck, 1966; Prance, 1978; 1979); though where the water is deeper and calm a dense mat floating aquatic and semi-aquatic macrophytes occurs (Junk, 1970; 1973; Fittkau *et al.*, 1975a).

It has been speculated that the origin of the suspended silt in the white water rivers was once thought to be in the Andean headwaters, but it is more probably derived from the foot zone of the Andes where the rivers leaving the mountains flow over soft tertiary sediments, and through their meanderings carry off much of this old alluvial land. Simple surface runoff apparently plays a minor role in the origin of the suspension load of these rivers. Only a portion of the suspended material is deposited in the present varzea (natural fertilization; Irion, 1976).

The term igapo is applied to the areas flooded by rivers carrying not turbid, but clear water. If such clear water contains dissolved humic substances it has a dark colour ('blackwater'; Klinge, 1967; Paolini, 1979; Zeichmann, 1976). The most impressive representative of blackwater rivers is the Rio Negro.

The areas flooded by these rivers do not carry a rich aquatic vegetation, and their forests are composed of species most of which do not occur in the varzea forest. Several species of the igapo forest (Hueck, 1966; Irmler, 1977; Mägdefrau and Wutz, 1961; Prance, 1978) tolerate floods to the extent that they maintain their green foliage, even if their crowns are submerged (Gessner, 1958; 1968).

The turbid and clear waters also differ in regard to their aquatic animal life, for the Blackwaters only support a sparse insect population (Stern, 1966; Stern *et al.*, 1970; Williams, 1968).

The waters also differ considerably in their chemical composition. The turbid waters are characterized both by their content of suspended inorganic matter, and also by their relatively high content of soluble salts of calcium, magnesium, potassium, and other cations, whilst the clear waters including the blackwaters contain extremely low contents of these elements but are richer in the salts of the alkali metals, and of aluminum and iron. (Anonymous, 1972a; Furch, 1976; 1978; Furch and Klinge, 1978; Furch *et al.*, in preparation; Furch and Junk, 1979; Sioli, 1964; 1967; 1976; and others). However, when compared to the world average for rivers (Bowen, 1966; Bayly and Williams, 1975; Meyback, 1977) almost all Amazonian waters are lower in almost all chemical elements (Figure 2.3). The turbid waters are nearer to that average than are the clear and blackwaters. It may be inferred from this observation that the biological differences between the waters are related to the differences in their chemical composition.

2.1.2.2. Subdivision of Amazonia

In attempting an interpretation of the different biological and chemical characteristics of Amazonian waters (Sioli, 1968a; Sioli and Klinge, 1961; Sioli *et al.*, 1969) related them to the topography and geological structure of the individual watershed (Tricart, 1978). In generalizing such observations Sioli (1975) put forward the idea that rivers are an expression of the terrestrial environment through which they pass. Furch and Junk (1979) made evident the fact that much of the chemical qualities of Amazonian waters may indeed be explained by petrographic differences between the respective catchments.

A good example of this generalization is that the blackwaters are typically derived from catchments where spodosols are the dominant soil type, i.e. coarse grained silicious rocks (Fittkau *et al.*, 1975b; Herrera, 1979a; 1979b; Klinge, 1965; 1966;

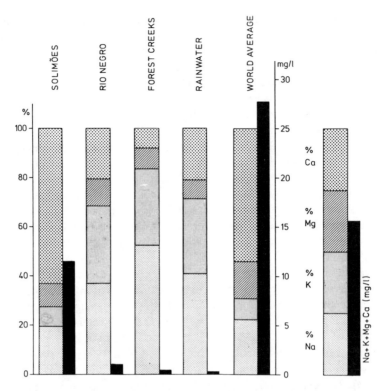

Figure 2.3 Sum of Na, K, Mg, and Ca (mg/l) and their proportions (w %) in selected Amazonian running waters, in comparison with Bowen's 1966 world average of freshwaters (After Furch, in preparation)

1967; Klinge *et al.*, 1977; Paolini, 1979; US Department of Agriculture, 1967), whereas the richer waters are from catchments with finer grained or more basic rocks.

Fittkau (1971; 1974) stratified all available information on the ecology of both terrestrial and aquatic environments in the Amazon region and produced an ecological map of the region.

In Western Amazonia, province III covers an extensive area of the foot-zone of the Andes. This province extends in the form of elongated strips along the courses of rivers carrying turbid water toward the Atlantic, thus splitting province IV of Central Amazonia in several 'islands'. The northern province I is developed on the Guiana shield, while province II covers the northern part of the Brazil shield.

While province III includes a rich and highly diverse flora and fauna, the other provinces are poorer in this respect. That these biotic differences are intimately related to the geochemistry of land and water is best examplified by molluscs bearing hard shells. These are extremely rare or even completely absent in Central

Amazonia where limestones or other rocks rich in calcium do not occur and the river water is extremely low in this element (Furch and Klinge, 1978). The hydrochemical data demonstrate that Central Amazonia includes waters ranking among the poorest in chemical elements, while the western province presents waters much richer in them. Provinces I and II are in this respect similar to Central Amazonia.

2.1.3. PERSPECTIVES FOR DEVELOPMENT

In the previous sections the Amazon region has been described as a landscape consisting of flooded and non-flooded portions generally characterized by low nutrient contents. It has also been shown that the region can be stratified according to the existing differences in the trophic conditions. In this section, the major units terra firme and varzea will be described in more detail and with respect to their development potential.

2.1.3.1. Terra Firme

The great majority of terra firme soils are acidic, heavily leached, and have a low base saturation, the dominant clay mineral being kaolinitic. Their natural fertility is therefore quite low (Alvim de Tarso, 1973b; 1976; 1978; Beek and Bramao, 1968; FAO, 1973; Irion, 1976a; 1978; 1979; Klinge, 1976a; 1976b; Sombroek, 1966; 1979; Van Wambeke, 1978), except for occasionally anthropogenic soils like 'terra preta do indio' (Hilbert, 1968; Sombroek, 1966) and for natural soils developed on chemically rich volcanic rocks (Alvim de Tarso, 1976; 1978). Among the least fertile soils are the already mentioned spodosols which occupy a great proportion of the region, particularly in its western part including the Rio Negro basin (Flores *et al.*, 1978; Herrera, 1979a; 1979b; Sombroek, 1979).

The low fertility of the terra firme soils is reflected in the structure and functioning of their natural plant cover, the Amazon rainforest (Golley *et al.*, in press; Schubart, 1978). Results of studies specifically undertaken to explore the nutrient question in the respective type of ecosystems (Medina *et al.*, 1977) have given substantial evidence that the system functions as a trap for nutrients supplied by the rainfall (Anonymous, 1972b; Herrera, 1979a; 1979b; Herrera *et al.*, 1978a; 1978b; Jordan, 1978; 1979; Jordan and Herrera, 1979; Jordan *et al.*, 1979a; 1979b; Klinge and Fittkau, 1972; Went and Stark, 1968; and others). Several of these authors have also attributed a particular importance to the detritus food chain for the functioning of both the rainforest ecosystem (Fittkau and Klinge, 1973) and the aquatic ecosystem embedded in it (Knöppel, 1970). In addition, several mechanisms for the conservation of nutrients have been described for the rainforest ecosystem (Herrera *et al.*, 1978a; Jordan *et al.*, 1979a; 1979b), among them the accumulation of high amounts of living phytomass both above ground and below ground (Klinge and Herrera, 1978; Klinge, 1976b).

In spite of the fragile nutrient situation of the terra firme rainforest ecosystem, symptoms of nutrient deficiency in its plant and animal species have not been reported so far. Symptoms of phosphorus, sodium, copper, and cobalt were however observed among cattle grazing on former rainforest soils (Sutmoeller *et al.*, 1966).

The productivity of the terra firme rainforest is only medium, as judged from the amount of litter produced (Klinge and Rodrigues, 1968; Klinge, 1977; Franken *et al.*, in press; Medina *et al.*, in preparation), but the forests are rich in animal species though the density of the exploitable mammalian populations is generally low (Smith, 1978b). Excessive hunting has endangered many of these species (Smith, 1974a; 1974b; 1976b; 1976c; 1978b). By opening-up the region, an opportunity was given to species previously unknown, such as the house sparrow (Smith, 1973), to invade it.

Low soil fertility of the terra firme is a serious obstacle to its development (Alvim de Tarso, 1973b; Schubart, 1978). The establishment of large scale monocultures including plantation is not recommended as a tool for the regional development (FAO, 1973). The traditional land-use system, shifting cultivation, will probably continue to be an important production system on the less valuable soils. The productivity of the system can be improved (Alvim de Tarso, 1976; Greenland, 1975; Smith, 1978a), particularly with respect to manioc and rice (Sanchez and Nureno, 1972). Medium and small scale planting of a variety of fruit trees and vegetables (oil palm, rubber, para-nut, guarana, black pepper, tomatoes, and others) is proposed by many individuals and development agencies (Alvim de Tarso, 1973a, 1976, 1978; FAO, 1973; Maas, 1969). Cattle raising on pastures established on former forest soils is now widely being carried out (Brücher, 1970; Falesi, 1976; Fearnside, 1979). Since the terra firme is naturally forest land, silviculture of both indigenous and exotic species, as well as management of the natural forests, is considered an economically important form of land use, particularly on the soils of very low fertility (Brünig, 1969).

Detailed soil mapping is considered essential for the selection of soils appropriate for any of these crop plants or land-use systems. In addition agronomic work will be needed for the control of pests and diseases, the fertilizers required, and the breeding of production cultures having low nutrient demands.

2.1.3.2. Varzea

The varzea wetland ranks high among the types of Amazon ecosystems having relatively good prospects for development (O'Reilly Sternberg, 1975; Tricart, 1978) since the rivers flooding these areas are rich in nutrients, both in solution and in the suspended inorganic and organic forms which are deposited on the soil. The agricultural utilization of these soils has already been recommended by Camargo (1958) and Sioli (1973).

Among additional observations pointing to favourable production conditions in the varzea may be cited:

(i) The natural primary production of aquatic and semi-aquatic grasses and herbs is due to the nutrient supply of the turbid waters (Fittkau *et al.*, 1975a; Howard-Williams and Junk, 1977 Junk, 1970; 1973; Siloi, 1968b). Results of studies of Venezuelan savannas inundated by turbid rivers confirm this observation (Briceno *et al.*, 1977; Bulla *et al.*, 1977; Escobar and Medina, 1977; Gonzales and Escobar, 1977; Pacheco *et al.*, 1977).

(ii) The natural secondary production of the aquatic fauna is also high. These include fishes, turtles, caimans, capybara, water fowl, and the even greater number of invertebrate species. Excessive hunting and fishing however has reduced the population of some of these species considerably (Junk, 1975; IUCN, 1975; Smith, 1974a). The high natural secondary production is based partly on the high amounts of plant matter available to the herbivores, and partly on the secondary production of animal species representing lower levels of the aquatic food chain that are available to the carnivores.

The considerable fluctuation of the water level in the varzea is obviously an obstacle to its permanent use under controlled hydrologic conditions. Such a control would require immense investments. However, certain forms of land-use avoiding such investments, are viable.

The present knowledge of the varzea ecosystem is, in many respects, still incomplete and much research is needed to bridge the respective gaps. It is not assumed that the varzea will be a future paradise. However, it is worthwhile, in our opinion, to study this particular ecosystem more intensively than in the past particularly with a view to its non-destructive use.

2.1.3.2.1. Plant Production

Many natural grass species of the varzea are, as mentioned, highly productive, being resistant to the local diseases and pests. They have, in addition, a moderately high content of proteins and nutrients, and most species are accepted by cattle. Pastures in these areas can be improved by replacing some of the less valuable with more valuable species. Species growing in deep water cannot be grazed but can only be utilized if they can be harvested mechanically.

In this context it is interesting to note that Mexican Indians have developed a system ('chinampa', Gomez-Pompa, 1977) in which aquatic plants growing in the ditches surrounding their fields are harvested and applied as green manure or compost to the crops grown on these fields. The green matter of varzea plants can similarly, after cropping, be carried to the nearby terra firme easily accessible at high water. There, the plant matter may be applied, as green manure or compost, to the vegetables and fruit trees grown there (Junk, 1979; Noda *et al.*, 1978).

The success of the Jari rice cultivation project which is yielding 12–14 t ha^{-1} yr^{-1}, in two harvests (Alvim de Tarso, 1976, 1978) should encourage the expansion of rice culture on varzea soils, particularly in areas with only moderate fluctuations in the water level.

Varzea forests occurring on higher ground have in the past supplied wood for construction purposes and fire wood. The stands have been heavily exploited, and may have disappeared. A study of remaining stands should be undertaken with respect to potentially merchantable species and their silivicultural management.

2.1.3.2.2. Animal Production

Among the varzea animals of commercial value, fishes are probably the most important ones. Amazon fisheries depend predominantly on the fishes of the turbid waters and the varzea, while the black and clear waters are much less important in this respect (Geisler *et al.*, 1971; Junk and Honda, 1976). A considerable increase in the landing of commercial fish species cannot be expected. Some species are probably already overexploited. The intensification of the fisheries by utilization of additional technology is not recommended (Chapman, 1977; Petrere, 1979).

In order to not overexploit the fish populations (about 200 different species occur in the Amazon rivers) fish farming should be developed (see also Ackefors and Rosen, 1979). The practical importance of such a programme would improve the fish supply to the economically less wealthy people, particularly those living in the poor terra firme. It would also allow the establishment of fish populations in existing and planned waters reservoirs. The culture of Amazon fishes would exclude the introduction of fish species native to other tropical areas.

In addition to fish farming, the utilization of animals like turtles, caimans, capybara, and water fowl should also be developed. Management systems for capybara, which are a large Amazon rodent (Hydrochoerus: Vos, 1979), have been developed in Venezuela (Ojasti, 1973; 1979). Experience in farming caimans is also available in these regions (Dickson, 1978). Caimans, being predatory, play an important role in controlling the size and composition of the fish population as well as the nutrient regime in the waters (Fittkau, 1973). The numbers of water buffalo should also be increased, for not only are they better adapted to the varzea conditions but their meat is also both palatable and nutritious.

2.1.4. SUMMARY

The knowledge of both the Amazon natural landscape units like terra firme and varzea and their development potentials is still limited, so it is vital to increase our understanding of the ecology of these units, including their rich and diverse biota, before any extensive development projects are undetaken. Already there is a programme for establishing ecological research stations within Brazil (Nogueira-Neto and de Melo-Carvalho, 1979).

The research programmes should be planned to allow the development of ecologically sound land use systems (Budowski, 1977; IUCN, 1975), particularly those which will be of the benefit to the people of Amazonia (Bousfield, 1979b; Breniere, 1979; Greenland, 1975).

In developing Amazonia, the conservation of selected areas in natural conditions should not be neglected, in order 'to conserve for present and future human use the diversity and integrity of biotic communities of plants and animals within natural ecosystems, and to safeguard the genetic diversity of species on which their continuing evolution depend' (Unesco, 1974).

The impact of the disappearance of the Amazon forests on both the climate and hydrology of the region should not be overlooked (Fränzle, 1976; Salati *et al.*, 1978).

Considering the fragile nutrient situation of a large proportion of Amazonia, major efforts should be directed to the development of the varzea wetlands where nutrients are supplied in relatively great quantities and at no cost by the floods of the turbid rivers.

REFERENCES

Ackefors, H., and Rosén, C. -G., 1979, Farming aquatic animals, *Ambio*, 8, 132–143.

Adis, J., 1979, Vergleichende ökologische Studien an der terrestrischen Arthropodenfauna zentralamazonischer Überschwemmungswälder, *Unpubl. Thesis*, University of Ulm, pp. 99.

Alvim de Tarso, P., 1973a, Los trópicos bajos de América Latina: recursos y ambiente para el desarrollo agricola, *Simposio Potencial Tropicos bajos América Latina,* Cali/Colombia 1973, pp. 43–61.

Alvim de Tarso, P., 1973b, Desafio agricola de regiao amazonica, *Ciencia e Cultura,* 24, 437–443.

Alvim de Tarso, P., 1976, Floresta amazonica: equilibrio entre utilizacao e conservacao, *Ciencia e Cultura,* 30, 9–16.

Alvim de Tarso, P., 1978, Perspectivas de producao agricola na regiao amazonica, *Interciencia,* 3, 243–251;

Anonymous, 1972a, Die Ionenfracht des Rio Negro, Staat Amazonas, Brasilien, nach Untersuchungen von Dr. Harald Ungemach, *Amaxoniana*, 3, 175–185.

Anonymous, 1972b, Regenwasseranalyseh aus Zentralamazonien, ausgeführt in Manaus, Amazonas, Brasilien, von Dr. Harald Ungemach, *Amazoniana*, 3, 186–198.

Bayly, I. A. E., and Williams, W. D., 1975, *Inland Waters and their Ecology*, Longman, Australia.

Beek, K. J., and Bramao, D. L., 1968, Nature and geography of South American soils, In *Biogeography and Ecology in South Ameruca* (Eds. E. J. Fittkau *et al.*), 18, 82–112.

Bousfield, D., 1979a, Scientists protest at US tycoon's Amazonian project, *Nature*, London, 280, 347–348.

Bousfield, D., 1979b, Urban deprivation in the heart of Amazon, *Nature*, London, 279, 752–753.

Bowen, H. J. M., 1966, *Trace Elements in Biogeochemistry*, Academic Press, London, New York.

Breniere, J., 1979, La recherche, une necessité pour la production des cultures vivrieres du petit paysan des pays en developement, *Agronomie tropical*, 33, 332–336.

Briceno, M. J. L. T. B., Lopez, D. H., Gonzalez, V., and Bulla, L., 1977, Ciclo biogeoquimico del Ca, Mg, Na, K y P en las sabanas inundables de Mantecal, Eco. Apure, Venezuela, In *Resumenes recibido para IV Simp. internacional de ecología tropical* (Ed. H. Wolda), Panama 1977, pp. 15–17.

Brücher, W., 1970, Rinderhaltung im amazonischen R. genwald, *Tübing. Geogr. Stud.*, **34**, 215–227.

Brünig, E., 1969, Forestry on tropical podzols and related soils, *Trop. Ecol.*, **10**, 45–58.

Brünig, E., and Klinge, H., 1975, Structure, functioning and productivity in humid tropical forest ecosystem in parts of the neotropics, *Mitt. BFH Reinbek*, **109**, 87–116.

Budowski, G., 1977, 'En busqueda de sistemas de uso de la tierra en los tropicos humedos', In *Resumenes recibido para IV Simp. internacional da ecologia tropical* (Ed. H. Wolda), Panama, 1977, pp. 13–22.

Bulla, L., Pacheco, J., and Miranda, R., 1977, Produccion, descomposicion y dinamica de una sabana bajo diferentes condiciones de inundacion, In *Resumenes recibido para IV Simp. internacional de ecologia tropical* (Ed. H. Wolda), Panama 1977, pp. 22–24.

Camargo, F. C., 1958, Report on the Amazon region, In *Problems of Humid Tropical Regions*, Paris, Unesco.

Chapman, M., 1977, Ecological management strategies for the fisheries of the Amazon basin, In *Resumenes recibido para IV Simp. internacional de ecologia tropical* (Ed. H. Wolda), Panama 1977, pp. 79–80.

Davis, L. C., 1964, The Amazon's rate of flow, *Nat. History*, **73**, 14–19.

Dickson, D., 1978, Brazil learns its ecological lesson–the hard way, *Nature*, London, **275**, 684–685.

Eden, M. J., 1979, Ecology and land development: the case of Amazonian rainforest, *Trans. Inst. British Geographers*, New Series, **3**, 444–463.

Eidt, R. C., 1968, The climatology of South America, In *Biogeography and Ecology in South America*, Monographiae Biologicae (Eds. E. J. Fittkau *et al.*), **18**, 54–81.

Escobar, A., and Medina, E., 1977, Estudio de las sabanas inundables de Paspalum fasciculatum, In *Resumenes recibido para IV Simp. internacional de ecologia tropical* (Ed. H. Wolda), Panama 1977, pp. 36–37.

Falesi, I. C., 1976, Ecosistema de pastagem cultivada na Amazonia bra, *Bolm téc EMBRAPA*, Belem, 1.

FAO, 1973, Evaluación y manejo de suelos en la región amazónica, *Boln latinoamer. fomento tierras y aquas*, **5**.

Fearnside, P. M. 1979, Cattle yield prediction for the Trans-amazon highway ot Brazil, *Interciencia*, **4**, 220–225.

Fittkau, E. J., 1970a, The role of caimans in the nutrient regime in the mouth-lakes of Amazon affluents, *Jl. Ecol.*, **58**, pp. 4.

Fittkau, E. J., 1970b, Role of caimans in the nutrient regime of mouth-lakes of Amazon affluents (an hypothesis), *Biotropica*, **2**, 138–142.

Fittkau, E. J., 1971, Ökologische Gliederung des Amazonasgebietes auf geochemischer Grundlage, *Münster. Forsch. Geol. Paläontol.*, **20/21**, 35–50.

Fittkau, E. J., 1973, Crocodiles and the nutrient metabolism of Amazonian waters, *Amazoniana*, **4**, 103–133.

Fittkau, E. J., 1974, 'Zur ökologischen Gliederung Amazoniens. I. Die erdgeschichtliche Entwicklung Amazoniens', *Amazoniana*, **5**, 77–134.

Fittkau, E, J., Irmler, U., Junk, W, J., Reiss, F., and Schmidt, G. W., 1975a, Productivity, biomass, and population dynamics in Amazonian water bodies, In *Tropical Ecological Systems*. Trends in terrestrial and aquatic research (Eds. F. B. Golley and E. Medina), pp. 289–311, Springer, New York–Berlin.

Fittkau, E. J., Junk, W. J., Klinge, H., and Sioli, H., 1975b, Substrate and vegetation in the Amazon region, *Ber. internat. Symp. Internat. Soc. Vegetation Sci.*, 75-90.

Fittkau, E. J., and Klinge, H., 1973, On biomass and trophic structure of the Central Amazonian rain forest ecosystem, *Biotropica*, **5**, 2-14.

Flores, P., Alvarado, S. A., and Bornemisza, E., 1978, Caracterización y clasificación de algunos suelos del bosque amazónico peruano, Iquitos, *Turrialba*, **28**, 99-103.

Franken, M., Irmler, U., and Klinge, H. (in press), Litterfall in the terra firme, riverine and inundation forests of Central Amazonia, *Trop. Ecol.*

Fränzle, O., 1976, Der Wasserhaushalt des amazonischen Regenwaldes und seine Besinflussung durch den Menschen, *Amazoniana*, **6**, 21-46.

Furch, K., 1976, Haupt- und Spurenmetallgenhalte zentralamazonischer Gewässertypen, *Biogeographica*, **7**, 27-43.

Furch, K., 1978, Limnochemische Untersuchungen an amazonischen Gewässwern. *Verhand. soc. Inter. Limnol.*, Karlsruhe 1978, 20-21.

Furch, K., and Junk, W. J., 1979, Water chemistry and macrophytes of creeks and rivers in southern Amazonia and the Central Brazilian Shield, *Paper presented to V. internat. Symp. Trop. Ecol.*, Kuala Lumpur 1979.

Furch, K., and Klinge, H., 1978, Towards a regional characterization of the biogeochemistry of alkali- and alkali-earth metals in northern South America, *Acta cient. venez.*, **29**, 434-444.

Geisler, R., Knöppel, H. -A., and Sioli, H., 1971, Ökologie der Sübwasserfische Amazoniens. Stand und Zukunftsaufgaben der forschung, *Naturwissenschaften*, **58**, 303-311.

Gessner, F., 1958, Igapo, eine Resie zu den Überschwemmungswäldern des Amazonas, *Orion*, **8**, 603-611.

Gessner, F., 1968, Zur ökologischen Problematik der Überschwemmungswälder des Amazonas, *Internat. Rev. ges. Hydrobiol.*, **53**, 525-547.

Gibbs, R. J., 1967, The geochemistry of the Amazon river system. Part I. The factors that control the salinity and the composition and concentrations of the suspended solids, *Bull. geol. Soc. Amer.*, **78**, 1203-1232.

Golley, F. B., Yantko, J., Richardson, R., and Klinge, H. (in press), Biogeochemistry of tropical forests. 1. The frequency distribution and mean concentration of selected elements in a forest near Manaus, *Trop. Ecol.*

Gomez-Pompa, A., 1977, Introduction to session I. In *Resumenes recibido para IV Simp. internacional de ecologia tropical* (Ed. H. Wolda), Panama 1977, p. 196.

Gonzales, J. E., and Escobar, A., 1977, Adaptación a las condiciones de inundación, productividad y valor nutritivo de las gramineas de la sabana inundable, In *Resumenes recibido para IV Simp. internacional de ecologia tropical* (Ed. H. Wolda), Panama 1977, pp. 43-44.

Goodland, R. J., and Irwin, H. S., 1975, *Amazon jungle: Green hell to red desert?*, Elsevier, Amsterdam, pp. 155.

Greenland, D. J., 1975, Bringing the Green revolutiom to the shifting cultivator, *Science*, **190**, 841-844.

Herrera, R., 1979a, *Unpubl. thesis*, University of Reading.

Herrera, R., 1979b, Nutrient distribution and cycling in Amazon Caatinga forest under groundwater spodosols in southern Venezuela, *Paper presented to Internat. Symp. Trop. Ecol.*, Kuala Lumpur 1979.

Herrera, R., Jordan, C., Klinge, H., and Medina, E, 1978a, Amazon ecosystems. Their structure and functioning with particular emphasis on nutrients, *Interciencia*, **3**, 223-232.

Herrera, R., Merida, T., Stark, N., and Jordan, C. F., 1978b, Direct phosphorus transfer from leaf litter to roots, *Naturwissenschaften*, **65**, 208–209.

Hilbert, P. O., 1968, Archäologische Untersuchungen am mittleren Amazonas, *Marburger Stud. Völkerkunde*, **1**, 1–337.

Howard-Williams, C., and Junk, W. J., 1977, The chemical composition of Central Amazonian aquatic macrophytes with special reference to their role in the ecosystem, *Arch. Hydrobiol.*, **79**, 446–464.

Hueck, K., 1966, *Die Wälder Südamerikas*, Fischer, Stuttgart, pp. 422.

Irion, G., 1976, Mineralogisch-geochemische Untersuchungen an der pelithischen Fraktion amazonischer Oberböden und Sedimente, *Biogeographica*, **7**, 7–25.

Irion, G., 1978, Soil infertility in the Amazonian rain forest, *Naturwissenschaften*, **65**, 515–519.

Irion, G., 1979, Jung-tertiär und Quartär im Tiefland Amazoniens, *Natur und Museum*, **109**, 120–127.

Irmler, U., 1976, An Essung von Eupera simoni Jousseaume (Bivaivia, Sphaeriidae) an den zentralamazonischen Ubergebiet.

Irmler, U., 1977, Inundation-forest types in the vicinity of Manaus, *Biogeographica*, **8**, 17–29.

Irmler, U., 1979, Uberlebensstrategien in dem saisonal überschwemmten Lebensraum 'zentralamazonischer Überschwemmungswald', Paper presented to Symposium *Charakteristische Faunen- und Floren- Elemente des Amazonasgebietes*, Vienna, November 1979.

IUCN, 1975, The use of ecological guidelines for development in the American tropics, *IUCN Publ. N.S.*, **31**, pp. 249.

Jahoda, J. C., and O'Hearn, D. L., 1975, The reluctant Amazon basin, *Environment*, **17**, 16–30.

Jordan, C. F., 1978, *Stem Flow and Nutrient Transfer in a Tropical Rain Forest*, Oikos.

Jordan, C. F., 1979, Nutrient loss from agro-ecosystems in the Amazon basin, In *Abstracts, V internat. Symp. Trop. Ecol.*, (Ed. J. I. Furtado), Kuala Lumpur, 1979, 88.

Jordan. C. F., and Herrera, R., (1979, Tropical rainforest: Are nutrients really critical?, *Amer. Naturalist*, (in press).

Jordan, C. F., Golley, F., Hall, J., and Hall, J., 1979a, Nutrient scavenging of rainfall by the canopy of an Amazonian rain forest, *Biotrophica*, (in press).

Jordan, C. F., Todd, R. L., and Escalante, G., 1979b, Nitrogen conservation in a tropical rain forest, *Oecologia*, **39**, 123–128.

Junk, W. J., 1970, Investigatiom on the ecology and production-biology of the 'floating meadows' (Paspalo-Echinochloetum) on the middle Amazon, *Amazoniana*, **2**, 449–495.

Junk, W. J., 1973, Investigation on the ecology and production-biology of the 'floating meadows' (Paspalo-Echinochloetum) on the middle Amazon, II, *Amazoniana*, **4**, 449–495.

Junk, W. J., 1975, Aquatic wildlife and fisheries, *IUCN Publ. N.S.*, **31**, 109–125.

Junk, W. J., 1979, Aquatic macrophytes: Ecology and use in Amazonian agriculture, In Abstracts, *V internat. Symp. Trop. Ecol.*, (Ed. J. I. Furtado), Kuala Lumpur, 1979, 90.

Junk, W. J., and Honda, E. M. S., 1976, A pesca na Amazonia. Aspectos ecologicos a economixos, Anais I encontro nac. limnologia, piscicultura e pesca continental, *Belo Horizonte*, 1975, 211–226.

Kleinpenning, J. M. C., 1977, A critical evaluation of the policy of the Brazilian government for the integration of the Amazon region, *Mimeo, Geogr. Planol. Inst. Katnol.* Universiteit Nijmegen/Netherlands, pp. 44.

Klinge, H., 1965, Podzol soils in the Amazon basin, *J. Soil Sci.*, **16**, 95–103.

Klinge, H., 1966, Podzol jubilee, *Nature*, London, **211**, 557–558.

Klinge, H., 1967, Podzol soils: A source of blackwater rivers in Amazonia, *Atas Simp. Biota Amazonica*, **3**, 117–125.

Klinge, H., 1976a, Nährstoffe, Wasser, und Durchwurzelung von Podsolen und Latosolen unter tropischem Regenwald bei Manaus/Amazonien, *Biogeographica*, **7**, 45–58.

Klinge, H., 1976b, Bilanzierung von Hauptnährstoffen in Ökosystem tropischer Regenwald, *Biogeographica*, **7**, 59–77.

Klinge, H., 1977, Fine litter production and nutrient return to the soil in three natural forest stands of eastern Amazonia, *Geo. Eco. Trop.*, **1**, 159–167.

Klinge, H., and Fittkau, E. J., 1972, Filterfunctionen im Ökosystem des zentralamazonischen Regenwaldes, *Mitt. deutsch. bodenkundl. Ges.*, **16**, 130–135.

Klinge, H., and Herrera, R., 1978, Biomass studies in Amazon Caatinga forest in southern Venezuela, 1., *Trop. Ecol.*, **19**, 93–110.

Klinge, H., Medina, E., and Herrera, R., 1977. Studies on the ecology of Amazon Caatinga forest in southern Venezuela, I, *Acta cient. venez.*, **28**, 270–276.

Klinge, H., and Rodrigues, W. A., 1968, Litter production in an area of Amazonian terra firme forest, 1., *Amazoniana*, **1**, 287–302.

Knöppel, H. -A., 1970, Food of Central Amazonian fishes. Contribution to the nutrient-ecology of Amazonian rain-forest–streams, *Amazoniana*, **2**, 257–352.

Kohlhepp, G., 1976, Stand und Problematik der brasilianischen Entwicklungsplanung, *Amazoniana*, **6**, 87–104.

Lovejoy, Th. E., 1973, The Transamazonica: Highway to extinction?, *Frontiers*, 1973, pp. 6.

Maas, A., 1969, Entwicklung und Perspektiven der wirtschaftlichen Erschließung des tropischen Waldlandes von Peru, unter besonderer Berücksichtigung der verkehrsgeographischen Problematik, *Tübinger Geogr. Stud.*, **31**, pp. 262.

Mägdefrau, K., and Wutz, A., 1961. Leichthölzer und Tonnenstämme in Schwarzwassergebieten und Dornbuschwäldern des tropischen Südamerika, *Forstwiss. Centralbl.*, 1961, 17–28.

Meade, R. H., Nordin, C. F., Curtis, W. F., Rodrigues, F. M. C., Vale, C. M. do, and Edmond, J. M., 1979, Sediment loads in the Amazon river, *Nature*, London, **278**, 161–163.

Medina, E., Herrera, R., Jordan, C., and Klinge, H., 1977, The Amazon project of the Venezuelan Institute for Scientific Research, *Nature and Resources*, **13**, 4–6.

Medina, E., Sobrado, M., and Herrera, R., 1978, Significance of leaf orientation for leaf temperature in an Amazonian sclerophyll vegetation, *Radiation and Environmental Biophysics*, **15**, 131–140.

Meggers, B., 1971, *Amazonia. Man and Culture in a Counterfeit Paradise*, Aldine/Atherton, Chicago–New York.

Meyback, M., 1977, Dissolved and suspended matter carried by rivers: composition, time and space variations, and world balance. In H. L. Golterman (Ed.), *Interactions Between Sediments and Fresh Water*, 25–32, Dr. W. Junk, B. V. Publishers, The Hague, Wageningen.

Noda, H., Junk, W. J., and v.d. Pahlen, A., 1978, Emprego de macrofitas aquatics ('matupa') como fonte de materia organica na cultura de feijao de asa (*Psophocarpus tetragonolobus*) em Manaus. *Acta Amazonica*, **8**, 107–109.

Nogueira-Neto, P., and de Melo Carvalho, J. C., 1979, A programme of ecological stations for Brazil, *Environmental Conservation*, **6**, 95–194.

Nye, P. H., and Greenland, D. J., 1960, The soil under shifting cultivation, *Techn. Comm. 51*, Commonwealth Bureau of Soils, Harpenden, England, 1–156.

Ojasti, J., 1973, Estudio biologico del chiguire o capibara, *Fondo Nacional de Investigaciones Agropecuarias, Caracas*, pp. 275.

Ojasti, J., 1979, Ecology of capybara raising on inundated savannas of Venezuela, In abstracts, *V Internat. Symp. Trop. Ecol.*, (Ed. J. I. Furtado), Kuala Lumpur, 1979, 137.

Oltman, R. E. (without year), Reconnaissance investigations of the discharge and water quality of the Amazon, Stencilled report.

Oltman, R. E., O'R. Sternberg, H., Ames, F. C., and Davis, L. C., 1964, Amazon river investigations. Reconnaissance measurements of July 1963, *Geol. Survey Circular*, **486**, 1–15.

O'Reilly Sternberg, H., 1975, The Amazon river of Brazil, *Geogr. Zeitschr.*, Beiheft, pp. 74.

Pacheco, J., Bulla, L., Preyra, E., Zoppy, E., Ramos, S., and Villarroel, G., 1977, Caracteristicas de la sabana tropical inundable en Venezuela, In *Resumenes recibido para IV Simp. internacional de ecologia tropical* (Ed. H. Wolda), Panama 1977, pp. 91–92.

Paolini, J., 1979, Huminstoffsystem in der Caatinga Amazonica bei San Carlos de Rio Negro, Venezuela, *Amazoniana*, **6**, 569–582.

Panero, R., 1967, *A South American 'Great Lakes' System*, HI-788/3-RR. Hudson Institute, Croton-on-Hudson, N.Y.

Panero. R., 1969, A dam across the Amazon, *Sci. Jb. 1969*, 59–60.

Petrere, J. M., 1979, Fisheries and fishing effort in the Amazonas State, Brazil, In Abstracts, *V internat. Symp. Trop. Ecol.*, (Ed. J. I. Furtado), Kuala Lumpur, 1979, 149–150.

Pires, J. M., 1973, *Tipos de Vegetacao de Amazonia*, Publ. avulsas, Museu Goeldo, Belém, **20**, 179–202.

Prance, G. T., 1978, The origin and evolution of the Amazon flora, *Interciencia*, **3**, 207–222.

Prance, G. T., 1979, Notes on the vegetation of Amazonia. III. The terminology of Amazonian forest types subject to inundation, *Brittonia*, **31**, 26–38.

Russell, J. A., 1942, Fordlandia and Belterra, rubber plantations on the Tapajos river, Brazil, *Economic Geography*, **18**, 125–145.

Rzóska, J., 1978, On the nature of rivers with case studies of Nile, Zaire, and Amazon, *Junk*, The Hague–Boston–London.

Salati, E., Marqués, J., and Molion, L. C. B., 1978, Origem e distribucao das chuvas na Amazonia, *Interciencia*, **3**, 200–206.

Sanchez, P. A., and Nureno, M. A., 1972, Upland rice improvement under shifting cultivation systems in the Amazon basin of Peru. *Techn. Bull. North Carolina Agric. Experim. Stn*, **210**, 1–20.

Schmidt, G. W., 1972, Amounts of suspended solids and dissolved substances in the middle reaches of the Amazon over the course of one year (August 1969–July 1970), *Amazonia*, **3**, 208–223.

Schubart, H. O. R., 1978, Criterios ecologicos para o desenvolvimento agricola das terras-firmes da Amazonia, *Acta Amazonica*, **7**, 559–567.

Schwerdtfeger, W., 1976, Climate of Central and South America, *World Survey in Climatology*, **12**, Elsevier, Amsterdam, Oxford, New York.

Sioli, H., 1957, Sedimentation in Amazonasgebiet, *Geol. Rundschau*, **45**, 608–633.

Sioli, H., 1964, General features of the limnology of Amazonia, *Verhandl. Internat. Verein. Limnol.*, **15**, 1053–1058.

Sioli, H., 1967, Studies in Amazonian waters, *Ates Simposio Biota Amazónica*, Belém 1964, **3**, 9–5.

Sioli, H., 1968a, Hydrochemistry and geology in the Brazilian Amazon region, *Amazoniana*, **1**, 267–277.

Sioli, H., 1968b, Principal biotopes of primary production in the waters of Amazonia, *Proc. Symposium Recent Advances Trop. Ecol.*, Varanasi 1968, 591–600.

Sioli, H., 1973, Recent human activities in the Brazilian Amazon region and their ecological effects, In *Tropical Forest Ecosystems in Africa and South America: A Comparative review*, (Eds. B. Meggers *et al.*), Washington, 321–334.

Sioli, H., 1975, Tropical rivers as expressions of their terrestrial environments, In *Tropical Ecological Systems: Trends in Terrestrial and Aquatic Research*, (Eds. F. B. Golley and E. Medina), Springer, New York–Berlin, 275–288.

Sioli, H., 1976, A limnologia na regiao amazonica brasileria. Anais I Encontro nacional limnologia, piscicultura e pesca continental, *Belo Horizonte 1975*, 153–169.

Sioli, H., 1977a, Amazonasgebiet–Zerstörung des ökologischen Gleichgewichtes?, *Geol. Rundschau*, **66**, 782–795.

Sioli, H., 1977b, Amazonien: Der Welt größter Wald in Gefahrl, *Umschau*, **77**, 147–150.

Sioli, H., 1979, Prospective effects of actual development schemes on the ecology of the Amazon basin, In Abstracts, *V internat. Symp. trop. Ecol.*, (Ed. J. I. Furtado), Kuala Lumpur, 1979.

Sioli, H., and Klinge, H., 1961, Über Gewässer und Böden des brazilianischen Amazonasgebietes, *Erde*, Berlin, 1961, 205–219.

Sioli, H., Schwabe, G. H., and Klinge, H., 1969, Limnological outlook on landscape ecology in Latin America, *Trop. Ecol.*, **10**, 72–82.

Smith, N. J. H., 1973, House sparrows (*Passer domesticus*) in the Amazon, *Condor*, **75**, 242–243.

Smith, N. J. H., 1974a, Destructive exploitation of the South American river turtle, *Yearbook Assoc. Pac. Coast Geographers*, **36**, 85–102.

Smith, N. J. H., 1974b, Agouti and babassu, *Oryx*, **12**, 581–582.

Smith, N. J. H., 1976a, Brazil's transamazon highway settlement scheme: Agrovilas, agropoli, and ruropoli, *Proc. Assoc. Amer. Geographers*, **8**, 129–152.

Smith, N. J. H., 1976b, Utilization of game along Brazil's transamazon highway, *Acta Amazonica*, **6**, 455–466.

Smith, N. J. H., 1976c, Spotted cats and the Amazon skin trade, *Oryx*, **13**, 362–371.

Smith, N. J. H., 1978a, Agricultural productivity along Brazil's transamazon highway, *Agro-Ecosystems*, **4**, 415–432.

Smith, N. J. H., 1978b, Human exploitation of terra firme fauna in Amazonia, *Ciencia e Cultura*, **30**, 17–23.

Sombroek, W., 1966, Amazon soils, *Agric. Res. Rep. PUDOC*, **672**, pp. 303.

Sombroek, W., 1979, Soils of the Amazon region. Paper presented to *Joint Meeting British Soc. Soil Sci. and Royal Congr. Soil.*

Stern, K., 1966, Über Weiß-Wasser und Schwarzwasser in den Tropen, *Ars medici*, **56**, 395–409.

Stern, K., Odenal. J., and Portillo, G., 1970, Über Schwarzwasser, *Ars medici*, **60**, 140–142.

Sutmoeller, P., Vahia de Abreu, A., v.d. Drift, J., and Sombroek, W. G., 1966, Mineral imbalances in cattle in the Amazon valley, *Commun. Depart. Agric. Res.*, Koninkl. Inst. Tropen, 53.

Tricart, J. L. F., 1978, Ecologie e développment: l'exemple amazonien, *Annales Géogr.*, **481**, 257–293.

Unesco, 1974, *MAB Report*, Series 22.

US Army Corps of Engineers, 1943, Report on Orinoco–Casiquiare–Negro-Waterway, Venezuela–Colombia–Brazil, Prepared for the Correlator of international Amazon Affairs, Washington, 4 vols.

US Department of Agriculture, 1967, *Supplement to Soil Classification System*, Washington, pp. 265.

Valverde, O., 1971, Shifting cultivation in Brazil—ideas on a new land policy, *Geogr. Arb.*, Heidelberg, **34**, 1–17.

Vos, A. de, 1979, The crocodile industry in Papua New Guinea: An example of eco-development, p. 40.

Van Wambeke, A., 1978, Properties and potential of soils in the Amazon basin, *Interciencia*, **3**, 233–242.

Walter, H., Harnickell, E., and Mueller-Dombois, D., 1977, Klimadiagramm-Karten der einzelnen Kontinente und die ökologische Klimagliederung der Erde, Map 2 Southamerica, Fischer, Stuttgart.

Went, F. W., and Stark, N., 1968, Mycorrhiza, *Bio. Science,* **18**, 1035–1039.

Wilhelmy, H., 1970, Amazonien als Lebens- und Wirtschaftsraum, *Staden-Jahrbuch*, **18**, 9–31.

Williams, P. M., 1968, Organic and inorganic constituents of the Amazon river, *Nature*, London, **218**, 937–938.

Zeichmann, W., 1976, Huminstoffe in südamerikanischen Flußsystemen, *Amazoniana*, **6**, 135–144.

Tropical Agricultural Hydrology
Edited by R. Lal and E. W. Russell
© 1981, John Wiley & Sons Ltd.

2.2

Seasonal Variations in the Hydrology of a Small Forested Catchment near Manaus, Amazonas, and the Implications for its Management

S. NORTCLIFF AND J. B. THORNES

2.2.1. INTRODUCTION

Despite the large body of literature on the effects of forest clearance on hydrology and soil conditions, the nature and timing of these effects is still the subject of much debate (see Rodda, 1976). This is partly because the term 'forest' refers to greatly differing vegetation types in different environments and because the word 'clearance' relates to many different types of practice. At one extreme is the 'chaparall' forest in the southwestern United States, and the other the dense secondary woodland of the cut-over tropical rainforest. In one case 'clearance' may refer to complete obliteration of the forest cover and its replacement with monoculture; in another the forest is allowed to recover quite rapidly as in shifting cultivation.

Another reason for the confusion arises from a difference in objectives of those who have examined the problem. Particular concern for flood effects has led to an emphasis on the impact of clearance on extreme discharge events often some distance down the main channel. In the soil erosion problem the main issue has been the changing energy of falling rain, while the pedologists have stressed the effects of interruptions to the nutrient cycle. The third reason for apparent disagreement is that the areas being affected must inevitably vary tremendously according to the geological, pedological, and climatic conditions. The wild speculations of some conservationists about the impact of clearance are not justified simply because the process of clearance, the environment in and on which it is carried out, and the nature of the response are almost infinitely varied and not necessarily uniformly bad.

Some of the most widely reported speculation has come from the forested tropical areas, especially Amazonas. This is not difficult to understand. As one of the largest remaining tropical rainforest stands in the world, and one which is under-going an intensive programme of development at the present, it is important that some appraisal should be made of the effects of such change. At the same time our knowledge of the dynamics of the vegetation cover and its interactions with the hydrological and pedological environment of this area are still very modest, despite the work of the Instituto Nacional da Pesquisas da Amazonia in Brazil (see for example Klinge, 1976), or the San Carlos Project of MAB (see for example Brünig *et al.*, 1979). The hydrology of this environment and the associated soil conditions are often assumed to be particularly simple. Large inputs of rainfall and uniformly high levels of soil moisture coupled with a closed nutrient cycle are supposed to lead naturally, in the event of clearance, to further impoverishment, extensive soil erosion and the eventual production of a 'red-desert'. Whilst in comparison with many areas of the world, where rainfall totals are generally smaller and intermittent and the range of soil moisture conditions are considerable, this may appear to be the case, in fact the seasonal climatic contrasts and the consequent hydrological contrasts are quite marked. This fact has implications for nutrient cycling, for pedogenesis, and, we believe, for the likely effects of forest clearance and agri-cultural practice.

This paper seeks to examine spatial and temporal inputs, total soil hydraulic head and water fluxes within different seasons from a series of sites across a single hillslope in Reserva Ducke near Manaus. Although the results strictly hold true only for this very limited domain, they may have wider implications, notably in under-standing the seasonal hydrological response of tropical forest environments.

2.2.2. ENVIRONMENTAL CONDITIONS

2.2.2.1. Site Characteristics

The field site chosen for this study is the catchment of the Barro Branco, a small forested catchment of approximately 1.5 km^2 within the National Forest Reserve, Reserva Ducke, 26 km north east of Manaus on the road to Itacoatiara (Figure 2.4). The area is underlain by the tertiary deposits of the Barreiras Series, which are sedimentary deposits of a wide textural range extensive in the basins of the Solimoes and Negro rivers to the west of Manaus. The catchment has three broad vegetation types described by Brinkman and Santos (1973) as (1) riverine forest, (2) carrasco forest, and (3) terra firme rain forest. The experimental site was located in the second of these types which is intermediate between the other two. It has canopy heights from 22–32 m, is quite heterogeneous, with an understorey of numerous seedlings and saplings and has some herbaceous plants and palms.

The soils found at the sites were investigated by a series of soil pits (shown in Figure 2.4) and correspond broadly with those described by Brinkmann and Santos (1973) as yellow latosols. In the nomenclature of Soil Taxonomy (USDA, 1975) they are probably plinthic haplorthox, being composed predominantly of

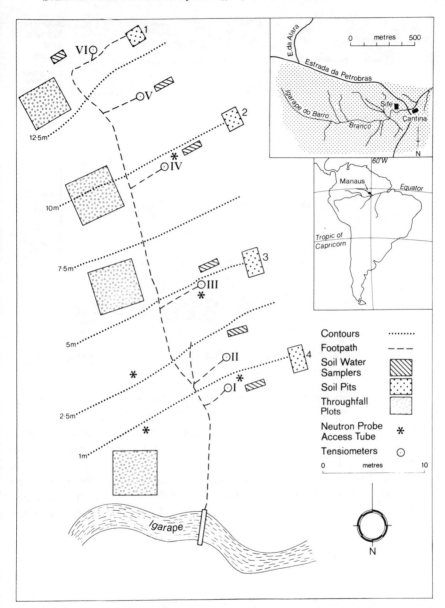

Figure 2.4 Map to show instrumentation of experimental slope. The inset shows the general location and the location within the reserve of the experimental site

kaolinitic clays and quartz, with few weatherable minerals and with hardened plinthitic nodules occurring at depth within the profile. Information on the profile is summarized in Figure 2.5. The dominant structural features are the strong root

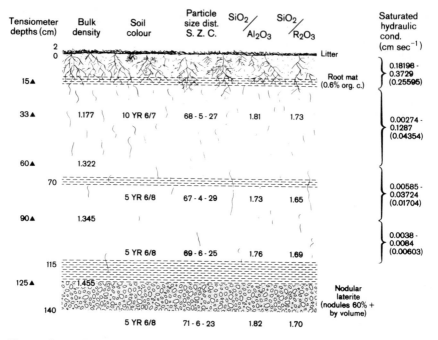

Figure 2.5 Schematic soil profile showing structural and hydrological parameters for the principal horizons

mat down to 15 cm and the persistent nodular plinthite layer found between 115 and 140 cm and running more or less parallel to the slope. The soils generally have sandy clay loam textures throughout their profiles but are often described as 'pseudosands' because of the presence of very stable microaggregates. The microaggregates proved difficult to disperse during particle size analysis and may account for the apparently very high porosity. Roots are present throughout the entire profile, though they are usually of very small size below about 20 cm. The average values of saturated hydraulic conductivity reveal quite a strong gradient between the uppermost and lowest parts of the profile, which is present throughout. At the very foot of the hillslope, at site I, white sands underlay a thin layer of downwashed latosolic material, but whether these constitute 'bedrock' or an alluvial fill in the valley bottom is not known. The general area is one of relatively flat interfluves bounded by slopes with average slope angles of about 20°, most of the slopes being relatively short (40–60 m length). The floodplains are generally 10–15 m wide in this part of the drainage system and 1–2 m above a sandy channel bed.

2.2.2.2. Climate

The temperature and precipitation regimes at Reserva Ducke have been the subject of an intensive study by Ribeiro (1976). In the period 1965–73 the range of temperature was from 14.3 to 37 °C, the average daily maximum being about 33 °C. During this period the average annual rainfall was about 2500 mm, a precipitation intensity of 60 mm hr^{-1} appears to occur with an annual return period, the wettest months are generally March (1965-73 average 350 mm) and April (1965-73 average 290 mm), the driest September (1965-73 average 70 mm). The monthly averages for 1910-1975 are plotted in Figure 2.6(a) for rainfall and temperature at Manaus from data of the Departamento Nacional de Meterologia. The strong contrast in precipitation between March, April, and May and July, August, and September was the basis of our choice of field observations. Although temperature begins to rise again in September, thermal conditions were fairly uniform throughout the experimental period. In any particular year the contrasts in precipitation may be much stronger than that shown in Figure 2.6(a). Rainfall for 1937 is shown in Figure 2.6(b) together with the annual harmonic estimated by Fourier analysis from the entire data set of 67 years. Analysis of the long term variations in precipitation does not show any significant trends in monthly rainfall totals but there has been an increase in the monthly maximum precipitations. This tends to concentrate the seasonal imbalance somewhat and the dry spells in the dry season are becoming longer.

The periods of our investigation were for March, April, and May, 1977 (centred on April) and for July, August, and September, 1978 (centred on August). Rainfall for the two periods is shown in Figure 2.7. The first period was characterized by regular and frequent rain with five storms of greater than 20 mm. The second period had little rain, with long gaps between the falls and only three storms of greater than 20 mm. In both periods the storms tended to be relatively concentrated occurring mainly in the early morning hours. Figures 2.8(a) and (b) show typical stage hydrographs for the Barro Branco during wet and dry seasons respectively. The upper diagram shows a typically very rapid response in the wet season, the short time to rise being a result of the large saturated area at the slope foot (Nortcliff and Thornes, 1978). In the dry season the hydrograph is less rapid in response and may, in very dry periods, be dominated by the diurnal evaporative cycle as shown in the lower diagram.

2.2.3. EXPERIMENTAL DESIGN

2.2.3.1. Observations

The overall experimental design covered flood plain, hillslope, and crest sections of the valley, but in this paper results are presented only for the slope hydrological analysis. Details of the entire experiment are to be found in Nortcliff and Thornes (1978). The hillslope was divided into six segments of approximately equal length (Figure 2.4) with instruments located near the middle of each segment. This design was based on the need to estimate vertical and downslope variations in water flux and is similar to those used in hillslope hydrology (e.g. Atkinson, 1978). At each

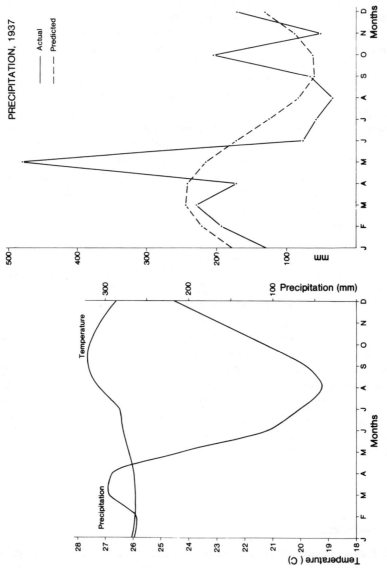

Figure 2.6 (a) Mean monthly temperature and precipitation at Manas for the period 1910–1970. (b) 1937 monthly rainfall and predicted rainfall (1910–1970)

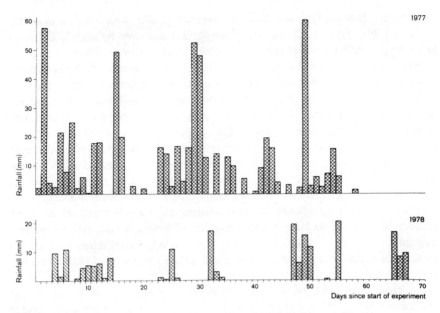

Figure 2.7 Rainfall totals (a) March, April, and May 1977, and (b) July, August, and September 1978

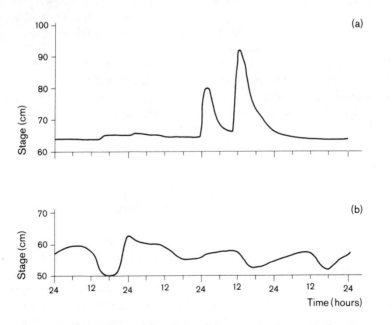

Figure 2.8 Typical wet (a) and dry (b) season hydrographs for the Barro Branco

site standard tensiometers, designed after Webster (1966), were installed to depths of 15, 33, 60, 90, and 125 cm below the surface and close by aluminium access tubes were installed to depths of two metres to enable observations to be made with neutron moisture and density probes. The neutron probe was calibrated against gravimetrically determined soil moisture values and standardized against the rate-scaler. It was possible to relate the neutron probe and soil tension data; tension (atmospheres)–soil moisture curves were drawn. In most cases a log/log regression provides a reasonably good fit to the scatter of points, and it is only under very dry conditions that underestimation of moisture and hence the unsaturated conductivity at that value, tends to occur. At the wet end the analysis is constrained so that unsaturated hydraulic conductivity calculated from the curve cannot exceed the saturated value. A problem which arises is that the neutron probe measures moisture over a relatively large volume, whereas tensions are measured over a very small volume. However, in so far as we are using the relationship to estimate unsaturated hydraulic conductivity which is spatially quite varied, a bulk rather than a point measure of moisture is almost certainly an advantage when applied to the computation of fluxes.

As mentioned above, the sampling design involved regular tensiometer readings throughout the working day and at intervals of 15 minutes or less during and in the period immediately following storm events. Such a sampling scheme invariably leads to a considerable bias (towards the wetter periods) in the data, but since this is the time when soil moisture changes occur most rapidly, this bias was justifiable. Observations of soil moisture using the neutron probe were taken only during the second period, and because of the time taken to complete a full set of readings for all sites, at a maximum frequency of twice per day.

Measurements of saturated hydraulic conductivity were obtained from undisturbed cores using a falling head permeameter. A large number of replicate samples were obtained and the mean values later used in the determination of the unsaturated conductivities. Estimates were also made from wetting front movements and from infiltration tests and all three sets show good agreement.

2.2.3.2. Calculations

Unsaturated conductivity is a function of soil moisture and we have used a method proposed by Campbell (1974) to derive estimates of unsaturated hydraulic conductivity.

At this point, for each sampled time the data available for each site and depth (a 6 × 5 matrix) includes values for soil moisture, hydraulic conductivities at that moisture, matric tensions, heights relative to the deepest tensionmeter at the lowest site and a vector of slopes between sites. The notation for this system is given in Figure 2.9. In referring to hydraulic head we have used the notation I to VI for sites, and A (15 cm) to E (125 cm) for depths. In referring to fluxes in what follows, the notation is of fluxes away from a point. Thus fluxes at site 1 depth 1 (point (1,1))

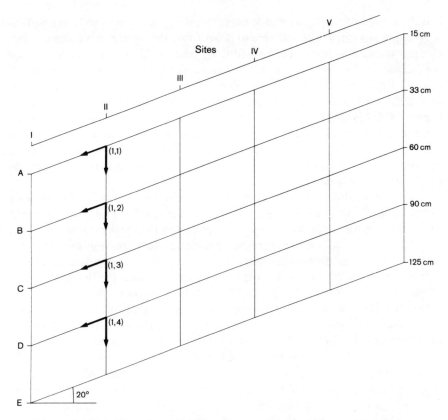

Figure 2.9 Notation of matrices for slope and vertical flux computations

on Figure 2.9 are the fluxes from IIA–IIB (vertical flux) and from IIA–IA (down-slope component of flux), i.e. away from IIA. Fluxes about the highest point are therefore annotated (5, 1). The data for matric tensions and heights are combined to give a matrix of hydraulic heads (Ψ) for each site and depth. The matrices of hydraulic heads, hydraulic conductivities and the slope vector are then used to compute the soil water fluxes using Darcy's Law.

2.2.4. RESULTS

2.2.4.1. Tension Variations

In so far as the total hydraulic head combines both gravitational head due to height on the slope, and matric pressure resulting from the sunction or pressure applied by the soil, i.e.

$$\Psi = x + \phi \tag{2.1}$$

it is the second of these, the matric potential, which varies seasonally in a particularly strong fashion. If the soil were to be saturated throughout, $\phi = 0$ and the lines of equal pressure with respect to an arbitrary base would run horizontally through the hillside. In the first season this was more or less the case as is illustrated by Figure 2.10(a). In this diagram total hydraulic head has been computed with respect to a depth of 125 cm from the surface, i.e. at the lowest tensionmeter. Thus in Figure 2.10(a) in all but one case there is an upward increase in head, more or less equal to the increase in depth (the line $\phi = 0$ shown in the figure). This is a sign of near saturation with poorly developed soil suction. In these circumstances the flux will be vertical and equal to the saturated hydraulic conductivity. Site 1 on the floodplain is the anomalous case.

A strongly contrasted case is shown for the dry season, Figure 2.10(b). Here the hydraulic head values are strongly negative even in the lowest layer of the soil, whilst they change suddenly to positive pressures in the uppermost horizon. This reflects the downward movement of water during a heavy storm in the dry season of 1978. Such a situation leads to very high fluxes, much higher in fact, at least for a short time, than the saturated hydraulic conductivities since the flux is now being driven by a very strong gradient in the uppermost horizons. Still, however, the lowest site (I) is quite wet, the pressures remaining positive above the lowest tensiometer. The generalized head distribution in this case is identical to that developed hypothetically by Weyman (1973) and confirms his model.

Although the tension remained consistently low throughout the wet period, in the dry period of July–August–September 1978 they rose consistently, especially at depth (Figure 2.10(c)). Near the surface the tensions were more immediately responsive to rainfall inputs resulting in the strong gradients illustrated in Figure 2.10(b). Tensions reached values higher than had been expected in such an overall wet environment. The rather rapid fluctuations observed in the near surface horizons obscured any results that might have risen from evaporative and transpirational fluxes, but these variations must be superimposed on the pattern shown in Figure 2.10(c), if the variations in stream level are to be accounted for. It is interesting to note, by way of comparison, that soil tensions in sandy soil of southeastern England (Lai, 1976) rarely reach the bubbling pressure of the instruments, which occurred towards the end of the dry season in Manaus. All the deeper sites in the experimental plot showed comparable build-up to a maximum at about day 48 (4/9/78), after which a series of storms tended to maintain the values more or less steady until the end of the observation period. We must assume that these high tensions during the whole of this period reflect both direct evaporation from the soil surface, and transpired water from the vegetation, though a direct feedback might occur to reduce transpirational losses, together with losses as deep percolation.

During the dry period the flood plain site (I) remained at much lower tensions than the other sites, the highest value being reached at the shallowest depth (A) on day 44 (1/9/78) when the vertical tension gradient was very steep. The highest values overall were reached in the mid-slope sites (III and IV) where values in excess of 650 cm of water were recorded several times at depths of 90 and 125 cm. During

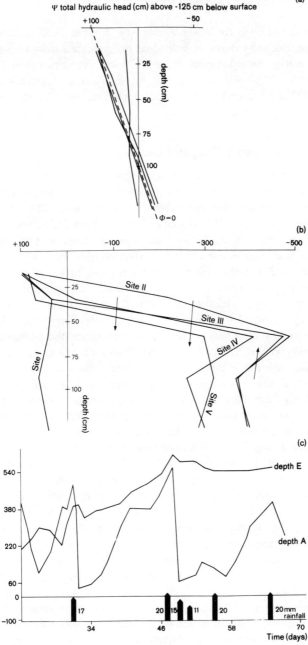

Figure 2.10 (a) Typical hydraulic head distributions in the wet season, ψ with respect to a depth of 125 cm. (b) Dry season hydraulic heads just after the onset of a large storm. (c) Trend in matric tensions at site II at the shallowest and deepest tensiometers over the dry-period observations. The arrows mark heavy rainfalls (in mm)

the wet period, although there was a slight rise in tensions generally towards the end of the period, the effects of individual storms were hardly recorded because tensions on average were between 15 and 125 cm. As storms passed, these were occasionally brought to positive pressures, inevitably at site I, but also at the lower two depths (90 and 125 cm) at site II reflecting the growth of a saturated wedge up the slope.

2.2.4.2. Spatial Variations in Computed Fluxes

Before describing the variations in computed fluxes, we stop to consider briefly the assumptions and other sources of error involved so that the magnitude of the variations can be held in their true perspective. Two main sources of error are involved. The first is the assumption of linear unique functions between unsaturated hydraulic conductivity and moisture and between moisture and tension. In fact in both cases the relationships are known to be hysteretic (Poulavassilis, 1962, 1969). The second rests in the errors associated with the regression procedures used to determine the slope of the log tension–log θ curve. Although the explained variance of these regressions are very high (consistently above 90 per cent), the flux rates computed are particularly sensitive to variations at the extremes. This is easily dealt with at the saturated end by constraining the value not to exceed the saturated conductivity. Occasionally at the 'dry' end the conductivities are *very* low when obtained by regression. Although this is infrequent, there is no simple correction which can be applied. The first source of error, due to hysteresis, will be smaller when considering fluxes arising from a fixed position with respect to the wetting or drying cycle. Fortunately in the drying period, when the hysteresis effect is likely to predominate, we are almost invariably observing the drying cycle, the situation for which most of the curves were developed. There are two other sources of potential error. First there are some locations, notably at the shallowest depths, where the inadequacy of our neutron probe practice made it impossible to obtain adequate calibration curves. In this case we have been obliged to use the regression curve for the 33 cm depths. Finally there are, inevitably, errors in the measurement of standard parameters, such as the saturated soil volume fraction and the saturated conductivities. In particular the variations in saturated conductivities are not systematic and the number of replications per site is low, so that average values for the different depths are adopted. Nonetheless these represent, as far as reasonably possible, the average conditions prevailing in the soils of this environment. The computed fluxes are particularly sensitive to variations in tension which, fortunately, is measured with a relatively low error. In summary the results are probably most sensitive to errors in the saturated hydraulic conductivities.

Assuming a saturated soil and a hydrological gradient of unity, then the vertical flux rate (in cm hr^{-1}) will be equivalent to the saturated hydraulic conductivity and, for flow parallel to the slope, it will be approximately proportional to the sine of slope angle. The corresponding values, based on the saturated conductivity

Table 2.1. Expected average saturated vertical flux components (cm hr^{-1})

	Depths		
A–B	B–C	C–D	D–E
92.14	15.67	6.13	2.17

Table 2.2. Expected average saturated downslope flux components (cm hr^{-1})

	Depths (between adjacent sites)		
A	B	C	D
31.51	5.35	2.09	0.74

Table 2.3. Average computed vertical flux components for the wet period, 1977 (cm hr^{-1})

Site	1 (from A–B) ·	Depth 2 (from B–C)	3 (from C–D)	4 (from D–E)
1	35.56	17.96	0.42	0.52
2	12.07	6.80	4.53	3.77
3	15.35	11.89	3.13	4.62
4	14.46	13.43	3.10	1.32
5	2.08	5.47	4.62	3.06

Table 2.4. Average computed downslope flux components for the wet period, 1977 (cm hr^{-1})

Site	1	Depth 2	3	4
1 (from II–I)	1.37	1.02	0.07	0.14
2 (from III–II)	3.51	2.63	1.66	1.44
3 (from IV–III)	4.68	4.21	1.03	1.63
4 (from V–IV)	5.51	5.51	0.97	0.56
5 (from VI–V)	1.09	2.30	1.17	0.94

values, are shown in Tables 2.1 and 2.2. Table 2.3 gives the mean values for *all* sites for the vertical wet-season (1977) fluxes. The mean slope fluxes are given in Table 2.4. These two tables demonstrate the overall dominance of vertical fluxes throughout the slope at rates reasonably close to the saturated values, except that

Table 2.5. Average computed vertical flux components for the dry period, 1978 (cm hr^{-1})

Site	1 (from A–B)	Depth 2 (from B–C)	3 (from C–D)	4 (from D–E)
1	5.59	186.35	0.012	0.0001
2	18.52	9.54	0.016	−0.0002
3	57.43	0.81	−0.013	0.38
4	16.66	18.66	0.35	−0.0009
5	0.87	6.92	0.41	−0.003

Table 2.6. Average computed downslope flux components for the dry period, 1978 (cm hr^{-1})

Site	1	Depth 2	3	4
1 (from II–I)	3.41	1.06	–	–
2 (from III–II)	1.94	0.24	–	–
3 (from IV–III)	0.82	0.02	–	–
4 (from V–IV)	2.10	1.32	–	–
5 (from VI–V)	0.29	0.71	–	–

at the shallowest depth of the highest site (5,1) which remained relatively dry even during the wet season. Whilst the reason for this is not fully understood it was recognized in the field that this site, which was on the flat crest, had thicker litter and surface organic horizons; this probably increased upper soil storage. The second feature is the reduction in vertical flux rates with depth for the mid-slope (2, 3, and 4) sites, and site 5 excluding the surface layer. At these sites both vertical flux at depth and lateral flux at the surface remain appreciable. At the lowest flux point both vertical and lateral fluxes are offset by the positive pressures which prevailed (Figure 2.10(b)) at the lowest (i.e. flood plain) site. The significant fall in vertical fluxes shown at depths 3 and 4, is a measure of the roughly layered structure of the soil as revealed by Figure 2.5. This layered structure is more evident when tracing the pattern of fluxes following a heavy rainstorm as shown below.

Tables 2.5 and 2.6 show the average computed fluxes for the dry season (downslope fluxes at depths 3 and 4 are insignificant and have not been included in Table 2.6). These tables are rather misleading since they are based on only 57 complete sets of observations (i.e. about 2000 observations) whereas the wet season tables are based on 173 complete sets (i.e. over 6000 observations). Second, as we shall show below, the distributions, especially for the near surface sites, are strongly bimodal. This occurs as a result of very high flux rates immediately following storms, separated by very low rates for the longer periods between. As mentioned earlier, there is also a bias in the sample towards the storm periods. This bimodal distribution is not found in the wet season since the potential gradients are only small then. Third, the very low rates are close to the limits posed by the methods of calculation used. What is evident, however, notwithstanding these caveats, is that the fluxes are much smaller (an order of magnitude at least in the lower horizons) than in the wet season and are somewhat closer to the published values (e.g. Harr, 1977; Dunne, 1978). Another feature is that lateral fluxes are proportionately much smaller, reflecting the overall dominance of the strong vertical gradient set up by evapotranspiration as revealed by the tension data in Figure 2.10(b).

2.2.4.3. Temporal Variations in Computed Fluxes

In Figure 2.11 the vertical flux components for site 3 depths 1 (A–B) and 4 (D–E) and for site 2 depths 1 and 4 are given for approximately twelve hours succeeding the large storm which occurred early on the morning of 6/5/77. The peak of the storm occurred between 0500 and 0600 when 59 mm fell. Light rain continued to fall until about 0900 hours but the period shown was almost entirely a 'drying' curve. Most remarkable is the relatively insignificant change in the vertical flux rates and even a fall in the lowest depth of site 2 later in the day. Basically the flux rates at this period are already close to the saturated value so that the input of even quite a heavy storm failed to raise the flux rates significantly, even though the tensions dropped from 68 cm to 0 for example at site 4. During the same period there was

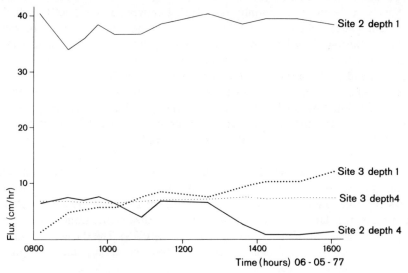

Figure 2.11 Vertical fluxes in the wet season at two sites and depths following a storm

an increase in the downslope flux component, though only by a relatively small amount.

This type of behaviour is likely to produce a relatively flat subsurface hydrograph with a recession limb which is sustained by inflow into the deeper horizons (the type B of Whipkey and Kirkby, 1978). Even in these conditions surface overland flow was only developed across the flood plain, presumably because of the higher gradients on the slopes compared with the flood plain proper (Nortcliff, Thornes, and Waylen, 1979).

The dry season fluxes present much stronger contrasts between storm and non-storm periods than during the wet season. In the dry season after a storm vertical fluxes at the shallowest depths are dominated by the strong pressure gradient between the advancing wetting front and the relatively dry soil beneath. This fact and the bias towards the storms explains some of the obvious anomalies in Tables 2.5 and 2.6. Figure 2.12(a) shows the vertical fluxes for two sites and depths during a dry season storm on 20/8/78. Although this storm was of relatively low intensity, only 17 mm of rain fell between 0850 and 1400 hr, the effects are quite characteristic. The sharp increase in vertical fluxes between levels A and B shown here for sites 2 and 4 was characteristic for all sites. At the second depth (between B and C) the vertical fluxes changed only slightly in the succeeding 40 hours. This indicates both the filling of the upper soil storage deficit coupled with high transpiration rates from the root mat, and the barrier effect of the very low conductivities between levels B and C (depth 2). This effect tended to promote an increase in the downslope component of fluxes at the upslope sites (Figure 2.12(b), site 4, depth 1), but again with little change at depth.

Fluxes at depth are uniformly low, ranging between about 0.0001–0.0025 cm hr^{-1}, and under some conditions turn negative indicating a tendency to vertical upward diffusion. Even here, during the major storms there were small increases in both vertical and slope flux components. As in the wet season fluxes at the base of the slope and in the surface horizons were greater than those upslope. In all cases fluxes from site 2, depth B (1, 2) seemed rather anomalous. This may be a reflection of the very low measured soil moisture saturation value (11 per cent) applied at site II depth B and the value is therefore somewhat suspect. Finally, we note that where the fluxes are low in both seasons, the ratio of the slope to the vertical flux component may increase appreciably as the strong vertical gradients due to wetting front action in the dry season and the overall control under saturated flow in the wet season give way to strong lateral gradients. However, in volumetric terms the total amount of water movement is small and we are able to confirm, from the entire data set, a conclusion that we reached earlier from examination of only two storms (Nortcliff, Thornes, and Waylen, 1979). The fluxes are dominated largely by vertical rather than downslope movement, though the latter may be quite significant at particular times.

To summarize, in the wet season even quite large storms produce little effect on the flux rates which are operating at or close to their saturated values. In the dry

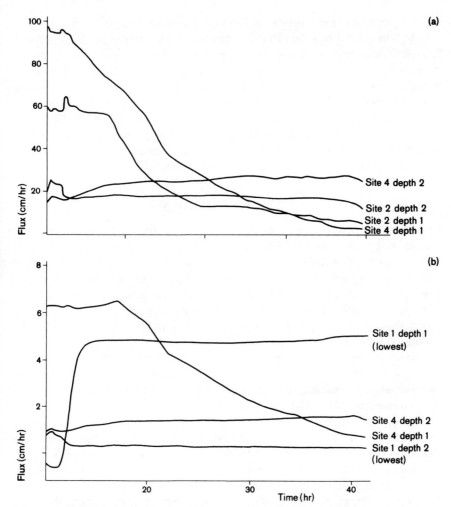

Figure 2.12 (a) Vertical fluxes in the wet season at two sites and depths following a storm. (b) Slope fluxes for dry season, also following a storm

season there are strong contrasts spatially and with depth for both vertical and
downslope components. Flux rates are generally high and sustained during the wet
season, and very low separated by sporadic very high levels in the few hours after
the onset of storms, especially in the uppermost horizons during the dry season. At
no time were intensities high enough to produce overland flow.

The results obtained here are for instantaneous fluxes and are, as we have said,
biased towards storm occasions. In this way there is a high-side bias. Nonetheless
the results are similar in pattern to those obtained by Roose and Godefroy (1977)
for the Orstom-Ivfa station near Abidjan, Ivory Coast. Proportionally, they too
found lateral flow to be small, a significant decrease in drainage rates with depth

under both forest and bananas, and demonstrated by balance computations the very high evapotranspiration rates. The latter is implied in our results through the large difference between upper and lower zone fluxes in the wet season, and the poor response of the stream discharge in the dry season. The overall values for instantaneous fluxes in our work suggest much higher vertical drainage in volumetric terms even given the bias towards storms. This is partly a result of the much greater total rainfall volume (1000 mm more) and the very high relative humidity in the canopy layers. Unfortunately we are not yet in a position to calculate a hydrological balance, but the relative magnitudes of the various components seem likely to be very close to those reported by Roose and Godefroy.

2.2.5. IMPLICATIONS

Because of the relatively low significance of lateral flux components, the direct contribution made from the hillslope to the stream hydrograph is mainly in terms of base flow, particularly since the gradient across the floodplain is very small. Only under very dry conditions is the channel hydrograph dominated by the hillslope components, as indicated by the lagged response of stage to the diurnal cycle of evapotranspiration (Figure 2.8(b)). Under normal conditions the lateral flux augments the slopefoot saturated wedge encouraging its growth back upslope. This can be observed from piezometers at the slope foot. However the dominant and very rapid response in the stream hydrograph comes essentially from saturated overland flow on the floodplain (Figure 2.8(a)). The water 'rises out' of the floodplain and discharges immediately into the channel or, in the higher reaches, flows down the floodplain parallel to the channel.

This dominance of flow generated on the floodplain, whose volume is largely controlled by the floodplain area, leads one to question the assumption that forest clearance would necessarily lead to significant changes in the pattern of floodwater generation. Even if the root mat were to be destroyed the upper horizons of the soil have sufficiently high hydraulic conductivities for vertical flow to dominate in the wet season. During the dry season, even with reduced transpiration, the stream hydrograph is still likely to be dominated by floodplain dynamics. Under these circumstances flood hydrology (as opposed to the total volumetric yield) need not differ substantially from that existing in the area today.

Since the concentrations of cations and anions in the soil water remain low, but effectively the same, in the wet and dry seasons, it follows that the flux of nutrients occurring through solution is much greater in the wetter season than in the drier season. On the other hand, as Klinge (1976) has convincingly shown, the major period of litter production and breakdown occurrs in the dry season, when water circulation is at a minimum. Breakdown occurs throughout the dry season culminating in an almost complete disappearance of fresh litter by September. The litter decomposition processes in the surface layers appear therefore to have little effect on the ionic concentration in the soil solution. In central Amazonas it is therefore difficult to see how deforestation must necessarily further impoverish the inorganic soils which are already extremely poor. On the other hand, given the small catchment hydrology described above, the ionic concentrations of water in small *igarapes* may offer a poor guide to the agricultural potential of the soils. Moreover, the idea of continued and heavy leaching of material throughout the year is at variance with the observations. There are times when the net flux tendency is reversed, and when soil moisture stress will limit the growth of shallow rooting crops.

It is difficult to imagine in tropical areas, how massive clearance would not inevitably lead to extensive splash, rill, and gully erosion. However, the evidence from Amazonas is not at all confirmatory on this point. Apart from road-cut sites, where there has been no attempt at replanting and slopes are extremely steep, there is little evidence of gully erosion. Moreover, even in streams draining cleared lands suspended sediment yields are extraordinarily low. Part of the explanation may lie in the fact that even in very large storms and under rains of very high intensity, the hydraulic conductivities of the latosols, near to saturation level, are quite high in the upper horizons, so that overland flow is quite rare. Unlike semi-arid areas, surface sealing of the soil and very low conductivities together with very high surface gradients in the upper soils appear only during the dry season and never in the floodplain and channel environments. The obvious conclusion, that the latter areas should be used to control flooding, is being acted upon by local planning agencies, who require that these lands be left under forest. Unfortunately we have not been able to observe the impact of heavy machinery on the land, which is likely to produce the most significant impact through the lowering of the surface conductivities, as it does everywhere else in the world. There is nothing inherent in these soils under natural circumstances which should give high susceptibility to soil erosion under clearance, though management practices will need to take careful account of the seasonal variability in subsurface moisture levels.

One further implication which may arise following forest clearance and subsequent use for agricultural crops relates to the relatively high soil tensions experienced during the drier part of the year. Such high tensions will probably give rise to plant stress and if such stress occurs at a crucial stage during the life cycle of the plant yields may be seriously reduced. Many of the crops grown in the humid tropics such as cocoa, oil palm and rubber are particularly susceptible to moisture stress, and even short periods of stress may reduce yields considerably. Sanchez (1976) reports briefly on the effects of drought on crops grown on an oxisol in Brasilia, Brazil. Because of aluminium toxicity at depth most crops root very shallowly, and for example, maize began to wilt after only six rainless days during the wet season. It becomes apparent therefore that in undertaking agricultural development of cleared areas considerable care must be taken with the timing of crop plantings, and that in order to maximize yields it may be necessary to irrigate crops during rainless periods.

2.2.6. CONCLUDING REMARKS

We have tried to show that the character of water movement in a small wet tropical catchment is in reality more complex than is commonly assumed. We have also attempted to indicate, however tentatively, the way in which this might be relevant to the question of forest clearance. What appears to be needed is greater recognition that environmental circumstances in the hot, wet tropics are, in themselves, enormously varied. In Amazonas there is an urgent need for further investigation of hillslope hydrology of cleared sites, linking up studies of plant water relationships on the one side with catchment hydrology on the other. Work on general catchment hydrology in the Manaus area has only been recently initiated. Such studies will indicate the broad effects of forest clearance but difficulties arise when several different effects are compounded, as we pointed out in our introduction. Further investigation of water budgets and fluxes on hillslopes are needed to compliment catchment studies because of the relative ease with which other controls may be held constant. Hillslope hydrological investigations are important in understanding the forest clearance's impact on soil erosion, nutrient movement and flood hazard. Such studies must also feed directly into the investigations of relationships between plant and soil water and hence be of practical use in agricultural development of the cleared areas, as they have been in other parts of the world.

ACKNOWLEDGEMENTS

The authors gratefully acknowledge the help they have received from several persons and institutions whilst carrying out this research. The work was initially sponsored in the first year by the Royal Society and in the second year by the Organization of American States Regional Scientific and Technological Development Program. In the field logistic support was provided by the Instituto Nacional de Pesquisas da Amazonia, Manaus, and the Max Planck Institute of Limnology. Our personal thanks go also to Dr. Wolfram Franken and especially to Dr. M. J. Waylen.

REFERENCES

Atkinson, T. C., 1978, Techniques for measuring subsurface flow on hillslopes, In *Hillslope Hydrology*, M. J. Kirkby (ed.), 73–121, John Wiley, London.
Brinkmann, W. L. F., and Santos, A., 1973, Natural waters in Amazonia, insoluble calcium properties, *Acta Amazonica*, **6(2)**, 33–40.
Brünig, E. F., Herrera, R., Heuveldop, J., Jordan, C., Klinge, H., and Medina, E., 1979, The international MAB rainforest ecosystem pilot-project at San Carlos de Rio Negro, In *Transactions of the Second International MAB-IUFRO workshop on Tropical Rainforest Ecosystems Research*, Adisormarto, S., and Brunig, E. F. (eds.), Hamburg: Reinbeck, 47–67.
Campbell, G. S., 1974, A simple method for determining unsaturated conductivity from moisture retention data, *Soil Science*, **117(6)**, 311–314.
Dunne, T., 1978, Field studies of hillslope flow processes, in *Hillslope Hydrology*, M. J. Kirkby (ed.), 227–293.
Harr, R. D., 1977, Water flux in soil and subsoil on a steep forested slope, *Journal of Hydrology*, **33**, 37–58.
Klinge, H., 1976, Bilanzierung von Hauptnährstoffen in Ökosystem tropischer Regenwald (Manaus), *Biogeographica*, **7**, 59–77.

Lai, P. W., 1976, Empirical evaluation of soil tension: with emphasis on variations during a drying period, *Geographical Discussion Paper*, **58**, 29 pp, London School of Economics.

Nortcliff, S., and Thornes, J. B., 1978, Water and cation movement in a tropical rainforest environment. I. Objectives, experimental design and preliminary results, *Acta Amazonica*, **8(2)**, 245–258.

Nortcliff, S., Thornes, J. B., and Waylen, M. J., 1979, Tropical forest systems: a hydrological approach, *Amazoniana*, **VI(4)**, 557–568.

Poulovassilis, A., 1962, Hysteresis of pore water, an application of the concept of independent domains, *Soil Science*, **93**, 405–412.

Poulovassilis, A., 1969, The effect of hysteresis of pore water on the hydraulic conductivity, *Journal of Soil Science*, **20**, 52–56.

Ribeiro, M. de N. G., 1976, Aspectos climatologicos de Manaus, *Acta Amazonica*, **6(2)**, 229–233.

Rodda, J. C., 1976, Basin studies, In *Facets of Hydrology*, Rodda, J. C. (ed.), 257–261, John Wiley: London.

Roose, E. J., and Godefroy, J., 1977, Pedogenes actuelle d'un sol ferrallitique remaine sur schist sour foret et sous bananerie fertilisée de basse Cote d'Ivoire, *Orstom-Irfa, BP. V51-Abidjan*, 116 pp.

Sanchez, P. A., 1976, *Properties and management of soils in the Tropics*, Wiley-Interscience, New York.

USDA, 1975, *Soil Taxonomy (Agricultural Handbook No. 436)*, Washington.

Webster, R., 1966, 'The measurement of soil water tension in the field', *New Phytologist*, **65**, 249–258.

Weyman, D. R., 1973, 'Measurement of a downslope flow of water in a soil', *Journal of Hydrology*, **20**, 267–288.

Whipkey, R. Z., and Kirkby, M. J., 1978, Flow within the soil, in *Hillslope Hydrology*, M. J. Kirkby (ed.), 121–144.

Tropical Agricultural Hydrology
Edited by R. Lal and E. W. Russell
© 1981, John Wiley & Sons Ltd.

2.3

The Hydrological Importance
of a Montane Cloud Forest Area
# of Costa Rica

F. ZADROGA

2.3.1. INTRODUCTION

Numerous watershed studies carried out in temperate regions influenced by normal atmospheric conditions have demonstrated that the elimination of forest cover generally gives rise to a temporary increase in stream-flow, principally due to reductions in evapotranspirational losses. The net increases in water yield are variable in amount, show largest relative increases in the dry season immediately following felling, and in general are approximately proportional to the basal area eliminated— up to a certain maximum value for total clearance (FAO, 1977; USFS, 1976). Also, under such conditions, the interception of moisture by forest vegetation normally constitutes a 'loss' through the vaporization 'in situ' of water that would otherwise have reached the ground.

These observations suggest that one of the best ways of increasing runoff would be to eliminate all or portions of the vegetative cover of a catchment—and indeed forest removal is used as an important watershed management technique in certain countries, such as the United States of America. However if the forest is a montane cloud forest, deforestation may cause a substantial decrease in water yield; a result of great hydrologic importance especially in the tropics.

Wherever clouds, mist, or fog impact forested areas and especially where these conditions persist in the form of orographic systems or advective sea fogs, moisture is intercepted by plant surfaces and precipitation occurs in the form of drip and stemflow, even though no rainfall occurs on adjacent open ground. This phenomena has been variously termed mist precipitation, fog drip, condensation drip, occult precipitation, and cloud moisture interception.

Several authors have commented on the importance of fog, mist, and other forms of cloud moisture in the expression of ecological relationships of high montane forest environments, including the modification and regulation of local climates, special plant physiological effects, variation in the composition and associations of plant communities, and contribution to the local water economies (Holdridge *et al.*, 1971; Kerfoot, 1969; Ekern, 1964). Cloudiness and the interception and subsequent redistribution of cloud moisture by vegetation is of special hydrologic importance, and may notably reduce evapotranspiration and increase net precipitation and run-off.

To ascertain more exactly the hydrologic role of cloud moisture interception in cloud forest environments, a five year experimental catchment research project has been proposed and funded to be carried out to the south of the Monteverde Cloud Forest Reserve in northern Costa Rica.

This paper gives some preliminary data supporting the foregoing observations on the hydrological importance of the cloud forest.

2.3.2. THE HYDROLOGIC IMPORTANCE OF CLOUD FORESTS

2.3.2.1. General Characteristics of Cloud Forests

In the tropics cloud forests normally occur in upland areas often with rugged topography where cloud belts are originated by moist ascending air masses. Thus, most cloud forests are found on volcanoes or high mountains, their average elevation being from 1,500 to 3,000 metres in Middle America and somewhat lower in the Caribbean (La Bastille and Pool, 1978). Such highland forests are generally a dense growth of trees and shrubs of varying but generally small stem diameters and heavily laden with epiphytic vegetation, including mosses, ferns, bromeliads, and leafy liverworts. They are not confined to regions with a continuously wet climate, as they are also found in regions with a sunny dry season. The average annual rainfall is usually well above 2500 mm and the humidity near the saturation point.

The daily temperature range is generally between 12 and 21 °C, depending upon factors such as latitude, altitude, aspect, and exposure. No frosts occur. Most of the water condensed or captured from the cloud reaches the ground as stemflow or drip, little being evaporated from the foliage due to the long periods of cloudiness.

Cloud forests can also occur under much drier conditions, such as those found in Lomas de Lachay in Peru and Bosque Fray San Jorge in Chile, on mountain slopes facing the Pacific Ocean, here the cloud moisture intercepted forms an important fraction of the local water supply as it may exceed the rainfall several fold.

2.3.2.2. Direct Hydrologic Benefits of Cloud Forest

Cloud forest ecosystems can be particularly valuable resources from a hydrologic point of view for three major reasons: (1) their effect in increasing net precipitation, (2) their regulation of flow regime, and (3) their low evapotranspiration rates.

2.3.2.2.1. Increasing Net Precipitation

The basic problem in the measurement of cloud or fog moisture input to natural ecosystems is the determination of the average rate of water delivery to the soil surfaces of those systems. Two variables control this rate of input: (1) the average rate of moisture collection per unit area of the natural vegetation surfaces, and (2) the average water storage capacity of those surfaces. Two general processes contribute to the moisture collection rate: (1) interception of liquid water droplets, and (2) vapour condensation.

To compute the magnitude of the interception process, it is necessary to know the volume rate of flow of moisture per drop size band width and the interception efficiency of the vegetation surface for each band width. The products of these two quantities must then be summed for all band widths represented in the moving cloud or fog.

$$I = \sum_{i=1}^{n} \epsilon_i \, \bar{v} \cdot X_i$$

where

ϵ_i = surface interception efficiency (%) for drop size band width i
i = drop size band width
$\bar{v}$ = mean velocity of the air cloud mass
X_i = moisture content of the cloud for band width i
I = total interception (volume of water)

In order to put fog interception on a comparable basis with other terms in the hydrologic balance equation, the amount intercepted must be expressed per unit horizontal area. The complex geometry of a natural vegetation surface with reference to the prevailing wind direction during a particular fog event is a serious complicating factor. Collection efficiency is probably not uniform over the vegetation canopy, and the area contributing to collection is very difficult to estimate relative to the projected horizontal area of the soil beneath the canopy. It is a commonplace observation that the windward sides of vegetation clumps are normally dripping wet, while the leeward sides may remain dry during periods of fog interception.

The commonly employed techniques for measuring the interception of liquid droplets of fog interception involves using artificial surfaces of different shapes and physical characteristics to simulate the effects of the natural vegetation, and these obviously produce erroneous estimates of intercepted cloud moisture. Even greater errors are introduced if these techniques are used to estimate the amount of water condensed by the vegetation. The practice of dividing the volume of moisture collected by the area of the collection surface for an estimate of the water collected by unit area of forest, amplifies the error to such a degree that it becomes difficult to assign any meaning to the observations. There is still no satisfactory solution to this problem.

It is also important to know how much of the total intercepted moisture reaches the soil, to play a role in soil storage and runoff processes. There is a critical length of time the cloud must persist before any water is added to the soil, depending on the concentration of the moisture in the cloud and its rate of flow. Moisture storage on vegetation surfaces is apt to be significantly greater than moisture storage on the artificial collection surfaces described in the literature, and this is especially so for small collection devices designed to fit over standard rain gauges. Because of lower resistances to flow, surface stored water is evaporated at a greater rate than transpired water. This means that traditional devices will overestimate interception.

2.3.2.2.2. Regulation of Flow Regime

It is logical to expect that cloud moisture interception have an important effect on the timing of runoff from catchments that have a substantial proportion of their headwaters under cloud forest. Depending upon the temporal distribution of cloud effects throughout the hydrologic year relative to the seasonal distribution of vertical precipitation, intercepted cloud water may or may not play a critical role in maintaining dry season flows. The typical positive effect occurs when cloud banks enshroud mountain slopes during dry season months when little or no rainfall (vertical) is recorded. Under these conditions fog-born moisture, mist, and other forms of cloud moisture may condense upon exposed vegetational surfaces, and drip or run down stems to the ground, thereby recharging soil and groundwater supplies and thus maintaining stream discharge.

It is commonly accepted in Costa Rica, among hydrologists, foresters, and other people involved in the use of water resources from highland catchments, that the existence of forest cover and particularly cloud forest in the headwaters of tropical rivers benefits runoff regulation. No quantitative analysis has been done, however, to demonstrate regime differences associated specifically with variations in forest cover.

2.3.2.2.3. Low Evapotranspiration Rates

There are no data on evapotranspiration losses from cloud forest areas in Costa Rica.

Weaver (1972) in his studies on 'Pico del Este' in the cloud forests of the Liquillo Mountains of Puerto Rico, noted that transpiration in elfin woodland and in montane rain forest was extremely low when compared to most other rates reported in the literature, regardless of habitat. Apparently, transpiration differences between the two habitats studied depended upon weather conditions. The extremely low saturation deficits of the air, frequent moisture deposition, and low insolation reaching the leaves because of frequent fog and closed canopy were reported to be the major causes of the slow transpiration. In-efficient xylem and effective scarcity of certain minerals may also be involved. A proportionately greater stomatal area reported for elfin forest leaves as compared to montane rainforest counterparts suggests that those cloud forest plants adapt to counteract this reduced transpiration.

Still other authors have speculated that the stunting of elfin woodland is directly relatable to reduced transpiration rates (Beard, 1944; Odum, 1968). However, data specifically on the transpiration rates of elfin woodland and other types of cloud forest is almost non-existant except for those estimates by Gates (1969), also in Puerto Rico.

2.3.2.3. Location and Physical Characteristics of the Study Area

The watersheds analysed in this paper constitute a major portion of the uplands of the Atlantic and Pacific slopes of the Cordillera de Tilaran, in north-central Costa Rica, Central America. Hydrographically, the study area extends over the headwaters portion of the Chiquito, Caño Negro, Peñas Blancas, and Jabillos watersheds of the San Carlos Basin, a majority tributary of the San Juan drainage system on the Atlantic slopes, and the Santa Rosa, Corobici, and Cañas watersheds on the Pacific slopes. The physiography, geomorphology, and soils of the area are described in detail by the author in a publication that can be obtained from him in Turrialba.

2.3.2.3.1. Climate and Special Atmospheric Effects

The climatic factors affecting Costa Rica have been described in general terms by several authors (including Portig, 1965; Holdridge *et al.*, 1971; and Dohrenwend, 1971), and in more specific terms by ICE (various).

The climate of the Cordillera de Tilaran resembles that of other tropical mountain ranges similarly exposed to the north-east trade winds and resultant orographic and cloud moisture effects (cf, Bayton, 1968, on the climate of the Luquillo Mountains of Puerto Rico, and Shreve, 1914, on that of the Blue Mountains of Jamaica). Due to the small land mass of Central America, the intertropical low-pressure trough is poorly defined over the Isthmus. As a consequence, the trade winds heavily laden with moisture from their passage over the warm Caribbean Sea are a predominant and important feature of the climate at higher elevations throughout the year (Dohrenwend, 1971).

The strong, cloud-bearing winds affecting the study area sweep from the Atlantic coastal plain up the river valleys and over mountain passes and gaps to eddy among the peaks, producing a spatially complex pattern of rainfall distribution and wind velocities. As the air mass descends the Pacific flank, the air heats adiabatically, and the cloud bank dissipates. As Lawton and Dryer (1978) report, the elevation of the lee edges of the cloud bank varies from 1300 to 1600 m.a.s.l. throughout the eastern facing Peñas Blancas watershed. Almost the whole of the four eastern watersheds studied may be enveloped by clouds and rain during certain times of the year, but the ridges and the headwalls of wind gaps are much more often exposed to blowing clouds and mist. Since much of the water input on these exposed cloud forest areas is due to cloud droplet catchment by the vegetation, these sites are perceptibly wetter than those to their lee. Annual rainfall, consisting almost entirely of normal vertical precipitation, at the community of Monteverde (1380 m a.s.l. on the lee slope) averages 2450 mm. (Instituto Meteorologico Nacional, 1977), while ICE reports rainfall maxima to the east of the Divide in the Peñas Blancas headwaters of up to 9000 mm.

Along the Atlantic slope of the Cordillera de Tilaran there is a definite dry season from January until the end of April. During this time there is much cloud cover but orographic rainstorms are rare. The trade winds blow quite strongly at this time, and give rise to an orographic cloud deck over the Caribbean side of the Cordillera. The dense mist and low cloud which covers the uplands often spills over to the Pacific side. As the dry season progresses, the strong winds and mist become less intense. There is increasing build-up of convective cumulus over the Pacific slope in the interspersed periods of calm weather.

By the middle of May, thunderstorms drift up the Pacific slope on local valley winds. Thunderstorm activity extends into October with a drier spell, the Veranillo occurring around the beginning of July. Strong trade winds continue to occur interspersed among the days of thunderstorm. In the earlier part of the rainy season, the winds are associated with dry periods, but by August they are once more carrying mist across the Divide. Beginning in September, thunderstorms are less frequent. November and December are dominated by strong winds, low cloud, and heavy mists.

2.3.2.3.2. Biologic Factors

Vegetation and life zones The forests of the study area are physiognomically complex, spatially variable and highly influenced by the special atmospheric conditions of high cloud mist and wind incidence under which they exist.

The upland cloud forests fall within the lower montane rainforest zone, and contain covex forest, leeward cloud forest, windward cloud forest, oak ridge forest, elfin forest, and swamp forest. Below these exposed ridges and their wet atmospheric cloud forest associations, are found in succession premontane rainforest (occurring between 2000 and 1000 m.a.s.l.) and tropical wet forest and transitional formations thereof (approximately between 1000 m.a.s.l. and sea level). The characteristics of the vegetation of these life zones are described in Holdridge *et al.*, (1971).

All the forests at the higher elevations are exposed to strong orographic rain and cloud moisture. These forests are always wet and perhaps with the exception of the forests of the tropical altitudinal belt (i.e. below 1000 m a.s.l.), do not have any effective dry months in their annual water balances. Leaf litter and epiphytes may dry out for brief periods of time on sharp ridges, but the mineral soil is wet in all associations throughout the year. Epiphytes are extremely abundant and play an important part in interception processes. Coverage of limbs and trunks occurs principally by bryophytes, ferns, and angiosperms.

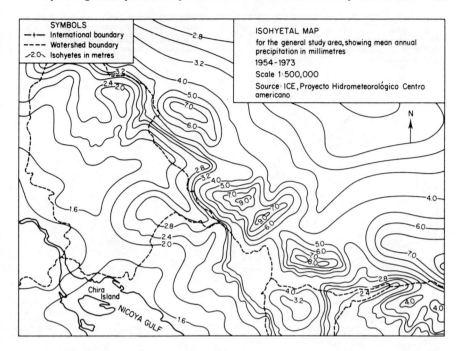

Figure 2.13 Isohyetal map for the general study area, showing mean annual precipitation in meters

2.3.2.3.3. Precipitation and Discharge Data for the Study Area

Precipitation data is available for a few selected stations within the study area but the spatial distribution of rainfall stations is still not adequate, to allow an accurate estimation of the rainfall over the watersheds to be made. Mean annual precipitation reaches its maximum values in the headwaters of the Rio Caño Negro and Peñas Blancas watersheds, with over 9000 mm, and in the headwaters of the Rio San Lorenzo and Balsa watersheds with over 8000 mm. Average daily, monthly, and yearly runoff data are also available for each of the four study watersheds. The precipitation and evapotranspiration map for the watersheds are shown in Figures 2.13 and 2.14.

2.3.2.4. Findings and Discussions

The percentage of lower montane forest associations occurring in each of the San Carlos sub-basins was similar in all four cases, although the relative proportions found for moist, wet, and rain forest varies considerably (Figure 2.15 and Table 2.7). The Chiquito and Caño Negro Rivers have 18.36 and 13.37 per cent of their headwaters in the lower montane moist forest life zone. The Jabillos River has 2.74 in moist, 5.95 in wet, and 4.85 rainforest in lower montane forest life zones; and the Peñas Blancas watershed, has only rainforest in its upland lower

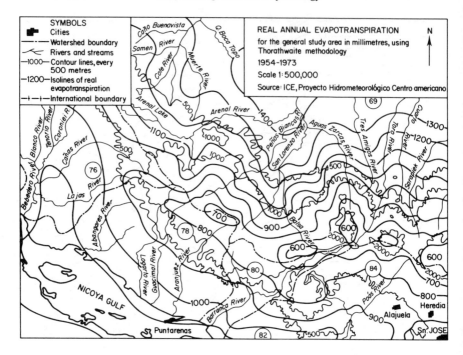

Figure 2.14 Real annual evapotranspiration for the general study area in millimetres, using Thorathwaite methodology

montane altitudinal belt, occupying 16.83 per cent of its total catchment surface area. Deforestation has been actively occurring in the Chiquito and Jabillos watersheds over the past ten years, and therefore the effective cover of lower montane forest is considerably less than a simple analysis of the ecological zones would indicate. Rough estimates based upon field observations and overflights show the Chiquito watershed has now only about 10 per cent forest cover left in the lower montane zone the Jabillos watershed has had a large part of its moist forest eliminated but retains the majority of its wet and rain forest associations essentially untouched. It is estimated that the Jabillos watershed has had its lower montane forest cover reduced to 10.85 per cent, distributed between moist (0.5 per cent), wet (5.5 per cent) and rain (4.85 per cent) provinces.

A comparison of the percentage total-record discharges that have occurred for each of these Atlantic slope watersheds for the five dry-season months of the year (i.e. January through May), during which period an estimated 23.4 per cent of the total annual precipitation falls (Table 2.8), gives the following results: As can be noted, differences in low flow discharges among these sub-basins show a general correlation with the lower montane forest associations data presented, but do not seem to be highly significant.

For comparison purposes, three adjacent watersheds on the Pacific slope of the

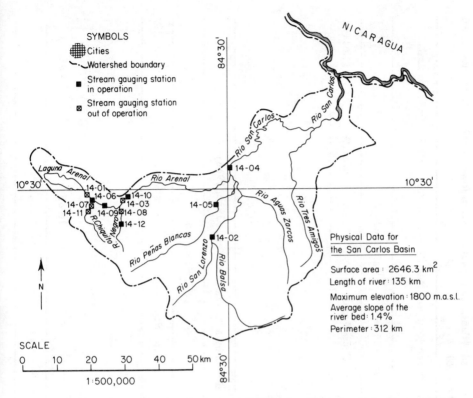

Figure 2.15 Basin No. 14: San Carlos and Sub-Basins Chiquito, Caño Negro, Peñas Blancas, and Jabillos

same Tilaran Mountain Range corresponding to the Santa Rosa, Corobici, and Cañas Rivers, were analysed in a similar manner for their low flow discharges (Table 2.9). These watersheds, of similar size, physiographic, and geomorphologic characteristics as those of the Atlantic slope, differ somewhat in soils but principally in climatic conditions and life zone distributions. The Santa Rosa, Corobici, and Cañas Rivers (Figure 2.16) are all affected by the dry Pacific climatic regime, and have essentially no lower montane forest vegetation occurring on them. An estimated 16.5 per cent of the total annual rainfall for the total record of all rainfall stations of the area fell during this analysis period.

The difference between the water yields of the Atlantic and Pacific slopes, compared with the monthly rainfall is shown in Figure 2.17. The results are startling for the Atlantic slope and quite predictable for the Pacific slope with respect to both water yield and regime. The annual measured runoff is 102 per cent of the annual estimated rainfall on the Atlantic side compared with 34.5 per cent on the Pacific, and the figure also shows that on the Atlantic side the monthly run-off exceeds monthly rainfall for about 7 months in the year, the excess being greater

Table 2.7. Percentage of total surface area of selected Atlantic slope watersheds in lower montane life zones, based upon theoretical distributions (a) and actual land use (b)

(a) Theoretical life zone distributions, based upon life zone map of Costa Rica (Data from Tosi, J., 1969, Tropical Science Centre)

Life zone	Watershed				
	Chiquito (%)	Caño Negro (%)	Peñas Blancas (%)	Jabillos (%)
Lower montane moist forest (bh-MB)	18.36	13.37	—	2.74
Lower montane wet forest (bmh-MB)	—	—	—	5.95
Lower montane rain forest (bp-MB)	—	—	16.83	4.85

(b) Estimates of current life zone distributions as altered by actual land use (as of June, 1968)

Life zone	Watershed				
	Chiquito (%)	Caño Negro (%)	Peñas Blancas (%)	Jabillos (%)
Lower montane moist forest (bh-MB)	10.0	13.37	—	0.5
Lower montane wet forest (bmh-MB)	—	—	—	5.5
Lower montane rain forest (bp-MB)	—	—	16.83	4.85
Total	10.0	13.37	16.83	10.85

Table 2.8. Percentage of total annual discharge for dry season months for the Atlantic slope watersheds, based upon mean monthly values in liters sec^{-1} km^{-2} for the entire record

Watershed	Jan	Feb	Mar	Apr	May	Total
			Month (%)			
Chiquito	7.0	5.7	4.1	3.0	3.1	22.9
Cano Negro	6.9	6.6	4.9	3.0	3.6	25.0
Penas Blancas	6.7	6.5	3.9	3.8	4.2	25.1
Jabillos	7.6	5.8	3.6	3.3	4.3	24.6
				Average value		24.4

Table 2.9. Percentage of total annual discharge for dry season months for the Pacific slope watersheds, based upon mean monthly values in liters sec^{-1} km^{-2} for the entire record

Watershed	Jan	Feb	Mar (%)	Apr	May	Total
Santa Rosa	5.8	4.2	3.5	3.1	3.8	19.8
Corobici	6.3	4.5	3.3	2.7	3.9	20.7
Canas	5.0	3.5	2.6	2.2	2.8	16.1
				Average value		18.9

in August, November, and March than in September, October, December, and April. On the Pacific side catchments, the stream flows follow that of the rainfall and only exceeds it during the dry season months of January to April when the flow is being fed by the run-down in basin storage.

The Atlantic slope pattern can be easily accounted for by two factors: (1) under-estimation of rainfall due to an inadequate rain gauge pluviometric installations, and (2) under-estimation of total catchment precipitation due to the occurrence and non-measurement of cloud moisture interception phenomena. The spatial distribution and density of rainfall stations over the study area is not adequate to permit an accurate picture of the rainfall distribution and especially so in the remote, forested headwaters where precipitation maxima are likely to occur. This undoubtedly accounts for a part of the discrepancy in the data between Atlantic slope runoff and rainfall, but it seems unlikely that this deficiency can be the sole cause; because there is a similar lack of rain gauge stations on the Pacific slope, where the Pacific rainfall runoff relationship was of the normal type.

The periods when the runoff is in excess of rainfall on the Atlantic side closely follow the periods when the north-east trade winds are dominant, and, since these are the principal agent causing cloud formation in the vegetation, it is reasonable to

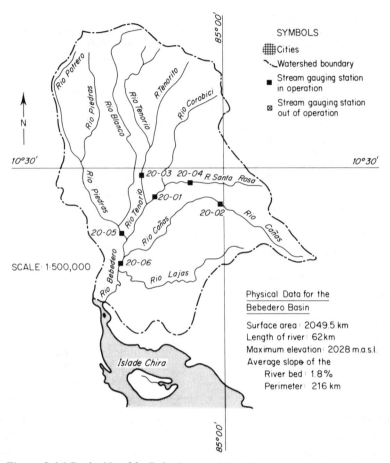

Figure 2.16 Basin No. 20: Bebedero and Sub-Basins Santa Rosa, Coro-
bicí and, Cañas

assume that the periods of excess runoff are also periods of extensive and continuous low cloud, which will also be the periods when the capture of 'occult' precipitation will be most significant. Further work is still needed before more reliable
estimates can be made of the magnitude of this occult precipitation over a watershed. It is not known, for example, how far it depends on the type of forest present
or how far it is reduced by the type of vegetation that succeeds deforestation.

2.3.3. SPATIAL DISTRIBUTION OF CLOUD FOREST AREAS
IN MIDDLE AMERICA AND THE CARIBBEAN

Cloud forests occur throughout Central America both from southern Mexico to
western Panama on the Isthmus, and from Cuba to Trinidad in the Caribbean. The

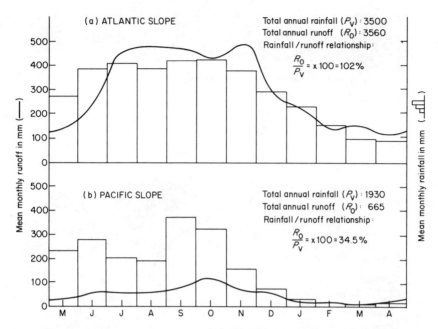

Figure 2.17 Comparison of rainfall/runoff proportions for the Atlantic and Pacific slopes of the Cordillera de Tilaran, Costa Rica (Note: Data summarized from the following sub-basins: (a) Atlantic Slope: Chiquito, Caño Negro, Peñas Blancas, and Jabillos Rivers of the San Carlos Basin (b) Pacific Slope: Santa Rosa, Corobici, and Cañas Rivers of the Bebedero Basin

total area of existing undisturbed cloud forest in this region is not known exactly but has been estimated to occupy between 7,000 to 15,000 km^2 (La Bastille and Pool, 1978). Currently only a small proportion of this cloud forest area is under, or destined to become under, a protected status in the form of national parks or reserves; yet they possess many rare endemic species of fauna and flora, have considerable tourist potential and they give general environmental protection to adjacent lowland areas.

The threats to cloud forest areas, however, are many and they are becoming one of the most rapidly disappearing forest ecosystems in the neotropics today. Due to rapidly increasing human populations in low lands adjacent to them, cloud forest land is continuously being deforested and converted to agricultural and other forms of land use.

Throughout the Central American region water resources are rapidly increasing in value, so it is increasingly important to have strictly enforced water management policies for the protection of river flows for hydro-electric power and irrigation and for the supply of potable water. The need for hydro-electric development is becoming particularly urgent due to shortages of other sources of energy and the high price of petroleum products, and this demands a new assessment of the role of soil and water conservation in Central America in light of future development needs.

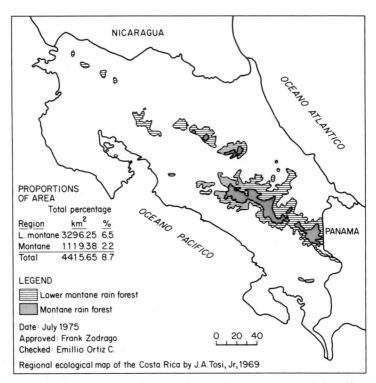

Figure 2.18 Costa Rica: Distribution of life zones potentially affected by cloud moisture interception

In Costa Rica, for example, those forest associations where cloud moisture originates significant increase in the water yield include the lower montane and montane
rain forest zones (Holdridge, 1967), which cover approximately 8.7 per cent of the
total surface area of the country (Figure 2.18). Costa Rica, the Central American
nation with by far the greatest hydrologic potential, should put emphasis in the
study and conservation of cloud forest areas.

REFERENCES

Bayton, H. W., 1968, The ecology of an elfin forest in Puerto Rico: 2. The micro-
climate of Pico del Oeste, *J. Arnold Arb.*, **49**, 419–430.

Beard, J. S., 1944, Climax vegetation in tropical America, *Ecology*, **25**, 127–158.

Dohrenwend, R., 1971, The energetic role of the Trade Wind Inversion in a tropical
subalpine ecosystem, *Ph.D. Thesis*, Syracuse University, Syracuse, New York.

Ekern, P. C., 1964, Direct interception of cloud water at Lanaihale, Hawaii, *Proc.
Soil. Sci. Amer.*, **28**, 419–21.

FAO, 1977, *Conservation Guide No. 1. Guidelines for Watershed Management*, 293
pp.

Gates, D. M., 1969, The ecology of an elfin forest in Puerto Rico, 4. Transpiration rates and temperatures of leaves in a cool humid environment. *J. Arnold Arboretum*, **50**, 93–98.

Holdridge, L. R., 1967, *Life Zone Ecology*, Tropical Science Center, San Jose, Costa Rica: 206 pp. illustr.

Holdridge, L. R., Grenke, W. C., Hatheway, W. H., Liang, T., and Tosi, L. R., Jr., 1971, *Forest Environments in Tropical Life Zones: A Pilot Study*, Pergamon Press New York, N.Y., 747 pp., illustr.

Instituto Meteorologico Nacional, Costa Rica, 1977, *Anuario Meteorologico*.

Kerfoot, O., 1969, Mist precipitation on vegetation, *Forestry Abstracts*, **29**, 8–20.

La Bastille, A., and Pool, D. J., 1978, On the need for a system of cloud forest parks in middle America and the Caribbean, *Environmental Conservation*, **5(3)**, 183–190.

Lawton, R., and Dryer, V., 1978, The vegetation of the Monteverde cloud forest reserve. 32p. 1 map. *Preliminary Draft of Paper*, unprinted.

Odum, H. T., 1968, Work circuits and system stress, In H. E. Young, Ed., *Primary Productivity and Mineral Cycling in Natural Ecosystems*, Univ. Maine Press, Orono, pp. 81–138;

Portig, H. W., 1965, Central American rainfall, *The Geographical Review*, **55(1)**, 68–90.

Shreve, F., 1914, *A Montane Rain-Forest*, Carnegie Inst. Publ. 199, 110 pp., Washington D.C.

USFS (United States Forest Service), 1976, *Food and Water-Effects of Forest Management on Floods, Sedimentation, and Water Supply*. Compiled by Anderson H. W., M. D. Hoover, and K. G. Reinhart. Pacific Southwest Forest and Range Experimental Station, USDA Forest Service General Technical Report. PSW-18/ 1976, 115 pp.

Weaver, P. L., 1972, Cloud Moisture Interception in the Luquillo Mountains on Puerto Rico, *Carib. J. Sci.*, **12(3–4)**, 129–144.

Tropical Agricultural Hydrology
Edited by R. Lal and E. W. Russell
© 1981, John Wiley & Sons Ltd.

2.4

Watershed Investigations for Development of Forest Resources of The Amazon Region in French Guyana

M. A. Roche

The economic progress of French Guyana is dependent on the development of its vast forest reserves (90,000 km^2) that cover 98 per cent of its area. There is an increase in industrial need for timber and for the production of food and fibre. The forest ecosystem, characterised by slowly permeable and highly weathered soils derived from crystalline rocks with slopes ranging from 15 to 50 per cent and receiving annual rainfall of 3,000 to 4,000 mm, poses a challenge to ecologists and agriculturists for its proper development. Oversight of some important factors may lead to irreversible degradation of this vast natural resource. Established under the framework of Unesco's MAB program is the ECEREX (Ecoulement d' Ecologie, Erosion, et Exploitation) project that is jointly coordinated by ORSTOM, GERDAT, and IRFA (Dubreuil, 1963; Heinz and Dubreuil, 1963; Hoepfener, 1974; Hoorelbeck and Lemaitais, 1972). This project was initiated in 1976 to investigate; (i) the ecological conditions under the natural forest ecosystem, (ii) the changes in soil and hydrologic balance, and shift in floral composition by deforestation, and (iii) the effects of different management systems on production from a range of farming systems including pastures, silviculture, and seasonal crops.

The hydrological investigations include the measurements of water runoff and soil erosion.

2.4.1. MATERIAL AND METHODS

These investigations were carried out on 10 watersheds of 1 to 1.5 ha each. Two watersheds were kept under natural forest cover as control. The layout of these watersheds is shown in Figure 2.19. Changes in soil and hydrological parameters were investigated for 8 systems shown in Figure 2.20. These were: (i) natural regrowth without burning existing vegetation, (ii) natural regrowth after burning

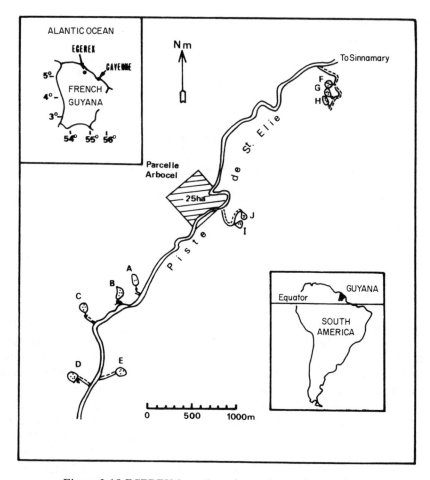

Figure 2.19 ECEREX Location of experimental watersheds

existing vegetation cover, (iii) Pinewood plantation, (iv) eucalyptus plantation, (v) citrus plantation, (vi) and (viii) two different pastures, and (viii) traditional farming. Prior to implementing these treatments, the initial baseline data under the forest cover was collected for two years. The observations on these watersheds were initiated over a period of two years from 1976 to 1978.

The soils of these experimental watersheds are representative of a large region. The hydrological and topographical characteristics of the soil are not homogenous among all watersheds. This variability should be kept in mind while interpreting the results.

Each of these 10 watersheds is equipped with a rain gauge and a V-notch or an H-flume and a water level recorder to monitor rainfall and runoff. Runoff samples are obtained for determining the sediment load.

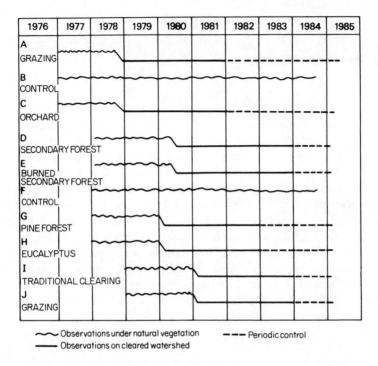

| 1976 | 1977 | 1978 | 1979 | 1980 | 1981 | 1982 | 1983 | 1984 | 1985 |

A
GRAZING

B
CONTROL

C
ORCHARD

D
SECONDARY FOREST

E
BURNED
SECONDARY FOREST

F
CONTROL

G
PINE FOREST

H
EUCALYPTUS

I
TRADITIONAL CLEARING

J
GRAZING

〜 Observations under natural vegetation　━ ━ ━ Periodic control
━━━ Observations on cleared watershed

Figure 2.20 ECEREX Experimental watersheds management project

Runoff plots (100 m²) have been established in the vicinity of these watersheds to monitor runoff and erosion on small plots. These plots are established on different soil types to investigate the magnitude of runoff and erosion from different soils of this region.

A methodology for comparative analysis of runoff and erosion from different watersheds has been reported earlier (Roche, 1978). The results of hydrological investigations were analysed for: (i) monthly and annual hydrologic balance, (ii) runoff/rainfall characteristics of individual storm events, and (iii) establishing cause-effect relationships among different parameters investigated. Simple correlation and regression were also computed between soil erosion and climatic erosivity factor, (R). Rainfall intensities computed on different time intervals were related to climatic erosivity factor (R), and to soil loss and runoff from each treatment.

Comparative analysis was also done for any specific hydrologic parameter for a given treatment with control, e.g. comparison of runoff characteristics from two watersheds for a single rainstorm event. These comparisons evaluate the effects of different types of rainfall events and of their temporal variability on hydrological characteristics. These effects are also evaluated for a range of farming systems at different stages of development.

2.4.2. RESULTS

2.4.2.1. Hydrologic Balance

2.4.2.1.1. Rainfall Characteristics

Rainfall amounts received over 1, 3, 5, and 10 minute intervals were not better correlated with runoff and soil loss than the total amount of rainfall received. Similarly, the multiple correlations of soil loss with rainfall erosivity (R), maximum intensity in 10 minutes, and an index of antecedent soil moisture regime did not improve the correlation compared with simple correlation with any of these variables.

Rains received in 1977 (3500 mm with 260 rainy days) were about normal, (Roche, 1977), although there were many high intensity rains with daily amounts exceeding 100 mm. One of the rainfall events totalled 350 mm in 54 hours and had a return period of 10 years. Runoff and erosion measurements were made for 415 storm events exceeding 3 mm. Rainfall events were distinguished from one another if they were separated by rains of intensity greater than 1.5 mm hr^{-1} for 1.5 hr. Rainstorm events of intensity ranging from 84 to 106 mm did not exceed their normal frequency.

2.4.2.1.2. Runoff

Runoff characteristics for watersheds A, B, and C, with slopes ranging from 15 to 25 per cent were monitored for 90 to 116 events, depending on the watershed. These watersheds have area of 1.5, 1.45, and 1.45 ha, respectively. The runoff was most frequently observed on watershed B.

Annual water yield was 925 mm, 845 mm, and 500 mm on watersheds A, B, and C with corresponding runoff coefficients of 27 per cent, 24 per cent, and 16 per cent respectively. The surface runoff from watersheds A, B, and C was 655 mm (19 per cent), 660 mm (19 per cent), and 230 mm (7 per cent) on watersheds A, B, and C, respectively (Figure 2.21). The highest monthly runoff coefficient of 32 per cent was observed on watersheds A and B, and one of about 8 per cent on watershed C.

2.4.2.1.3. Evapotranspiration and Groundwater Recharge

The difference between the rainfall and water yield from watersheds A, B, and C was estimated to be 2550, 2600, and 3000 mm yr^{-1} respectively. These values are more than 1600 mm yr^{-1} of evapotranspiration reported by ORSTOM for large Guyanese watersheds. This implies that the ground water recharge on watersheds A, B, and C was 950 mm, 1000 m, and 1400 mm, respectively. Interflow seepage, that was frequently observed in watershed C, and is a source of perennial flow in watersheds F, G, and H is caused by the groundwater recharge. Nevertheless, the seepage

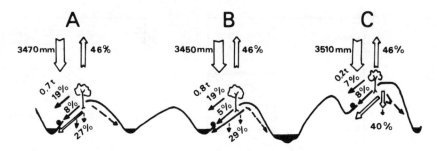

Figure 2.21 ECEREX Distribution of various terms of hydric balance of A, B, C, experimental watersheds, under forest. Example: rainfall on A basin was 3470 mm of which about 46 per cent was evapotranspiration, 19 per cent surface runoff, 8 per cent subsurface runoff, and 27 per cent infiltration. Soil erosion was 0.7 t ha^{-1} yr^{-1}

losses are difficult to quantify because the acquifer boundaries do not always coincide with the surface waterdivide.

The results presented indicate similarities in watersheds A and B, although runoff losses were slightly higher on watershed B. The runoff coefficient of watershed C is the lowest of the watersheds.

2.4.2.1.4. Runoff-Rainfall Relationship

Data in Figures 2.22 to 2.24 shows that total water yield from watershed A is more than from B, although the surface runoff was more from B than A. This difference between the two watersheds are attributed to water retention and transmission characteristics of the subsurface layers. The relationship between mean monthly runoff and rainfall are shown in Figures 2.22 to 2.24.

2.4.2.1.5. Soil Erosion

Under forest cover, the bed load was estimated to be 0.4 t ha^{-1} yr^{-1} from watersheds A and B, and 0.1 t ha^{-1} yr^{-1} from watershed C. The correlation between suspended sediment load and climatic erosivity factor (R) were computed to assess the soil erosion from unsampled storms. By using these regression equations, soil loss from suspended sediments was computed to be 0.34, 0.35, and 0.04 t ha^{-1} yr^{-1} with corresponding sediment density of 46, 36, and 12 mg l^{-1} for watersheds A, B, and C respectively. The soil erosion was also similar from watersheds A and B (0.7 and 0.8 t ha^{-1} yr^{-1}), but was significantly less from watershed C (0.2 t ha^{-1} yr^{-1}). In general, soil erosion under forest cover was low.

The concentration of dissolved constituents in water runoff ranged from 10 to 13 mg l^{-1}. This level of dissolved elements is lower than those reported for major Guyanese rivers. In addition to these elements, the concentration of dissolved silica was found to be 2 to 4 mg l^{-1}. The dissolved concentration in the ground

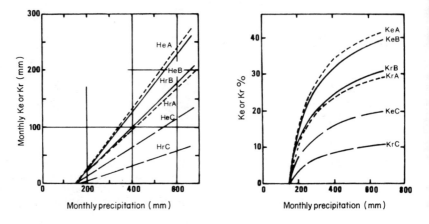

Figure 2.22 ECEREX (a) Correlation between the monthly water yield *He* or the monthly surface runoff Hr and the monthly rain intensity *P* on A, B, C, watersheds under forest. (b) Correlations between mean monthly water yield coefficients *Ke* or surface runoff coefficients *Kr* and the monthly rain intensity *P* on A, B, C, watersheds, under forest

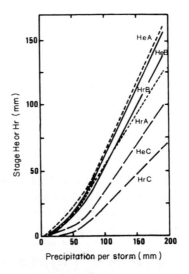

Figure 2.23 ECEREX Correlations between water yield *He* or surface runoff *Hr* for different storm events and the amount of rainfall per storm *P*, on A, B, C, watersheds under forest

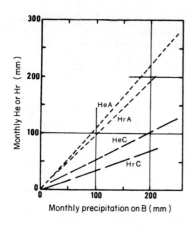 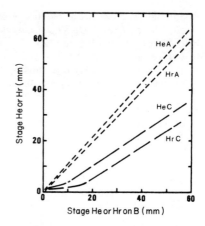

Figure 2.24 ECEREX (a) Correlation between monthly water yield *He* or surface runoff on A and C basins and the monthly water yield and surface runoff on the B check watersheds, under forest (b) Correlations between water yield *He* or surface runoff *Hr* for different storm events on A, C basins and the corresponding water yield or surface runoff on check watershed B under forest

water may be higher than in the surface water. Considering both surface and sub-surface flow, about 0.5 t ha⁻¹ yr⁻¹ of dissolved nutrient elements may be removed from the basin as a whole.

2.4.3. CONCLUSIONS

A methodology has been developed to monitor runoff and soil erosion from water-sheds under different land use. A computer program enables comparative evaluation of hydrologic phenomenon for different storm events, and for different watersheds. The preliminary results obtained indicate that soil erosion and runoff are influenced by hydrologic characteristics of the soil. For example, high vertical drainage of soils in watershed C is the reason for low surface runoff. Consequently, the soil erosion from watershed C is also low. On the contrary slowly permeable subsoil character-istics of watersheds A and B is responsible for 60 to 70 per cent runoff coefficient for intense rainstorms and for high erosion of about 1 t ha⁻¹ yr⁻¹. Small soil erosion under forest cover is attributed to the protective effects of vegetation cover, and to the binding effects of roots and leaf litter.

REFERENCES

Dubreuil, P., 1963, Le bassin versant de la Crique Virgile (Guyane Francaise). Rapport preliminaire, *Sect. Hydrol. ORSTOM Guyane,* 73 pp.
Heinz, G., and Dubreuil, P., 1963, Les regimes hydrologiques de Guyane francaise. *Mem ORSTOM,* 119 pp., 49 figures.

Hoepfener, M., 1974, Les bassins versants de la Crique Gregoire. *Rapport preliminaire: Sect. Hydrol. ORSTOM Guyane,* 107 pp.

Hoorelbeck, J., and Lemaitais, L., 1972, Le bassin versant representatif de la Crique Cacao, *Sect. Hydrol. ORSTOM,* 33 pp.

Roche, M. A., 1977, Hydrodynamique et evaluation du risque de pollution dans un estuaire a marees (Guyane francaise), *Cah. ORSTOM, Ser. Hydrol.,* **XIV,** 4, 345–382, 21 figures.

Roche, M. A., 1978, Objectifs et methodologie d'etude comparative sur l'hydrologie et l'erosion des bassins versants experimentaux Ecerex. *Bull. liaison DGRST.* 8 pp. 2 figures.

Tropical Agricultural Hydrology
Edited by R. Lal and E. W. Russell
© 1981, John Wiley & Sons Ltd.

2.5

Estimating Potential Evapotranspiration from a Watershed in the Loweo Region of Zaire

N. Sengele

Estimates of the evapotranspiration of a forested catchment in the humid tropics are highly variable, depending on the method employed. Pan evaporation may be less than the evapotranspiration of a forested catchment by as much as 20 to 30 per cent (Ringoet *et al.*, 1961). Dupriez (1959) obtained satisfactory results with the energy balance method that compared favourably with the evapotranspiration measured by the lysimetric technique. Field hydrologic balance method, though expensive and time consuming, may yield better estimates of evapotranspiration than the Pan evaporation of the lysimetric technique.

2.5.1. EXPERIMENTAL SITE

The Loweo catchment is located in the Yangambi flora reservation in the Congo basin at approximately $0° 50'$ N and $24° 30'$ E and at an altitude of approximately 450 m (Bultot and Dupriez, 1968). Out of the total area of 46.3 km², perennial crops occupy about 40 per cent and the remaining 60 per cent is the forest reserve. Topographically this region is a large plateau indented with steep valleys. The texture ranges from predominantly coarse textured soils in the valleys to those with high clay content in the plateau (Table 2.10). The natural forest reserve is mostly semi-deciduous dominated by *Scordophloeus zenkeri* with islets of evergreen forest colonized by *Brachystegia laurentii* and *Gilbertiodendron dewevrei*. The valley bottom and hydromorphic regions are covered with soft wood and climbers.

2.5.2. METHODS

Daily evaporation was measured with Class A Pan. The evapotranspiration measurements from bare soil and under vegetation cover were made with 4m² lysimetric

83

Table 2.10. Soil characteristics in the catchment

Characteristics	Plateau Soil (1)	Plateau Soil (1)	Alluvial Soil (1)
Clay (%)	30.2	30.2	64.5
Silt (%)	4.8	1.8	26.1
Porosity (%)	37 to 40	37 to 40	+50
Field capacity (%)	16.9	18.0	48.6
Permanent wilting point (%)	10.8	11.6	26.8
Apparent density (g cm^{-3})	1.61	1.61	1.24

tanks. These observations were made at the meteorological station located in the vicinity of the Loweo catchment.

Rainfall was monitored with 15 non-recording (4 dm^2 opening) and two recording rain gauges (1 dm^2 opening) distributed over the catchment (Figure 2.25). The effective rainfall over the catchment (weighted average) was computed using Thiessen's method. Water yield from the catchment was monitored by a Parshall flume, and river gauging was facilitated by the use of a Leupold and Stevens water level recorder. A current meter was used for instantaneous flow measurements.

Modified Penman's formula was used to calculate the evaporation from a free water surface. The original formula developed at Rothamsted (Penman, 1948) was adapted for climatic conditions at Yangambi by Bernard and Frere (1956). These formulae are presented in equations (2.1) and (2.2) respectively.

$$V_w = 0.26(1 + 0.146\ U_2)\ (E_s - e) \tag{2.1}$$

$$V_w = f(u)\ (E_s - e) \tag{2.2}$$

where

$$f(u) = 0.16(1 + 0.22\ U_2)$$

Substituting $f(u)$ in equation (2.1) gives the modified Penman's formula (equation (2.3)).

$$V_w\ (\text{mm day}^{-1}) = 0.16\ (1 + 0.22\ U_2)\ (E_s - e) \tag{2.3}$$

where

E_s = saturation vapour pressure,
e = actual vapour pressure, and
V_w = potential evaporation from free water surface.

The empirical relation between the potential evapotranspiration (V_p) and the evaporation from free water surface (V_w) for Yangambi is shown by equation (2.4).

$$V_p\ (\text{mm month}^{-1}) = 0.91\ V_w + 2.5 \tag{2.4}$$

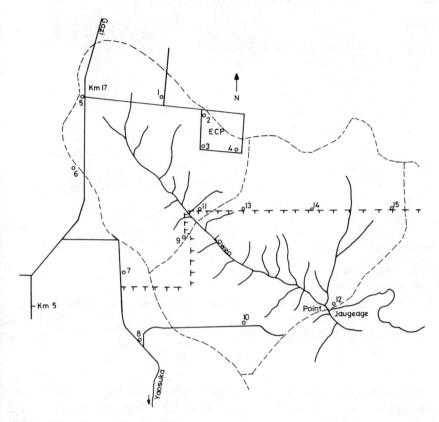

Figure 2.25 Map of the Loweo watershed. The location of the raingauges is shown in numbers 1 to 15

In spite of the variations in monthly estimates, equation (2.4) can be used to obtain reliable estimates of annual potential evapotranspiration. Nevertheless, it is important to realize that potential evapotranspiration is the quantity of water evaporated from a free water surface that receives the same energy load and has identical Bowen ratio as the vegetative cover being considered. The Bowen ratio, which depends on the microclimate within the multi-storey canopy, can be considered uniform over the whole region. With this assumption, the potential evapotranspiration for a natural vegetation cover can be computed using equation (2.5) (Bultot and Dupriez, 1974).

$$V_p = f \cdot E_{0T} \tag{2.5}$$

and f is the transfer factor given by equation (2.6)

$$f = \frac{(1 - \alpha_0) G - N - \Delta Q_T}{(1 - \alpha_T) G - N - \Delta Q_T} \tag{2.6}$$

where

α_T = albedo over the vegetative surface,
α_0 = albedo over the free-water surface,
G = global solar radiation,
N = terrestrial radiation,
ΔQ_T = soil heat flux, and
E_{0T} = evaporation from free water surface.

The annual computed evapotranspiration form a vegetative cover by this formula is generally more than the lysimetric measurements (Table 2.11).

Total evapotranspiration over the basin was estimated by equation (2.7) (Lahaye, 1978).

$$V = P - D \tag{2.7}$$

where

V = evapotranspiration
P = rainfall, and
D = river's flow at the gauging point that includes seepage and surface runoff.

2.5.3. RESULTS

An example of the step-wise water balance according to the method of Thornwaite (Bernard and Frere, 1956) is shown in Table 2.12. The computation of water balance began when the soil-water deficit was replenished by rains from April to May and August to December. The permanent wilting point (PWP) was observed on monthly basis only once in February during the period of Harmattan (Sengele and Crabbe, 1973), although there were many 5-to-10 day periods when soil moisture in the surface reached the permanent wilting point. From the point of view of plant-water requirements, it is advisable to use a short duration of 5 to 10 days. The use of shorter time duration is particularly important during the critical stages of plant growth (Talla and Sengele, 1977).

The data in Table 2.13 indicates that variations in the annual soil-water reserves are not significant. Nevertheless, the water deficit observed from January to March had a significant adverse effect on plant growth. These results are in conformity with observations made earlier by Dupriez (1964). The second period of water deficit observed between June and July is serious for those crops planted late in the season, because the grain development stage may coincide with this period of water deficit. An analysis of data in Figure 2.26 indicates that the return period of a dry season lasting from November to April is 10 years and of the permanent wilting point to exceed 20 days is 5 years. The second short season in June is of less significance for perennial crops.

Table 2.11. Comparison of mean monthly measured (V_{pm}) and calculated (V_{pc}) potential evapotranspiration (period 1974–1978)

Evapotranspiration	J	F	M	A	M	J	Jl	Au	S	O	N	D
Measured (V_{pc})	3.9	4.5	4.2	4.5	4.3	4.0	3.9	4.1	3.9	4.0	3.5	3.8
Calculated (V_{pm})	3.1	3.5	3.3	3.7	2.9	2.9	3.1	3.1	3.1	3.1	3.0	2.8
Deviation	0.8	1.0	0.9	0.8	1.4	1.1	0.8	1.0	0.8	0.9	0.5	1.0

Table 2.12. Monthly values of the components of the water balance at Yangambi, 1977

Components	J	F	M	A	M	J	Jl	Au	S	O	N	D
V_p	90	96	107	103	101	87	72	80	90	94	96	79
P	8	19	144	294	120	49	105	84	205	149	267	109
$P - V_p$	−82	−77	+37	+191	+19	−38	+33	+4	+115	+55	+171	+30
W_a	−82	−159	−122	—	—	−38	−5	—	—	—	—	—
ΔS	−82	−124	−87	—	—	−46	−3	—	—	—	—	—
ΔW_a	228	176	213	300	300	254	297	300	300	300	300	300
Θ_g	39.9	33.4	38.0	48.8	48.8	44.3	48.4	48.8	48.8	48.8	48.8	48.8

V_p = Potential evapotranspiration
P = Precipitation
ΔW_a = Cumulative water deficit
Θ_a = Present soil-water reserve
Θ_g = Gravimetric soil moisture content
ΔS = Change in soil-water reserve

where

$\Theta_g = \Theta_0 + h\, W_a$

Θ_0 = average soil moisture in the profile (per cent) over and above the permanent wilting point

h = depth of the layer multiplied by its bulk density.

Table 2.13. Inter-annual variations in monthly values of water storage (W_a) during the period 1971–1978, assuming the field capacity for these soils to be at 300 cm of water suction

Year	J	F	M	A	M	J	Jl	Au	S	O	N	D
1971	300	247	300	300	300	300	300	300	300	300	300	300
1972	261	270	300	300	300	300	300	300	300	300	300	300
1973	278	296	265	300	300	300	300	300	300	300	300	291
1974	300	268	285	300	300	289	300	300	300	300	300	275
1975	260	248	260	300	300	268	300	300	300	300	300	300
1976	240	228	191	256	300	300	300	300	300	300	300	300
1977	228	176	213	300	300	264	297	300	300	300	300	300
1978	252	237	300	300	300	300	300	300	300	300	300	300

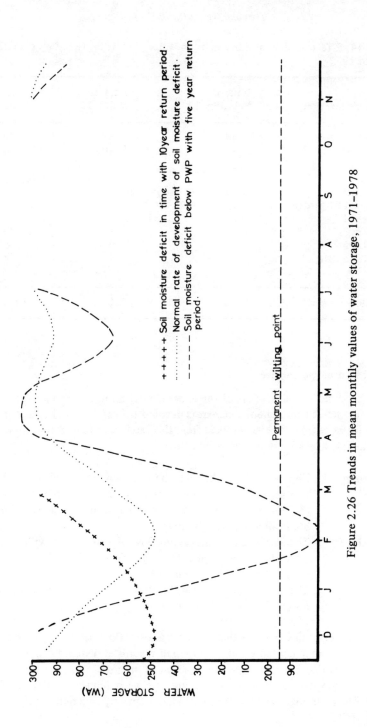

Figure 2.26 Trends in mean monthly values of water storage, 1971–1978

Table 2.14. Effect of high intensity rains on water yield as measured by river discharge in 1977

Rain gauge location	Rain of February 13	Rain of March 15	Rain of March 31
Km 17	44	41	54
Coffee tree	46	40	54
Singili	55	36	51
Moitondi	69	37	53
Wayangonde	68	43	50
Agrostolic test garden	75	67	47
P.C.T. km 1.5	–	32	49
P.C.T. km 3.3	–	42	51
P.C.T. km 5.1	–	40	59
Reservation road	–	69	63
Loweo	–	72	43
Flood flow (1 sec^{-1})	1250	1200	1250

2.5.3.1. Hydrologic Balance

Loweo catchment drains a part of the experimental farm, the oil palm plantation located in the central region, some areas devoted to seasonal food crops, and the natural forest reserve. The water yield from this catchment for 3 dates during 1977 shown in Table 2.14 is briefly described below:

February 13 : Five hours after the rain, the river flow increased to 960 1 sec^{-1} The second peak in this composite rainstorm resulted in a flow of 1250 1 sec^{-1}. The flow decreased to the normal rate about 24 hours after the peak discharge.

March 15 : This rain caused a maximum flow of 1200 1 sec^{-1} within 7 hours after the storm event. The flow decreased to the normal rate about 24 hours after the peak discharge.

March : The maximum flow was 1250 1 sec^{-1}, and it decreased to the normal rate about 24 hours after the peak discharge.

These observations indicate that a flow rate of 1500 1 sec^{-1} may be the maximum for a 50-mm rainfall event. The rainfall records at Yangambi from 1966 to 1978 indicate that 30 rainfall events exceeded 50 mm day^{-1}, out of which 17 exceeded 80 mm day^{-1}. Similar observations have been reported earlier by Bultot (1956). The hydrologic equipment installed has a capacity to monitor a flow rate of 3500 1 sec^{-1}.

Table 2.15. Monthly mean rainfall at various locations in the Loweo watershed

Rain gauge location	J	F	M	A	M	J	Jl	Au	S	O	N	D	Total	Weighting coefficient
1. P.C.T. – FP	128.9	65.5	169.9	152.4	125.4	104.9	64.6	155.0	226.0	149.7	177.0	129.5	1578.8	3.85
2. P.C.T. – NW	128.5	76.3	102.5	154.9	103.8	101.7	66.0	125.7	174.0	185.1	140.7	94.4	1353.6	0.75
3. P.C.T. – SW	132.6	86.5	152.6	152.4	120.2	100.9	72.9	143.0	226.8	216.9	186.5	116.5	1607.5	1.86
4. P.C.T. – SE	127.9	76.7	153.3	159.9	114.9	94.9	82.7	128.7	214.5	184.5	91.3	108.0	1437.3	1.96
5. Km 17	161.2	78.7	136.0	142.1	116.6	140.7	93.5	207.7	202.9	164.9	169.9	102.8	1607.0	2.95
6. Coffee program	160.9	97.9	134.1	158.7	95.1	147.3	98.6	176.6	226.7	173.9	164.2	137.3	1671.3	1.77
7. Singili plot	157.5	95.5	125.7	157.3	117.2	157.9	141.7	154.6	233.1	179.5	151.9	106.4	1678.3	1.06
8. Trial garden	154.5	99.6	120.3	165.1	117.9	128.3	100.5	130.3	217.3	154.6	148.4	141.7	1678.3	0.59
9. Moitondi plot	137.3	67.8	127.2	174.7	119.6	147.0	106.4	159.9	228.9	256.8	180.5	92.5	1703.3	3.78
10. Reservation	148.6	68.5	119.5	163.8	136.5	150.2	101.9	155.3	243.6	274.4	186.6	85.5	1734.4	5.42
11. Wayangonde plot	139.7	70.5	127.9	163.5	106.5	143.1	182.0	181.5	242.1	220.5	159.9	85.3	1846.5	3.86
12. Loweo	130.7	84.0	110.6	171.9	126.3	129.7	105.4	137.1	258.9	231.1	184.6	89.4	2304.9	3.22
13. Km 1.5	142.0	71.2	140.1	180.1	125.1	137.9	118.9	190.8	236.6	260.6	179.4	99.7	1782.4	2.71
14. Km 3.3	140.3	71.2	140.1	180.1	125.1	137.9	118.9	190.8	236.6	260.6	179.4	100.9	1791.5	4.75
15. Km 5.1	144.4	76.9	157.8	171.9	142.1	138.0	105.6	182.8	238.6	239.7	173.7	101.5	1773.0	7.74

2.5.3.2. Evapotranspiration

Mean river discharge was calculated assuming the flood flow of 1500 1 sec^{-1} and the normal flow of 500 1 sec^{-1}. The drainage component (D) in equation (2.7) was estimated to be 341 mm or 21 per cent of the 1600 mm rainfall received over the catchment. Therefore the evapotranspiration over the catchment amounted to 79 per cent of the rainfall received.

2.5.4. DISCUSSION

The results of this investigation are summarized in Table 2.16. A comparison of the evapotranspiration over the catchment with the Pan evaporation indicates that the Pan evaporation was less than evapotranspiration only during a short period of May and June. Similar results were reported by Bernard and Frère (1956) and Ringoet *et al.* (1961). An analysis of the daily fluctuations in water yield record indicated a sinusoidal function with a peak flow rate recorded at about midnight and the low flow during mid-day, corresponding with periods of the least and the maximum evapotranspiration, respectively. However, the correlation of the basin evapotranspiration with potential evapotranspiration measured over grassed surface by lysimetric method was low (equation (2.8)).

$$V_b = 62.1 + 0.25 \ V_g, \qquad r = 0.54 \tag{2.8}$$

where

V_b = evapotranspiration over the catchment, and
V_g = evapotranspiration over the grassed surface.

The analysis of runoff hydrograph indicated that water deficit was more on sandy than on clayey soils. This argument is also supported by the data in Table 2.13, and by the peak flow rate in the stream. In sandy soils, the gravitational water is drained out of the profile within a short period after the rain. Similar observations were reported by Dupriez (1964).

The effects of mixed-cropping on the available soil moisture reserve is shown in Figure 2.27. The available water reserves are related to the root system of the component crops. From these results, it is difficult to assess the adverse effects of drought without the necessary information on soil characteristics and crops to be grown.

There exists a linear correlation between the catchment evaporation and the potential evapotranspiration measured over grass, although the correlation coefficient is rather low ($r = 0.50$). This low correlation may partly be attributed to discrepancies in water balance obtained with draining lysimeters, where the drainage due to a rainfall received towards the end of a month will be completed and accounted for in the following month.

Error caused by delayed rains are negligible during the dry season, when the

Table 2.16. Comparative values of the major components of the water balance (mm)

Month	Rains in the basin	Loweo flow	Percentage drainage	Evapotranspiration	Rain (kms)	Potential evapotranspiration (Paspalum)	Actual evapotranspiration (Paspalum)	Drainage actual evapotranspiration (Paspalum)
January								
February								
March	149.1	23.6	15.9	125.3	127.5	129.6	101.1	26.4
April	243.1	26.3	10.8	216.8	172.1	103.4	130.7	41.4
May	100.5	21.3	21.2	79.6	92.6	89.5	89.0	3.6
June	97.5	20.2	20.7	77.3	149.3	104.0	104.2	45.1
July	118.1	19.9	16.9	98.2	118.6	63.3	63.1	55.5
August	159.5	19.8	12.4	139.7	112.5	76.2	76.8	35.7
September	213.8	21.9	10.2	191.9	104.8	93.5	97.1	107.7
October	167.1	24.5	14.7	142.6	243.0	104.3	110.6	132.4
November	121.2	19.9	16.4	101.3	103.7	98.5	63.1	40.6
December	127.1	20.4	16.1	106.7	119.8	77.8	102.0	17.9
Total	1497.0	218.0	14.6	1279.0	1444.3	945.1	937.7	506.3

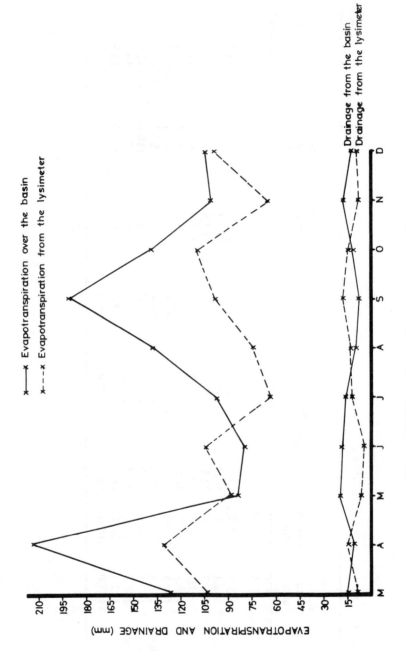

Figure 2.27 Trends in monthly evapotranspiration and drainage

correlation between the catchment evapotranspiration and the evapotranspiration over grass is generally high (Dupriez, 1959).

2.5.5. CONCLUSIONS

1. Evapotranspiration over a catchment can be estimated by the water balance method. Assuming certain biological and climatic conditions, evapotranspiration was estimated to be 79 per cent of the precipitation received.
2. Pan evaporation gives an overestimation of evapotranspiration during humid periods, and is equal or slightly lower than evapotranspiration in dry months.
3. Investigations of water balance from climatic data can enable the assessment of soil moisture reserve. It is necessary to adopt the 5 or 10 days scale in order to better understand the water balance. With the erratic distribution of rainfall and the desiccating winds in dry periods, it is important to consider soil and edaphic factors.
4. This method of water balances is far from being perfect and gives only approximate measures of evapotranspiration.

REFERENCES

Bernard F., and Frere, M., 1956, Expression pratique de l'evapotranspiration potentielle d'une surface naturelle en climat tropicale. *Muscllanea geofisica.* Lunda (Angola)

Bultot, F., 1956, Etude statistique des pluies intenses en un point et sur une aire au Congo. *Publication INEAC.*

Bultot, F., and Dupriez, G. L., 1968, Le bac evaporatoire en usage dans le reseau hydrometeorologique Belge, *Extrait Bull. d'AIHS*, XIIIe annee, No. 2.

Bultot, F., and Dupriez, G. L., 1974, L'evapotranspiration potentielle des bassins hydrographiques en Belgique, *Publ. serie A No. 25*, Int. Roy. Met. Belge.

Dupriez, G. L., 1959, La cuve lysimetrique de Thornwaite, comme instrument de mesure de l'evapotranspiration, en regions equatoriales, *Extrait Bull, d'AIHS*, No. 49.

Dupriez, G. L., 1964, L'evaporation et les besoins en eau de differentes cultures dans la Region de M'Vuazi (Bas-Congo). *Serie Scient.*, No. 106, Publ. INEAC.

Lahaye, J. P., 1978, Etudes hydrologiques du barrage de Bagre sur la Volta blanche. *Publication CIEH Rep. Haute-Volta.*

Penman, H., 1948, Natural evaporation from open water, bare soil, and grass proceedings. *Roy. Soc., Series A*, CXCIII London.

Ringoet, M., Molle, A. L., and Myttenaire, C. O., 1961, L'evapotranspiration et la croissance des vegetaux dans le cadre du bilan energitique serie. *Scient. No. 92 INEAC.*

Sengele, N., and Crabbe, M., 1973, *Etude elimatique de Yangambi.*, (unpublished) INERA.

Talla, J., and Sengele, N., 1977, Contribution a l'etude de la Cryptodesertification au Zaire. *Conference des Nations Unies sur la Desertification.* Nairobi (Kenya) 29 August–9 September, 1977.

PART 3

Change in Land-use and
Hydrological Conditions

Tropical Agricultural Hydrology
Edited by R. Lal and E. W. Russell
© 1981, John Wiley & Sons Ltd.

3.1

The Effects of Forest Clearing on Soils and Sedimentation

S. H. KUNKLE AND A. J. DYE

3.1.1. INTRODUCTION

In many countries deforestation is a major cause of soil degradation, erosion, and sedimentation. As population pressures mount, forests are cleared for fields, grazing space, or new farms, but primarily trees are felled for fuel-wood—the main source of energy in most of the world. Deforestation is such a common problem that globally the area under forests is decreasing by about 20 million hectares each year.

Many steep, arid, or other marginal sites would formerly have been left protected by trees or brush, but today land and wood shortages force farmers to clear these marginal lands. Even shrubs are cut for fuel in many places, leaving soil exposed to erosion. Another tragic loss also occurs as remote forested areas are cleared, because unique species of plants or animals may be destroyed, losing forever whatever special contributions these organisms might someday have made to mankind.

In many areas farmers practise shifting cultivation, also called swidden, slash-and-burn agriculture, or milpa, where a small peice of land is cleared and burned, then crops are grown for one or more years, before moving on to repeat the cycle on another patch of ground. During the first few years after clearing, the soil is relatively fertile and crops are good, but gradually the soil becomes worn out, and usually weed competition increases. Then the farmer moves on to a new site, allowing trees to reclaim the area once more and rebuild the soil during a fallow period. Traditionally, a shifting cultivator would not return to clear the same piece of land for 15 to 20 years, when the soil has recovered most of its fertility, structure, and organic content. But because of population pressures, these farmers are reducing the fallow period to a shorter time. Crop yields decline, erosion becomes a more serious concern, and some lands are overgrown with noxious grasses and other weeds tolerant to soils of low fertility. In more arid situations, forest or shrub cutting may lead to wind erosion, dust storms, blowing sand, and other down-wind impacts.

Besides the erosion or degradation of soil on a cleared site itself, deforestation can lead to damages downstream, such as reservoir siltation, sediment damage to irrigation works, higher flood peak flows during brief storms and debris problems (logs, branches, etc.) which cause serious local flooding, reducing dry-season stream flows, and the drying up of springs and wells. It should be recognized that the impact of deforestation on streams is much greater if the cutting occurs near stream channels. Therefore, channels can be protected by leaving strips of trees to act as buffer zones against sediment loads and channel erosion.

Recent studies in India at 17 major reservoirs, for example, show that the reservoirs are filling at about three times the expected rate, because 'very vast areas of forest have been deforested' (Tejwani, 1977). Likewise, high rates of siltation were measured by Rapp (1976) in Tanzania at three reservoirs (up to 729 m^3 km^{-2} yr^{-1}), and one other catchment he studied was yielding a massive 13,500 m^3 km^{-2} yr^{-1}.

One of the most dramatic examples of the impact of deforestation is occurring in Panama. The watersheds that feed runoff into the Panama Canal originally were covered with dense rain forests. But in recent years, over one-third of the forests have been cut down and the land cleared for cultivation or for pastures. The impact on the canal's operations is becoming critical during dry seasons, since the reduced area of forest cover is less able to maintain the necessary steady water flow into the canal. In 1977, for example, a serious drought imparied the canal's operations. Each ship passing through the channel requires 192,000 m^3 of stored fresh water and, during the drought, water to operate the locks was in short supply. There is no doubt that the deforestation had aggravated these droughty conditions. Deforestation has also caused serious sedimentation effects. Because of sediment, ship channels are blocked and there is also a reduced storage capacity of Lake Alajuela which supplies water to the locks. The future of the canal is at stake, and for this reason a major watershed protection program including soil conservation and reforestation, is to be implemented in the near future (Wadsworth, 1978).

3.1.2. SOME COMMON CAUSES AND EFFECTS OF DEFORESTATION

3.1.2.1. Cutting Forests for Cultivation

Historically, shifting cultivation, like nomadic use of grazing lands, has not necessarily been a cause of high erosion. But in recent practice, shifting cultivation is usually detrimental for one or more of the following reasons: (a) the tree growth (fallow) period is too short compared with the cultivation period, (b) the clearings are so large that the surrounding forest cannot reseed the land, (c) slopes are too steep to prevent erosion, and (d) soil fertility is depleted to the extent the soils are nearly sterile (Watters, 1971).

Increased ecological understanding of tropical forests should contribute to more effective management. Much of the research to date has been fragmented and often with conflicting results. A recent study in Peru (Seubert, Sanchez, and Valverde,

1977) was established to compare the effects of slash-and-burn clearing with those observed from bulldozer clearing. Crop yields were higher and soil properties were more favourable in slash-and-burn clearings than on lands cleared by the tractors. Bulldozer clearing resulted in increased bulk densities and reduced infiltration rates. The exchangeable calcium and magnesium available for plants in the topsoil more than doubled in the slash-and-burn area while there was little change in the availability of the nutrients in the bulldozed plots. Even if the area is to go into permanent cultivation, the traditional slash-and-burn method of land clearing appears to be more advantageous in terms of nutrients than tree removal by tractor. Clearing costs also were less expensive with the slash-and-burn method.

While it superficially appears that tropical forested ecosystems are very productive and therefore should be suitable areas for conversion into croplands, the results from such actions have not always been successful. Yields normally decline rapidly, and even slash-and-burn agriculture can disrupt those nutrient-conserving mechanisms which maintain higher productivity in undisturbed forests. It has become apparent that the recycling of nutrients within the ecosystems is dependent upon humus layers and root mats. This process has been verified by a recent field study (Stark and Jordan, 1978) which showed that nutrients can be held in the root mat and humus layer, above the soil surface, until absorbed by roots. According to some observers mycorrhizae in these upper layers help digest dead organic matter and pass nutrients through their hyphae to the root cells of plants (Benge, 1978).

3.1.2.2. How Long Under Fallow

As Ofori (1973) points out, in dense forest areas in the tropics, with 10 to 15 years under fallow, there is a large accumulation of organic matter through leaf litter and the fine-root remains of the vegetation, and the nutrient level of the soils is generally high. But unfortunately the fallow period is often reduced to a much shorter time. As a result, fertility is insufficiently restored and soon the land must be abandoned because crop yields are low. A study in Venezuela (Watters, 1971) indicates that a crop:fallow time ratio of at least 1:5 is needed, for example, 3 years of cropping and then 15 years of fallow. Other researchers likewise have calculated that perhaps 15 years is the minimum time period required for organic matter, nutrients, porosity, and other soil factors to return to a level adequate to make cropping reasonably productive; the exact time, of course, depends on the particular location. Therefore, theoretically, the practice of shifting cultivation is acceptable if the forest clearing is restricted to gentle slopes, if an adequate cropping:fallow ratio is observed, and if care is taken that organic matter and essential microorganisms are not destroyed by erosion or exceptionally severe burning.

3.1.2.3. Cutting Trees for Fuel

In arid zones, deforestation is typically the result of the search for fuelwood or wood for charcoal as much as for production. In fact, over 80 per cent of wood cut in developing countries is used for fuel (Arnold, 1978). Wind erosion can be a common problem following tree cutting in the more arid areas. For example, near Bara Village in the Kordofan Provinces of the Sudan, lands were formerly used with a cycle as follows: 4 to 5 years cultivation; 3 to 5 years of fallow, while a new stand of gum arabic trees sprouted from the stumps; 7 to 10 years of tapping the trees for gum; and finally a harvest of the trees for making charcoal and then the return to the cultivation phases to complete the cycle. The mean total length of this traditional cycle was about 17 years. But in recent times, the farmers have compressed the cycle to only 9 years, cultivating 2 to 3 years, leaving 1 to 3 years under fallow, and tapping the trees only 3 to 5 years. This shorter cycle leads to decline in soil fertility. The land becomes exhausted, and farmers in the area now abandon plots permanently after about three such short cycles. The land is also exposed to wind erosion, and in the traditional village cultivation area it is now common to see soil blown away and drifting sand covering fields (Hammer-Diggerness, 1977).

It also must be recognized that recent increased prices of kerosene or other fuels exert even greater pressures on forests. The impact in some countries is extreme, such as seen along the northern coast of Haiti, where the forests are being destroyed and turned into eroded badlands in the quest for charcoal (Kunkle, field observations, 1976).

3.1.2.4. Clearing Land for Grazing

In many tropical countries, erosion, sedimentation, and hydrologic changes often are the result of the conversion of forests into grazing lands in the upper portions of watersheds. This is especially the case in Latin America, where pressures to produce beef for export has lead to increased clearing of marginal lands. The impact of converting forests to grasslands is well documented by various studies. For example, in California, USA, steep catchments were converted from essentially deep rooting small trees and brush cover (with roots down to 20 metres) to shallow-rooted grasses and herbaceous vegetation. Grazing was carefully controlled, yet after the land was cleared the incidences of landslides, soil creep, and mudflow began to occur in one catchment at a rate about 10 times as great as before clearance, and runoff was increased by 50 per cent (Murphy, 1976).

3.1.2.5. Hauling Timber and Building Roads

In some of the humid tropical countries, commercial forest harvesting can have a major impact on watersheds, particularly because of the effect of building the roads required for the wood extraction. The roads have a secondary effect as well, in that often shifting cultivators or others then gain access to an area that formerly was isolated from their potential impacts. Several studies have shown that the roads frequently become the number one source of sediment, and dirt roads frequently serve as the source of sediment-laden runoff causing severe gully formation. The problem is that often roads are neither properly laid out nor maintained. In brief, roads built into a forest area can sometimes have a greater impact, especially downstream, than the deforestation as such.

3.1.3. SOME EFFORTS TO TACKLE THE PROBLEM

In contrast to the gloom described above, there are good opportunities to make use of trees to protect watersheds and also to improve crop production. Application of known conservation techniques, such as the use of tree windbreaks, can protect against wind erosion and also enhance crop production. Trees also can help rehabilitate soils, by improving soil structure, by adding organic matter, and by bringing up nutrients from the deeper subsoil. In some cases, trees and crops may be grown in succession or in combination, to help protect or improve soils. Finally, marginal lands, for example those with salty or steep slopes, may be reforested with certain types of trees to provide fuelwood. Whereas such land may be of no value for farming given the growing shortage and increasing value of firewood, even areas formerly used for the marginal production of crops may now be profitably used for fuelwood or charcoal production. Fuel can sometimes cost more than food. It is essential to develop systems in which trees can best be filled into schemes for rural development and river basin planning.

3.1.3.1. Integrating Forestry with Agriculture

How should soil and water conservation be fitted into a rural development plan? The starting point is to classify each piece of land according to its inherent capability, to identify those sites which are best suited for farming, forestry, or other uses, including identification of conservation works needed. Some workers prefer the term 'land suitability', to that of land capability, recognizing that a piece of land is capable of being used for many purposes, but is more beneficially put to a particular use. For example, even a relatively flat piece of land may best be used to grow fuelwood, if fuelwood fetches a better income than, say, yams. At today's high fuelwood costs in many countries, and considering the low investment cost for forestry as compared to cultivation, areas are likely to be rated suitable for tree growing which by a physical classification would be capable of cultivation.

Forestry and various agricultural uses of land have been integrated successfully in a number of cases. In some areas in China, for example, tree windbreaks and other forestry conservation works have been credited with doubling agricultural yields in the last 10 years, with thousands of kilometres of shelterbelts established to protect fields (FAO, 1978).

A current project by the Honduran Forest Development Corporation in cooperation with the Food and Agriculture Organization (FAO), German aid, and the US Peace Corps is designed to apply good conservation practices to watersheds in two mountaineous regions. The strategy of improving land use combines both social and technical aspects through four activities, working with groups or cooperatives of small farmers:

(a) forest extension work in existing forests;
(b) soil conservation works, such as terraces, for the small farmers, using food for work as an incentive;
(c) agricultural extension for the introduction of new cultivation techniques on the land which is terraced;
(d) reforestation of the cutover or marginal lands, such as on the steeper slopes above the terraces (Tschinkel, 1978; and personal communications).

The advantages of integrating these activities have been demonstrated and this project model appears to be a good way to encourage the small farmers to increase their productivity while also protecting the watersheds.

Farmers need incentives to carry out conservation. The Government of Indonesia/ FAO project in Indonesia is a good example of combining conservation, largely erosion control, with benefits (profits) for the farmers. Although conceived originally as a costly land rescue operation, the rehabilitation project is, in fact, leading to an economically sound investment. Erosion control is achieved by using a suitable mixture of trees interplanted with fodder crops on slopes of over 50 per cent. This offered some quick returns for the farmers. The combined package was: (1) *Pinus merkussi* for permanent cover, by-product resin, and eventually wood; (2) *Albizia falcata* to improve soil quality; (3) underplanting of elephant grass (*Pennisetum purpureum*) and legumes, to provide income on a short-term basis (FAO, 1976).

An interesting and apparently successful forestry project for farmers is being developed in the Sudan. In the Nile and northern province, valuable agricultural land has been buried by sand dunes, with the crops damaged and the irrigation canals clogged. Therefore, a project has been implemented to establish shelter-belts (i.e. tree windbreaks), which will serve several purposes:

(a) protect the fields from the sand and help crop production;
(b) provide fuelwood and construction poles;
(c) provide fodder for livestock;
(d) ameliorate the living conditions and microclimate in the area.

Trees were raised in nurseries, mainly *Prospis chilensis*, but also *Eucalyptus camaldulensis*, *E. microtheca*, and *E. conocarpus*. A number of shelterbelt species from abroad also are being tested (Khalifa, 1978; and personal communications, 1979).

An important Sudanese government directive supports the above efforts. This states that no plan for any new agricultural scheme will be accepted unless the position of forestry is clearly defined. It is proposed in the directive that at least 5 per cent of the area under agricultural development in arid and semi-arid zones of the country be put into trees. At first it was difficult to convince local farmers of the need to set aside strips of their land for shelter-belts. The Sudanese Farmers Association took on this task, and the project has demonstrated the value of shelter-belts so well that farmers are now requesting tree seedlings and advice for establishing their own windbreaks.

In many countries where shifting cultivation is indigenous various methods have been developed to increase the productivity of the area. One well-known forestry system is the taungya where migrant farmers live temporarily on the land that is being planted with a commercial forest crop such as teak (*Gmelina arbore*), caring for the growing trees and growing their food crops between them until they became so large they shade out the crops. Another system is to plant trees at a wide spacing so allowing the continuous cultivation of food crops which may include cereals, beans, peas, and cabbages. The trees may produce food for sale or consumption, such as cashew nuts or various tropical fruits, or they may be chosen to help raise the soil fertility. Some tropical trees have remarkable root systems, and their deep roots may serve to bring nutrients from deep in the soil to upper layers where the crops are planted. For example, *Prosopis cineraria* sends roots down to 30 m or more while *Acacia tortilis* roots may spread out 40 m more laterally (Poulsen, 1978).

In the Sahelian zone of Senegal, *Acacia albida*, a leguminous tree which adds nitrogen to the soil, is planted at 10 m X 10 m spacing to rehabilitate worn out soil and enrich the savanna for grazing purposes. Forest guards are used to protect the new trees. *Acacia albida* is also interplanted in millet fields there to add nutrients to the soil, and 44 trees per hectare are said to add the equivalent of 50 to 60 metric tons of manure per hectare to the top 10 cm of soil if the leaves and pods are allowed to fall and remain on the ground (personal communications from T. Greathouse in 1977).

Some research using other leguminous trees for improving soil looks promising. *Leucaena leucocephala* for example not only adds nitrogen to the soil, but also may have mycorrhizae which are able to metabolize phosphorus and other minerals, take them up in the mycorrhizal mass, then slowly release these nutrients for use by plants. (Went and Stark, 1968). The tree itself has other useful characteristics. It can sprout from stump can tolerate alkaline soils, grows rapidly, and is useful for both fodder and fuelwood. Studies underway in the Philippines have shown the common local *L. leucocephala* will yield as much as 88 m^3 ha^{-1} yr^{-1} of wood on some fertile deep soils or several times the yield of many commonly-used tree species (Benge, 1978; and Benge and Curran, 1976). The yields, however, are much less on poor soils, especially on previously eroded sites.

There are other interesting possibilities with these various 'agri-silvicultural' or 'agro-forestry' systems, as they are sometimes called, with trees and crops grown together or in succession. To gain knowledge on these various systems the International Council for Research in Agro-forestry (ICRAF) was recently set up in Nairobi. Its objective, briefly, is to study traditional agro-forestry systems and techniques and to develop improved agro-forestry measures which can be used for the prevention of land degradation, or for land rehabilitation, while helping to maintain or improve the productivity of the land. Much remains to be learned in this ancient but poorly understood field.

3.1.3.2. Using Trees for Rehabilitation Work

On eroded areas, innovations are sometimes necessary if we are to pick trees or shrubs that meet as many of the following pre-requisites as are feasible:

(a) appeal of the tree or shrub to local farmers or local use, i.e. benefits in addition to erosion control *per se*; for example, some species are good for fuel, some produce good fodder, etc.:

(b) good survival of the young trees and fast growth on impoverished sites;

(c) ability to produce a large amount of litter to protect the soil from raindrop splash;

(d) a strong and wide-spreading root system with numerous fibrous roots is desirable if the trees are to be used for soil conservation, but such a root system is undesirable if food crops are to be interplanted; in landslide zones, deep mechanically-strong roots are usually essential;

(e) ease of establishment of trees or shrubs and little need for maintenance; for example, with some species simple plant cuttings can be used;

(f) capacity to retain foliage at least through the rainy season (in flatter areas this point is less critical);

(g) resistance to insects, disease, or rodents.

Few species would meet all the above criteria, but a species is normally selected because it meets a few of these points very well.

Appeal of the tree or shrub to local people is critical. The first point above can be the most critical, since often the main problem is to pick tree species that the local farmers like, for example, a tree that yields a nut, oil, seed, fodder, good quality fuelwood, or other needed products. Everyone takes better care of something that is highly valued. For instance, bamboo used to check soil erosion also can be a source of fodder. In south-eastern Nigeria, cashew trees (*Anacardium occidentale*) were first planted as an anti-erosion measure, but now the fruits and seeds are also important by-products used for both food and commercial purposes (personal communication from Dr. B. N. Okigbo, Nigeria, in 1978).

Trees for erosion control also can be used for livestock fodder. In Nepal, for example, fodder plantations can be planted at 1100 to 1500 trees per hectare, using *Ficus* spp., *Albizia* spp., *Caetanopsis* spp., *Leucaena* spp., and other types. Starting at 5 years, the fodder can be harvested from the trees at any time, year after year, but especially during the critical dry season when the grass supply is scarce. One mature fodder tree supplies a month's supplementary feed for a cow or a buffalo, with the trees yielding 5 to 12.5 metric tons of leaves per hectare annually after about 10 years of age (Harcharik and Kunkle, 1978). However, in some reforestation sites it may be essential to use non-palatable species, to discourage animals from destroying the trees. For example, *Prosopis juliflo* foliage generally is not appealing to livestock, yet the tree grows in poor soils, can tolerate drought well and produces fuelwood fairly rapidly on sites that otherwise are useless.

Time is also critical, and most poor rural people have no interest in tree products that are available only a generation from now. One innovation to save time that has met with success in Indonesia is the planting of *Calliandra calothyrus*. This vigorously coppicing and branching tree shrub is cut back at age 8 to 12 months to encourage coppicing (stump sprouting), and it produces usable fuelwood within the first year. Mixing species also can help to provide some early benefits. For example, certain shrubby *Lespedeza*, which yield fodder and fuel (woody stems) as early as the first year, can be interplanted with tree species that produce both fuelwood and industrial wood over the longer period (Arnold, 1978).

Trees can be used to put salt-affected lands back into productivity. Work in India has shown that certain eucalyptus species, *Prosopis juliflora, Acacia Arabica,* as well as other trees are useful to rehabilitate salt-affected lands which have been lying barren for decades. These lands can once more become sources of fuel, fodder, or building posts (Yadav, 1977).

In areas where sand dunes have covered former fields, trees have been used for rehabilitation with considerable success in Libya, Morrocco, Algeria, Tunisia, Iran, and other countries (Kunkle, 1976; Lal, 1973). Studies by Jensen (1976) demonstrated that tree roots are often able to reach down to the original soil layers under the sand where they can tap moisture and nutrients and this allows them to survive and grow on the dunes.

3.1.4. CONCLUSIONS

There are many reasons why the rate of deforestation has increased in recent years. Forests are cleared for farming, grazing, fuelwood, or other purposes, but usually the clearing stems from pressures of expanding populations and shortages of land, food, and resources. While it appears that lands supporting tropical forest ecosystems are very productive, in fact these areas often are not capable of supporting cultivation, and attempts to do so frequently result in eroded wastelands or low marginal farming at best. Traditional shifting cultivators accepted the productivity constraints of forest lands and allowed the land to rest for many years to replenish the humus layer and permit soil regeneration between the periods of cropping. However, land shortages have forced many slash-and-burn farmers to shorten traditional fallow periods, often with land degradation occuring in the form of disrupted nutrient cycles, hydrologic, and erosion problems, reservoir siltation, and local downstream flooding.

In some countries shifting cultivation is only an intermediate stage in the conversion of forested ecosystems into large-scale cattle production systems. This conversion of forest to grazing land can however lead to very serious erosion if the forest is essential for watershed protection, for if this ecosystem is fragile not even shifting cultivation should be permitted. Unfortunately not only has shifting cultivation taken place in these areas but also the cultivators have often been making a cash income by selling their rights to the land they have cleared to cattle ranchers and moved on to clear more forest.

Rehabilitation of eroded areas via reforestation with locally desirable species not only halts the erosion, but also can produce fodder, food, or other valuable benefits for rural people.

Attention is now being focused on ways to develop ecologically viable systems to use tropical forest areas while protecting fragile soils. Some of these systems are slight modifications of approaches used by shifting cultivators. Forestry conservation activities coupled with agronomic uses have served to increase yields and actually improve soil fertility.

From a socioeconomic viewpoint, a pattern of land use should be designed to incorporate production systems which permit rural families to sustain an acceptable standard of living while protecting resources.

More integrated efforts are needed in future research activities. More accurate knowledge of a region, in conjunction with well-developed extension systems and adequate training plans, will permit wiser use of forest resources, and help prevent large scale destruction of tropical forests.

REFERENCES

Arnold. J. E. M., 1978, Wood energy and rural communities, *Position Paper, Eighth World Forestry Congress*, Jakarta, 16–28 Oct. 1978, Forestry Department, FAO, Rome, 31 pp.

Benge, M. D., 1978, Renewable energy and charcoal production, *Agency for International Development*, Washington, D.C. 29 pp.

Benge, M. D., and Curran, H., 1976, Bayani (Giant Ipil-ipil, *Leucaena leucocephala*) —A source of fertilizer, feed, and energy for the Philippines. *USAID Agriculture Development Series*, USAID Magsaysay Center, Manila, Philippines, 26 pp.

FAO, 1976, Food and agriculture organization of the United Nations, *Upper Solo Watershed Management and Upland Development: Indonesia*, United Nations Development Programme, Rome, 48 pp.

FAO, 1978, *Forestry and Rural Communities*, Forestry Department, Food and Agriculture Organization of the United Nations, 56 pp.

Hammer-Diggerness, T., 1977, *Wood for fuel-energy crisis implying desertification the case of Bara, the Sudan*, Geografisk Institut, University of Bergen, Bergen, *Master's thesis*, 128 pp. plus *Addendum*.

Harcharik, D. A., and Kunkle, S. H., 1978, Forest plantation for rehabilitating eroded lands, pp. 81–101, In *FAO Conservation Guide No. 4*, Forestry Department FAO, Rome.

Jensen, A. M., 1976, A review of some dune afforestation procedures, In *FAO Conservation Guide No. 3*, Forestry Department FAO, Rome, 125 pp.

Khalifa, K. O., 1978, *Agri-Silviculture and Desert Encroachment Control in the Nile Province—A Sudanese Experience Project Report*, Sudan Council of Churches, P.O. Box 460, Khartoum, 8 pp.

Kunkle, S. H., 1976, Forestry support for agriculture through watershed management, windbreaks and other conservation measures, *Position Paper, Eighth World Forestry Congress*, Jakarta, 16–28 Oct. 1978. FAO Forestry Department, FAO, Rome, 28 pp.

Lal, R., 1973, Soil erosion and shifting agriculture, In *Proceedings, Regional*

Seminar on Shifting Cultivation and Soil Conservation in Africa, Ibadan, Nigeria, 2–21 July 1973. FAO Soils Bulletin No. 24, 248 pp.

Murphy, A. H., 1976, Watershed management increases rangeland productivity, *California Agriculture*, **30(7)**, 16–21.

Ofori, C. S., 1973, Shifting Cultivation—Reasons Underlying Its Practice. In *Proceedings, Regional Seminar on Shifting Cultivation and Soil Conservation in Africa*, Ibadan, Nigeria 2–21 July 1973. FAO Soils Bulletin No. 24, 248 pp.

Poulsen, G., 1978, Man and tree in Tropical Africa, *IDRC-Paper No. 101e*, International Development Research Centre, Ottawa, Canada, 31 pp.

Rapp, A., 1976, *Soil Erosion and Reservoir Sedimentation—Case Studies in Tanzania*, Expert Consultation on Soil Conservation and Management in Developing Countries, 22–26 November 1976. FAO, Rome, 11pp.

Seubert, C. E., Sanchez. P. A., and Valverde, C., 1977, Effects of land clearing methods on soil properties of an Ultisol and crop performance in the Amazon jungle of Peru, *Tropical Agriculture (Trinidad)*, **54(4)**, 307–321.

Stark, N. M., and Jordan, C. F., 1978, Nutrient retention by the root mat of an Amazonian rain forest, *Ecology*, **59(3)**, 434–437.

Tejwani, K. G., 1977, Trees reduce floods, *Indian Farming*, **26(11)**, 57.

Tschinkel, H., 1978, Watershed management in Honduras, In *Final Report, 12th Session of the Working Party on the Management of Mountain Watersheds*, Forestry Department, FAO, Rome, 29 pp.

Wadsworth, F., 1978, Deforestation—death to the Panama Canal, pp. 22–25, In *Proceedings of the US Strategy Conference on Tropical Deforestation*, US Agency for International Development, Washington, D.C. 78 pp.

Watters, R. F., 1971, Shifting cultivation in Latin America, *FAO Forestry Development Paper No. 17*, Food and Agriculture Organization of the United Nations, Rome, 305 pp.

Went, F. W., and Stark, N., 1968, Mycorrhiza, *Bioscience*, **18**, 1035–1039.

Yadav, J. S. P., 1977, Tree growth on salt-affected lands, *Indian Farming*, **26(11)**, 43–45.

Tropical Agricultural Hydrology
Edited by R. Lal and E. W. Russell
© 1981, John Wiley & Sons Ltd.

3.2

An Evaluation of Land-clearing Methods for Forest Plantations in Nigeria

J. B. BALL

3.2.1. INTRODUCTION

The demand for wood products in Nigeria is increasing so rapidly that the natural forests are unlikely to be able to supply the quantities of industrial roundwood required in fifteen years' time. Plantations of fast-growing indigenous and exotic trees could provide one method of meeting the demand, but the programme of conversion of logged high forest will have to be rapidly increased from about 20,000 ha yr^{-1} in the moist lowland forest and from small pilot schemes in the Guinea savanna at present. Table 3.1 shows the programme that will be needed if present demand predictions are correct.

Clearing for and weeding of the plantations are two operations that must be carried out to a high standard to ensure successful establishment, but they can be constrained due to lack of labour at critical seasons. Mechanized techniques have therefore been developed for both processes, and this paper discusses the criteria for selection of these techniques that have been used in planning projects in the moist lowland forest and Guinea savanna zones.

3.2.2. SITE CHARACTERISTICS

Topography and drainage of the Nigerian forest reserves are generally not constraints to mechanical clearing, since steep slopes should not be cleared because of the risk of erosion, and seasonally flooded valley bottoms are unsuited to most plantation species. Some sites have more suitable topography than others; for instance, most of the reserves in the Guinea and derived savanna are fairly flat and intersected by fewer valleys than reserves on basement complex rocks in the moist lowland forest. Where there is high rainfall in the moist lowland forest, areas lying on sedimentary deposits are level, but are not, however, suitable for all plantation species due to the very rapid loss of fertility on clearing.

Table 3.1. A suggested 20-yr forest plantation development programme for Nigeria (from FAO, 1979b)

	Area (in 1,000 ha)				
	1981–85	1986–90	1991–95	1996–2000	Total
Moist Forest and Derived savanna	150	200	250	300	900
Guinea savanna	30	100	250	400	780

Table 3.2. Climatic data for some vegetation zones of Nigeria (from FAO, 1965)

	Lowland moist forest	Derived savanna	S. Guinea savanna	N. Guinea savanna
Rainfall (mm)	1300–2500	1100–1700	1100–1600	900–1300
Length of dry season (months)	1–3	3–4	4–5	5–6
Potential evapotranspiration (mm)	1000–1200	1200–1400	1500–1800	1900–2200
Soil moisture deficit[a] (mm)	180–500	600–800	900–1100	1200–1800

[a]Soil moisture deficit is the difference between the potential evapotranspiration and the rainfall in the dry season.

The soil type was generally not considered to be a constraint because on the ferruginous soils derived from the basement complex, on both the moist lowland forest and the savanna zone, the upper horizons are well drained. The gravel layer underlying these horizons and the infertile subsoil should not be brought to the surface if possible, as tree growth appears to be reduced where this occurs. The ferralitic soils, derived from the basement complex or from sedimentary deposits, were often more clayey in texture but based upon experience in Kenya where mechanical clearing and cultivation was carried out throughout the year on very heavy clays, the clay texture is not expected to be a constraint. All of the soil types are highly erodible with most of the soil loss appearing to occur in the heavy storms at the beginning of the rainy season when soil is fully exposed after burning the debris from the clearing.

Climate varies widely over the potential plantation sites. Some illustrative figures are given in Table 3.2.

Climate affects the choice of clearing method in as much that the longer the dry season lasts the better is the burning of the debris. Even stumps may be burned and completion of the clearing early in the dry season is less critical. Climate also determines the weeding method for many plantation species, and the weeding method determines the clearing method, as discussed in Sections 3.2.3 and 3.2.4.

Table 3.3. Characteristics of forest and woodland in some vegetation zones of Nigeria

	Stem/ha		B.h. diameter Class (cm)		Basal area (m^2/ha)
	5 cm+	15 cm+	Mode	Upper	
(1) Lowland moist forest	Variable	80–160	20–30	150+	14–20
Derived savanna					
(2) S. Guinea savanna	750	100	10	50	11–14
(3) N. Guinea savanna	980	140	10	50	8–10

Source: (1) FAO, 1979a
(2) Allan and Akwada, 1974
(3) Allan and Akwada, 1973
B.h. = breast height

Vegetation of the potential plantation sites varies greatly: i.e. from the closed high forest in the moist lowland and through the forest/woodland/grassland mosaic of the derived savanna to the woodland of varying density of the Guinea savanna. The structure, basal area, and diameter of the individual trees are important in deciding on clearing methods.

The clearing technique chosen for moist lowland forest must be able to deal with a large volume of material as well as the large sizes of the individual trees.

3.2.3. CHARACTERISTICS OF THE PLANTATION SPECIES

The main plantation species are broad leaved in the moist lowland forest and the forested parts of the derived savanna. They are *Gmelina arborea, Tectona grandis, Terminalia ivorensis, Nauclea diderrichii*, and mahoganies. The first three are usually deciduous in the dry season and therefore have low moisture requirements; the other two are evergreen and require more moisture. Thus they are generally planted where the dry season is shortest. Weed growth is vigorous, but, although there is some competition for moisture and nutrients, the main damage from the weeds is mechanical. There is therefore no need to do any more than cut or crush the weeds and exposure of the soil should be discouraged in order to prevent erosion.

In the savanna zone and the grassland of the derived savanna, the main species for industrial roundwood will be *Pinus caribaea, P. oocarpa, Eucalyptus camaldulensis, E. tereticornis*, and other eucalyptus species. Competition must be eliminated to promote rapid extension of the root system as soon as possible in the short growing season and to conserve soil moisture in the dry season. This is done by cultivation with plough or harrow.

3.2.4. THE CHOICE OF TECHNIQUE

The mechanized technique available may be summarized thus:

Knock-down	Uprooting	Individual tractor with blade.
		Team of 2 tractors with chain and
		1 tractor with blade.
	Shearing	Individual tractor with blade.
Wind-row		Rakes, blades, etc.
Burning		Rakes, blades, fans, etc.

In the moist lowland forest the weeding method will be above ground so there is no need to remove stumps. Shearing by V- or angle-blade was therefore chosen because it should be cheaper since the stumps are not removed and there is less debris to burn. Erosion may also be reduced by leaving stumps in the ground.

In the savanna region the weeding technique cultivates the soil so stumps must be removed. Output using a chaining technique is highest but due to the high capital cost it is suitable only for large plantation programmes. A smaller programme of a few hundred hectares a year would use a single tractor and a dozer blade.

3.2.5. OUTPUTS

A comparison of the outputs of proposed methods was made from the literature. Figures obtained in Nigeria were available for the Guinea savanna but experience elsewhere had to be used for data relevant to the moist lowland forest. The literature from which this information was obtained is given under the references below.

The techniques evaluated were:

(1) Knockdown (stumping) by one 224 kW tractor with tree pusher.
(2) Knockdown (shearing) by one 224 kW tractor with cutting blade.
(3) Windrowing by one 140 kW tractor into lines 50 m apart.
(4) Burning of windrows (moist lowland forest) assisted by one 224 kW tractor with rake.
(5) Manual clearing of moist lowland forest using chain saws. No stumping or windrowing, but includes burning.
(6) Knockdown (stumping) by one 60 kW tractor with tree pusher.
(7) Knockdown (stumping) by two 140 kW tractors with anchor chain, assisted by one 60 kW tractor with tree pusher.
(8) Windrowing by one 100 kW (southern Guinea) or one 60 kW tractor (northern Guinea) into lines 50 m apart.
(9) Burning of windrows (savanna) assisted by one 100 kW tractor with rake (estimated output).
(10) Manual clearing of savanna woodland and with hand tools, with stumping and heaping. Burning output was estimated.

Contract outputs were converted from costs by the use of Government rates and working hours. Actual outputs are probably higher. It is significant that manual outputs in the moist lowland forest are not comparable to these in the savanna because the specification for the job does not include stumping in the former.

Table 3.4. Estimated outputs of some land clearing techniques

Technique	Knockdown	hr ha^{-1} Windrow	Burn
(A) Moist lowland forest		(3)	(4)
(1)	5	5	2.5
(2)	5	5	1.5
(5)		600–1000(d)	
(B) Guinea savanna		(8)	(9)
(6) Southern	n.a.	1.3	(0.7)
Northern	1.7	1.5	(0.5)
(7) Southern	0.26	1.3	(0.7)
Northern	0.13	1.5	(0.5)
(10) Southern	800(c)	600(d)	(100)
Northern	420(d)	430(c)	(70)

Source: Allan and Jackson, 1972; Allan, 1973a; Allan, 1973b; Allan and Akwada, 1973; Allan and Akwada, 1974; Caterpillar Tractor Co., 1974; Dibbitts, 1976; FAO, 1974; Letourneux 1960. n.a. = Not available; (c) = contract; (d) = direct labour; Value in parentheses = estimate.

Finally, it must be stressed that these are estimated outputs for planning and project preparation only. They must be revised with further experience. The source literature in some cases gave widely differing figures, often with inadequate details for comparison. The importance of basal area in determining output, stressed by Allan and Jackson (1972) is confirmed.

3.2.6. EFFECTS OF THE CLEARING METHOD ON THE SOIL

It may be assumed that complete clearing, even by manual methods, is detrimental to soil fertility, structure, and moisture-retaining properties. The soil is generally fairly well covered by weed growth within four months after burning the moist lowland forest region, but extensive sheet erosion occurs in that time, particularly where there is little debris left from the burning. Canopy closure of *Gmelina arborea*, the main plantation species of this zone, occurs by the end of the second year. This species will be grown on a 7 to 8 year rotation for pulpwood or 15 year rotation for sawtimber. Whether the organic matter in the soil will build up sufficiently to prevent site deterioration at the end of this period is under investigation by forest research organizations in Nigeria and elsewhere.

Mechanical clearing is likely to increase sheet erosion because clearing will be more complete, although some debris may remain in windrows. Heavy equipment is also likely to compact the soil, and if uprooting is done, it will bring subsoil to the surface.

A programme of research and development will therefore have to be started at the same time as a large-scale mechanical land clearing. There must be rapid feedback of research results to the field to prevent harmful techniques being continued and to introduce improved methods as soon as possible. One of the main research projects must be to carry out long term studies into the effects of mechanical clearing on the soil, including possible interactions with plantation species, soil type, rotation length, etc.

Research in the short term should concentrate on means of maintaining soil cover. This may be done in various ways:

(a) *Agricultural crops.* Maize, for example has grown 50 cm or more before the trees are planted, but to provide sufficient soil cover to reduce erosion it may have to be planted so closely that it interferes with the tree crop.

(b) *Other cover crops.* Several other cover crops exist, such as *Stylosanthes guianensis,* but generally establishment is slow. *Eupatorium divinorum,* the dominant weed of much of the plantation area in the moist lowland forest zone, might be used if its growth could be kept down by frequent weeding.

(c) *Herbicides.* In moist lowland forest zone the main weeds and the plantation species are all broad leaved so wide spectrum herbicides acting on the foliage are unsuitable. Triazine herbicides, that are active in the soil, are suitable, particularly as maize is resistant to them and a combination of maize and trees should be possible. In the savanna zone the main weeds are grasses, and selective weedkillers such as Dalapon which is specific to grasses do not harm, at normal dosages, *Pinus caribaea* or *Eucalyptus spp.* (Ball, 1971). Furthermore, it was found in trials in Kenya savanna conditions that *Pinus patula* had survived significantly better after the dry season where herbicides had been applied than where harrowing was done (Ball, 1974). The potential of herbicides to retain soil cover is considerable, but problems such as these of cost, toxicity, and transport of diluent remain.

3.2.7. CONCLUSION

Mechanical clearing methods are necessary to create plantations to meet the requirements of the country in future, but they will have even more deleterious effects on soil properties than the present manual methods have. A choice of different methods is available so the least harmful ones can be selected, and while it is not possible to await definitive research results before starting an expanded plantation scheme, a research programme with rapid feedback can ensure that the harmful effects are minimized.

REFERENCES

Allan, T. G., and Jackson, J. K., 1972, Land clearing trials at Afaka forest reserve (1971), *Paper to the 3rd Annual Conference of the Forestry Association of Nigeria,* Forest Research Institute of Nigeria, Ibadan.

Allan, T. G., 1973a, Land clearing and preparation trials using Caterpillar, Fleco, and Rome equipment (1972), *Research Paper No. 17 (Savanna Series),* Forest Research Institute of Nigeria, Ibadan.

Allan, T. G., 1973b, *Trip to the Ivory Coast,* Internal report, Savanna Forestry Research station, Zaria.

Allan, T. G., and Akwada, E. C. C., 1973, Land clearing and preparation trials at Afaka forest reserve, *Project Working Document FO,* NIR/64/516, FAO, Rome.

Allan, T. G., and Akwada, E. C. C., 1974, Land clearing trials at Mokwa forest reserve, *Paper to the 5th Annual Conference of the Forestry Association of Nigeria,* Forest Research Institute of Nigeria, Ibadan.

Ball, J. B., 1971, Herbicides in Uganda forestry, *Technical Note 173/71,* Uganda Forest Department, Entebbe.

Ball, J. B., 1974, Weed control on the Turbo Afforestation Scheme, Kenya, *Proceedings of the 5th East African Weed Control Conference,* Tropical Pesticides Research Institute, Arusha, Tanzania.

Caterpillar Tractor Co., 1974, *The Clearing of Land For Development.*

Dibbits, H. J., 1976, Trials on methods of land clearing in a tropical rain forest at IITA, *Internal Paper,* IITA, Ibadan.

FAO, 1965, *Crop Ecologic Survey in West Africa,* prepared by J. Papadakis, FAO, Rome.

FAO, 1974, Essai de presentation uniformisee des conditions, d'execution, des resultats et des couts des reboisements, *FO: MISC/74/3,* FAO, Rome.

FAO, 1979a, The indicative inventory of reserved high forest in Southern Nigeria, based on the work of H. Sutter, *FO: SF/NIR/71/546 Technical Report 1,* FAO, Ibadan.

FAO, 1979b, Plantations, based on the work of J. B. Ball, *FO: SF/NIR/71/546 Technical Report 3,* FAO, Ibadan.

Letourneux, C., 1960, Utilisation de materiel mechanique pour les plantations forestieres tropicales, *Bois et Forêts des Tropiques,* **71,** 19–40.

Tropical Agricultural Hydrology
Edited by R. Lal and E. W. Russell

3.3

Land Clearing and Development for Agricultural Purposes in Western Nigeria

D. C. COUPER, R. LAL, AND S. L. CLAASSEN

3.3.1. INTRODUCTION

Approximately five billion hectares of land are available for development in the tropics but only an estimated 25 per cent of this area is considered suitable for agriculture. Much of the potentially cultivable land will be utilized for large scale development schemes which may be successfully accomplished if appropriate clearing technology is used to minimize soil degradation. Productivity may then be sustained through judicious post-clearing soil management, suitable cropping sequences and crop combinations.

If, however, land clearing is tackled as an engineering problem, disregarding soil and other agro-climatic factors, soil degradation may be such that expensive remedial measures are incapable of reversing the damage done.

Until recently, much of the forest land cleared for cultivation in tropical Africa was cleared by shifting cultivators using traditional hand tools with minimal damage to the soil. Increasingly, however, as labour costs rise and mechanization of land clearing increases, heavy equipment is being used for forest clearing, very often with most damaging results to the soil. This may be partially attributed to the use of construction type heavy equipment or bulldozers which are designed to move soil rather than clear forest. With the wrong equipment and inexperienced operators, very often the thin layer of top soil, built up over many years under forest conditions, is removed together with the forest debris in the clearing process.

A considerable area of potentially cultivable land in the humid and sub-humid tropics is now under primary or secondary forest cover. Approximately ten million hectares are cleared annually by shifting cultivators alone and it is estimated that because of ever increasing demographic pressure, the arable land area in tropical countries may have to be increased from 737 million ha in 1970 to 890 million ha in 1985 with an annual rate of 6 to 10 million ha of new land development (Boerma, 1975; Boer, 1977; Thijsse, 1977a, b; Silva and Laurence, 1977).

Proper planning, with due consideration to all factors concerned is the key to the successful development of land and water resources in tropical environments. The objectives of this investigation were therefore to evaluate the efficiency of various land-clearing methods, cost and inputs in post-clearing land-development, and the effects of clearing methods on crop yields.

The effects of land-clearing on soil and environments have been investigated by many researchers. The problem arising from large-scale forest-clearing by heavy machinery have been emphasized by many (Bunting, 1955; Cunningham, 1963; Nye and Greenland, 1964; Ahn, 1968; Scotter, 1970; Webb, 1975; Daniel and Da Kulasingam, 1975; Seubert *et al.*, 1977; Lal and Cummings, 1979). With recent advances in the development of suitable machinery for land-clearing purposes (Anon, 1956, 1959, 1975; Kumar, 1963; Caterpillar Tractor Co., 1974), mechanized land-clearing is being widely adopted for development of land and water resources in the tropics. Feasibility of mechanized land-clearing methods have been investigated in the savanna region of Nigeria (Allan, 1975; Allan and Akwada, 1976), in the humid tropics of Malaysia (Ibrahim, 1978), in tropical Latin America (Weert and Lenselink, 1972; Weert, 1974; Setzer, 1967), and in the Ivory Coast (Martin, 1970, 1976).

Though there can be drastic changes in soil and environments by deforestation, many workers have reported that soil properties can be restored by proper management. Mature dry forest does produce a considerable amount of biomass and adds plant nutrients to the soil surface. Dommergues (1963) observed that total dry weight of the vegetation in a mature forest may be as much as 3,400 t ha^{-1} contributing annually 200 kg N, 100 kg P, 900 kg K, 2700 kg Ca, and 250 kg Mg to the soil. Though these quantities may vary depending upon the density and maturity of the forest cover, improper removal of this vegetation may deplete the nutrient reserves of the soil. In addition, soil fauna may also be adversely affected. Drift (1963) reported that clearing the forest in Surinam resulted in the disappearance of most constituents of the forest soil fauna. The density of desirable macro-arthropods in a recently cleared soil was only 10 to 30 per cent of that in the forest soil. Land-clearing may result in a more drastic reduction in soil organic matter and other chemicals for soils under forest cover than those supporting savanna vegetation (Blic, 1976). The objective of subsequent soil management systems should be to minimize these alterations (Feller and Milleville, 1977; Ollagnier *et al.*, 1978).

3.3.2. MATERIALS AND METHODS

Land-clearing experiments were conducted at the research farm of the International Institute of Tropical Agriculture (IITA) located in south-west Nigeria, approximately 30 km south of the northern limit of the lowland rainforest zone of the West African tropics.

This western region of Nigeria lies roughly between the longitudes 3° and 6° east and latitudes 6° and 8° north. The bimodal character of rainfall distribution leads to two distinct growing seasons: the first season from late March to late July, ending in a short dry period of approximately one month, and the second shorter season from late August to early November.

The clearing experiments were conducted in the November–March dry season of 1978–1979. The secondary forest vegetation at the time of clearing was approximately 15 years old. Tree density of all sizes and ages was approximately 200 ha^{-1}. Most of the large trees consisted of oil palm (*Elaeis guianeensis*), Kolanut (*Cola nitide*), Cacao (*Theobroma cacao*), *Albizia glaberrima*, *Holarrhina floribunda*, *Newbouldia laevis*, *Antiaris africana*, *Ficus exasperata*, and *Leconiodiscus cupanioides*. In addition to large trees, the underbrush consisted of woody climbers and small, erect, woody plants. The predominant species were *Combretum spp.*, *Hippocratea spp.*, *Cnestis ferruginea*, *Paullinia pinnata*, *Secomone afzelii*, *Alchornea laxiflora*, and *Chassalia kolly*. A complete survey of the floral species was done by the Forestry Department of the University of Ibadan, prior to land-clearing.

The soil of the trial area had been under shifting cultivation until 1969, and is highly variable. The soil belongs to Egbeda/Iwo association on the Upper slopes of the catena and to Ibadan and Apomu series towards the lower concave section of the terrain. Physical and chemical characteristics of these soils have been described in an earlier report (Moormann, Lal, and Juo, 1975). The soils of these associations are medium to light textured near the surface, with sandy clay to clay subsoil, and a layer of angular and subangular quartz gravel immediately below the surface. The soils at the lower end of the catena have soft plinthite at 15 to 20 cm below the surface that hardens to lateritic concretionary material on exposure by clearing and subsequent cultivation. In general, the soil is a clayey skeletal kaolinitic iso-hyperthermic oxic paleustalf. The mean slope is about 5 per cent with a range of 2 to 10 per cent in the upper and lower part of the catena.

Land clearing methods consisted of the following:

(1) *Traditional clearing:* Clearing, using traditional methods, was carried out by hand labour, using machete, axe, hoe, and other native tools. After the underbrush operation, which removes woody climbers and small erect woody plants, all trees exceeding 45 cm in girth one metre above ground were not cut. Those less than 45 cm in girth were cut one metre above the ground surface. About 3 to 4 weeks after clearing, the biomass was burnt *in situ*. Man hours required for each operation were carefully monitored.

(2) *Manual Clearing:* The manual clearing was also carried out using native tools and manually operated chain saws (Plate 1). After underbrushing, all trees were dug out to a depth of approximately 20 cm. All trees and underbrush material were then burnt *in situ* about 3 to 4 weeks after felling (Plate 2). Unburnt trees, particularly the trunks of oil palms, were removed to the plot boundaries. Time required (man days) for each operation was recorded separately.

Both manual and traditional clearing was done by contract labour which was paid according to the work done rather than on a daily basis. The time estimates obtained are therefore realistic.

(3) *Crawler tractor with Shear Blade Attachment:** The shear blade is a flat bottomed cutting blade (Rockland SA-4) with a stinger (Plate 3) which is front mounted on a 21C Fiat Allis crawler tractor. The total horsepower of the tractor is applied to the sharp angled cutting edge. This treatment involved cutting all trees off at ground level, windrowing, and burning in windrows. Windrows were approximately 100 m apart. Unburnt trunks were removed to the plot boundaries.

The removal of tree stumps is normally required during the clearing process to prevent future damage to cultivation equipment. This process however inevitably results in considerable soil disturbance, erosion, and degradation. Where no-till farming will be practised, stumps need not be removed, and therefore shear blade clearing becomes attractive. Tractor-drawn no-till planters are commercially available which will ride over stumps without sustaining damage.

No-till farming is the term used to describe the establishment of a crop without seed-bed preparation, using chemicals only for weed control.

(4) *Crawler Tractor with Tree Pusher/Root rake Attachment:* The front mounted tree pusher (7B-3) attachment extends above and forward from the tractor (21C Fiat Allis) providing the tractor with added leverage in felling the trees (Plate 4). The tree pusher is mounted above a tined root-rake which travels through the soil to a depth of 50 cm removing tree roots, stumps, and debris. Considerable soil disturbance takes place during this operation. Both root-rake and tree pusher were front-mounted on the crawler tractor.

Traditional farming will be practised on the traditionally cleared plots and in the first year, maize and cassava were planted through the standing and felled trees with manually operated equipment recently developed at IITA for the small farmer. Both no-tillage and conventional tillage were used for crop establishment on land cleared either manually or with the tree pusher, and only no-tillage was used on land cleared with the shear blade because roots and stumps were not removed. Conventional tillage for the first maize crop consisted solely of a disc harrowing to a depth of 15 cm as a preliminary disc ploughing was not considered necessary on this land which had just been cleared. No-tillage plots were sprayed twice with 2.5 l ha^{-1} of paraquat (1-l'-dimethyl-4,4'-bipyridinium ion). The first spray was made 1 week before planting and the second (together with the pre-emergent herbicide) just after seeding. Both no-till and conventionally tilled plots received the pre-emergent herbicide atrazine (2-chloro-4-ethylamino-isopropylamino-1,3,5-triazine) applied at 2.5 kg ha^{-1}. No herbicides were applied to the traditionally cleared treatments.

All six clearing and post-clearing management treatments were replicated twice and their positions were completely randomized over the area. The field layout is shown in Figure 3.1. The plot size ranged from 2.5 to 4 ha. Plots designated for conventional tillage were surveyed for one-metre contours and graded channel

---

*(The mention of commercial names does not necessarily imply their endorsement by IITA).

Plate 1 **Manual** clearing with traditional tools

Plate 2

Plate 3 A flat-bottomed front-mounted shear blade

Plate 4 Front-mounted tree pusher/root rake attachments

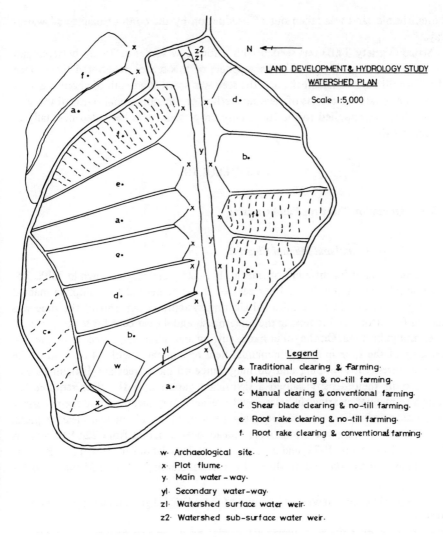

Figure 3.1 Watershed plan for land development and hydrology study

terraces were then constructed at a vertical interval calculated by the following equation:

$$\text{Vertical Interval (feet)} = \frac{\text{Percentage slope}}{3} + 2$$

The channel gradient was 0.5 per cent and the channels from each plot drained into a graded and grassed waterway. Depending on the slope, about 12 to 15 per cent of the cultivable land was taken out of production by the contour banks and water-ways.

Maize (Variety TZB) was seeded with a four row seeder at 75 cm between and 25 cm within rows. The double disc openers of this seeder penetrated the surface of the no-till plots satisfactorily and the seed was sown at a depth of about 4 cm. In order to evaluate the effects of clearing method on the nutritional status of the soil, no fertilizer was applied to the first maize crop. Maize was harvested manually in all treatments.

3.3.3. RESULTS

3.3.3.1. Operation Time for Land Clearing

3.3.3.1.1. Traditional and Manual Clearing

The time required for different operations in manual clearing is shown in Table 3.5. Traditional clearing required 33 per cent of the time needed for complete manual clearing operation. In the traditional clearing system, 37 per cent of the time was needed for cutting and removing the underbrush and 47 per cent for stumping small trees and in burning. On the other hand, cutting and felling large trees consumed 54 per cent of the time in complete manual clearing compared with 34 per cent of the time required for stumping and burning. Since oil palm trees are difficult to cut, variation in their density caused a significant variation in the time required for complete manual clearing. In spite of all precautionary measures and the necessary supervision, stumping was inadequately performed in all manual clearing plots. The manual clearing operations were spread over a period from 22nd December, 1978, to 30th April, 1979, and involved a total daily labour force of 12 to 50 men. Liefstingh (1965) observed in Ghana that manual clearing required 35 man days for

Table 3.5. Time (man days ha^{-1}) required for traditional and manual clearing oper-ations

Operation	Complete manual clearing[a]	Traditional clearing[b]
Underbrush[c]	21 ± 5	21 ± 5
Cutting and felling trees	96 ± 32	5 ± 1
Stumping and burning	60 ± 26	27 ± 6
Total	177 ± 9	57 ± 21

[a] Mean of 5 replications
[b] Mean of 3 replications
[c] Mean of 8 replications
(1) Additional work was required for removing unburnt palm trees from the plot.
(2) One working day is 8 hours.

Table 3.6. Operating time required for mechanized clearing and windrowing with front mounted root-rake/tree-pusher combination and front mounted shear blade

Clearing treatment	Machine time (ha hr^{-1})	Man days ha^{-1} [a]	Fuel consumption (l hr^{-1})
Root-rake/tree-pusher	0.37 ± 0.11	24.8 ± 11.5	45.8 ± 9.7
Shear blade	0.52 ± 0.11	29.1 ± 3.1	40.8 ± 7.1

Each figure is a mean of 10 replications. The experiment was conducted for clearing operations while windrowing uphill as well as downhill on representative slopes of the experimental site.
[a] Additional man days were required for removing roots and stems not cleared by the machine.

underbrushing, 84 man days for felling trees, and 52 man days per hectare for stumping and burning. These results are comparable to those obtained in this investigation.

3.3.3.1.2. Mechanical Clearing

The operation time required for mechanized clearing by the two systems adopted is shown in Table 3.6. On average, it required 1.94 working-hours ha^{-1} to clear with the shear blade attachment compared with 2.70 working-hours ha^{-1} required with the tree-pusher/root-rake combination. Accordingly, the fuel consumption was also less for the shear blade than for the tree pusher treatment: 98 and 123 l ha^{-1} respectively. Since roots and stumps were not removed, there were more branches and twigs left on the soil surface of the shear blade cleared plot compared with that cleared by the tree-pusher attachment. Consequently, the man days required for picking up this debris were more in the former than in the latter treatment (Table 3.6).

Since the shear blade is flat-bottomed, roots and stumps are left in the soil. Burning in windrows was therefore easier in the case of the shear blade clearing than with the tree-pusher/root-rake combination. Moreover, there was considerably less soil removed to the windrows in shear blade clearing than with the tree-pusher/root-rake attachment. However, the efficiency of clearing by the shear blade method depends considerably on the skill and experience of the operator. An inexperienced operator may shear the trees 10 to 15 cm above the ground surface (Plate 3), which then necessitates a second clearing operation.

3.3.3.2. Post-Clearing Land Development

Terracing, the post-clearing land development, was done only for the conventionally tilled plots but no additional land development was necessary for the no-till treatments. The construction of graded contour banks and grass waterways for the conventional tillage plots required time of both skilled and unskilled workers and the use of heavy equipment such as a grader. The time required for these operations is

Table 3.7. Post-clearing development for conventionally tilled plots

Operation	Professional services (hr ha^{-1})	Unskilled labour (man days ha^{-1})	Equipment (hr ha^{-1})
Grid survey	0.92	1.2	—
Mapping	0.8	—	—
Surveying and placing contour banks	0.4	1.6	—
Grading contour banks	1.25	—	1.25
Roads and Waterway construction using D-7 caterpillar	1.50	—	1.50
Grassing	—	14.4	1.34

Grader was rented at ₦75.00 hr^{-1}
Caterpillar D7 available at ₦120 hr^{-1} ₦1 – US$1.80
These values are average at 4 plots totalling 14.15 ha in area and with soil slopes ranging from 3 to 10 per cent.

Table 3.8. Effects of clearing methods and tillage systems on maize grain yield

Clearing treatment	Tillage system	Grain Yield (t ha^{-1})
Traditional clearing	Traditional seeding	0.50[a]
Manual clearing	No-tillage	1.56[b]
Manual clearing	Conventional tillage	1.59[b]
Crawler tractor/shear blade	No-tillage	1.98[b]
Crawler tractor/tree-pusher	No-tillage	1.36[b]
Crawler tractor/tree-pusher	Conventional tillage	1.75[b]

[a] Planting of maize in the traditional plots was carried out by Messrs. Garman and Pedoline.
[b] Planting carried out by Gaspardo Leonard planter.

shown in Table 3.7. Post-clearing development costs may be as high as US $450/ha (Couper, Lal, and Claassen, 1979).

The operational time for post-clearing management was also different for the conventionally tilled compared with the unploughed plots. These estimates were, however, not obtained for this study and have been reported earlier (Couper, Lal, and Claassen, 1979).

3.3.3.3. Crop Performance and Grain Yield

Maize growth was extremely uneven in the tree pusher treatment. In general, growth was vigorous in the windrow regions and relatively poor in the adjacent areas from which surface soil was removed. Removal of roots and stumps also resulted in exposure of the subsoil which caused poor growth. Growth was generally uniform in the manually cleared plots, though symptoms of P deficiency were commonly observed in all treatments. In the traditionally cleared treatments, maize growth was drastically suppressed by the shading effects of trees which were not cut. Maize growth was the poorest in traditionally cleared treatments, except where the tree density was low.

As was expected, there was considerable bush regrowth in the shear-blade treatment, but it is debatable if maize growth was adversely affected due to competition for light, soil moisture, or nutrients. Bush regrowth was less in the tree-pusher and manually-cleared treatments.

Maize grain yield was the lowest in traditionally cleared plots, and the highest yield was obtained in shear blade clearing followed by no-tillage farming (Table 3.8). With the exception of traditionally cleared plots, tree-pusher clearing followed by no-tillage system also produced low yields. This low yield may be attributed to high soil compaction caused by the tree pusher treatment with passages of heavy machinery (Lal and Cummings, 1979). A low mean yield of about 1.5 t ha^{-1} is attributed to the fact that no fertilizer applications were made to the first maize crop immediately after land clearing.

3.3.4. GENERAL DISCUSSION

3.3.4.1. Relative Performance of Different Land Clearing Methods

Clearing with shear blade attachment was the most rapid and economic method of land clearing. Martin (1976) also observed from his investigations in Colombia, Peru, and the Ivory Coast that clearing with the shear blade was the most economical method. The manual clearing was the slowest and the most expensive of the methods investigated. The application of the shear blade method, however, may be limited to land development for either establishment of plantation crops or for those arable crops that may be grown by a no-till system of soil management. In addition to the skill and experience of the operator which is necessary for proper use of the shear blade, precautionary measures are necessary to ensure that the cutting edge is sharpened, or the efficiency deteriorates and stumps are left torn and projecting above the ground level. The sharpening operation with angle grinder and portable generator lasted 20 to 30 minutes after every 4 to 6 hour operation. The shear blade was also not effective in felling thin woody plants and herbaceous growth. The blade simply rode over them without cutting. Two passes in opposite directions were required to cut thin springy trees which tended to bend under the shear blade. However, oil palm trees were more easily removed by the shear blade than the tree-pusher/root-rake attachments.

3.3.4.2. Desirability of Mechanized Land Clearing Operations

Bringing new land under cultivation for increasing agricultural production should not be done with the objective of carrying out as much as possible, as cheaply as possible. This unthinking philosophy could result in short-term gains at the expense of irreversible damage to the natural-resource base. Land and water resources should be developed so that productivity can be sustained economically and indefinitely. There are many examples of failures of large-scale land development schemes in the tropics that have led to the loss of vast tracts of arable land.

The development of land for agricultural purposes should be based on the principle of the removal of existing plant cover in such a way that the existing equilibrium between soil, vegetation, and climate is minimally disturbed. If the topography and soil characteristics are such that erosion hazards and degradation of soil quality are minimal, all operations may be carried out mechanically. If the area to be cleared is small and the labour is cheap and readily available, manual clearing is preferred. The indiscriminate use of heavy equipment, though quick and economical, can cause havoc to soil and environments.

Where primary or secondary forest has to be cleared for plantation or annual crops, for which efficient selective herbicides exist, no-tillage farming should be considered in order to reduce the risk of rapid soil degeneration. Stumps and roots need not then be removed, and the more economical and safer shear blade method of forest clearing may be adopted.

Care must be taken to carry out clearing when soil moisture is minimal; experienced operators should be used; the distance between windrows should not be excessive and the removal of stumps should be avoided.

Shear blade clearing, if properly carried out, may be expected to minimize post-clearing soil losses and provide a suitable base for sustained yields in the future (Table 3.8).

3.3.5. ACKNOWLEDGEMENTS

The authors wish to record their appreciation to the Scientists of IITA for assistance and advice, and also to the SCOATRAC Company of Nigeria and the Fiat-Allis Company for supplying the equipment for mechanized clearing.

REFERENCES

Ahn, P. M., 1968, The effects of large scale mechanized agriculture on the physical properties of west African soils, *Ghana J. Agr. Sci,* **1(1),** 36–40.

Allan, T. G., 1975, Studies of mechanized land development in Nigeria and possible effects of such work on employment, *FAO/AGS/MPR/75/14,* 14 pp.

Allan, T. G., and Akwada, E. C. C., 1976, Land clearing and site preparation in the Nigerian savanna. Savanna afforestation in Africa, *FAO/DANIDA Training Courses on Forest Nursery and establishment techniques for African savanna and papers for symposium on savanna Afforestation Kaduna Nigeria,* **1976,** 123–138.

Anon, 1956, Advanced designs in land clearing equipment, *World crops,* **8(4),** 137–141.

Anon, 1959, Reclamation of tropical rain forest in south Sumatra, *Way Ahead,* **7(4),** 11–15.

Anon, 1975, Putting idle land to work, *World Farming,* **17,** 20–21.

Blic, P. De, 1976, ORSTOM: The behaviour of Ivory Coast ferrallitic soils after mechanical clearing and cultivation: role of inherited characteristics of the natural environment, *Cahiers ORSTOM, Serie Pedologie,* **14(2),** 113–130.

Boer, I. J..De, 1977, The relation between agriculture and protection of nature in Asia. 2. Shifting cultivation in north Thailand, *Tijdschrift-Koninklijke Mederlandse Heide Maatschappij,* **88,** 103–111.

Boerma, A. H., 1975, The world could be fed, *J. Soil and Water Conserv.,* **30,** 4–11.

Bunting, A. H., 1955, A review of land clearing, *Mem. Res. Div. Min. Agr. Sudan,* **58,** 121–132.

Caterpillar Tractor Co., 1974, The clearing of land for development, *Caterpillar Tractor Co. Publication,* USA, pp. 111.

Couper, D. C., Lal, R., and Claassen, S., 1979, Mechanized no-tillage maize production on an alfisol in tropical Africa. In R. Lal (ed.), *Soil Tillage And Crop Production,* IITA Ibadan, Nigeria, 147–160.

Cunningham, R. K., 1963, The effects of clearing a tropical forest soil, *J. Soil Sci.,* **14(2),** 334.

Daniel, J. G., and Da Kulasingam, A., 1975, Problems arising from large scale forest clearing for agricultural use. The Malaysian experience, *Planter,* **51,** 250–257.

Dommergues, Y., 1963, Biogeo-chemical cycles of mineral elements in tropical formations, *Bois Forests Trop.,* **87,** 9–23.

Drift, J. Van Der, 1963, A comparative study of the soil fauna in forests and cultivated land on sandy soils in Surinam. Uitg. Natuurwet, *Studierkring Suriname en Ned. Antillen,* **32**, 1–42.

Feller, C. and Milleville, P., 1977, Evolution of recently cleared soils in the region of Terres Neuves (eastern Senegal) (1) Study and evolution of the prinicpal morphological and physico-chemical characteristics, *Cahiers ORSTOM Serie Biologie,* **12**, 199–211.

Ibrahim, Bin Haji Ismail, 1978, *Mechanical land-clearing,* Planter (Malaysia), **54 (629),** 475–498.

Kumar, H., 1963, Land-clearing techniques, *Ghana Farmer,* **7(2),** 43–51.

Lal, R., and Cummings, D. J., 1979, Clearing a tropical forest. 1. Effects on soil and micro-climate, *Field Crops Res.,* **2(2),** 91–107.

Liefstingh, G., 1965, Chemical clearing, a possibility, *Ghana Farmer,* **9(1),** 8–14.

Martin, G., 1970, Mechanized land clearing for the establishment of industrial oil palm plantations, *Oleagineux,* **25,** 11, 575–80.

Martin, G., 1976, Method of estimating the time required for land clearing and mechanical windrowing of an industrial oil palm plantation, *Oleagineux,* **31,** 59–62.

Moormann, F. R., Lal, R., and Juo, A. S. R., 1975, Soils of IITA, *IITA Tech. Bull.,* **3,** 38 pp.

Nye, P. H., and Greenland, D. J., 1964, Changes in soil after clearing Tropical Forest Plant, *Soil,* **21(7),** 101–112.

Ollagnier, M., Lauzeral, A., Olivin, J., and Ochs, R., 1978, Development of soils under oil palm deforestation, *Olegineux,* **33,** 537–547.

Scotter, D. R., 1970, Soil temperatures under grass fire, *Aust. J. Soil Res.,* **8(3),** 273–279.

Setzer, J., 1967, The impossibility of rational land use in the region of the upper xingu, Mato Grosse (Brazil), *Rev. Brasiliera Geogr.,* **29(1),** 102–109.

Seubert, C. E., Sanchez, P. A., and Valvarde, C., 1977, Effects of land clearing methods on soil properties on an ultisol and crop performance in the Amazon jungle of Peru, *Tropical Agriculture (Trinidad),* **54(4),** 307–321.

Silva, F. R. D., and Laurence, M., 1977, The importance of the utilization of bulldozers in the development of cerrado soils in Brazil, *Agricultural Services Bulletin,* **FAO 28,** 165–177.

Thijsse, J. P., 1977a, Soil erosion in the humid tropics, *Landbouwkundig Tijdschrift,* **89,** 408–441.

Thijsse, J. P., 1977b, Deforestation and erosion in Java, *Landbouwkundig Tijdshcrift* (Netherlands), **89,** 443–447.

Webb, B. H., 1975, Objective planning of land clearing for mechanised agricultural development, *Planter,* **51,** 231–249.

Weert, R. Van Der, 1974, The influence of mechanical forest clearing on soil conditions and resulting effects on root growth, *Trop. Agric.,* **51,** 325–331.

Weert, R. Van Der, and Lenselink, K. J., 1972, The influence of mechanical clearing of forest on some physical and chemical soil properties, *Surinamise Landbouw,* **20(3),** 2–14.

Tropical Agricultural Hydrology
Edited by R. Lal and E. W. Russell
© 1981, John Wiley & Sons Ltd.

3.4

Deforestation of Tropical Rainforest and Hydrological Problems

R. LAL

3.4.1. INTRODUCTION

Tropical rainforest is one of the ancient, complex, self-supporting, and stable eco-systems. If the rainfall is adequate, tropical rainforest occurs in a region approximately 10 degrees north and south of the equator. In this closed ecosystem, trees and other plant species are in equilibrium with soil and their environment. Most plant nutrients are tied up in vegetation, and there is an effective nutrient cycling. Rainfall interception, surface detention, evaoptranspiration, and soil-water storage effectively decrease water runoff to a minimum. Multi-storey canopy and leaf-litter protect the soil against raindrop impact and prevent soil detachment. Leaf litter and other organic residues rapidly decompose thereby enhancing the activities of soil fauna.

However, this vast natural resource and a reservoir of genetic diversity is rapidly shrinking in favour of arable land use. In Africa and other tropical regions, this closed ecosystem was always being upset by shifting cultivation. The fragmented belt of existing rainforest in South America, Africa, Borneo, and New Guinea (Figure 3.2) is rapidly retreating. Large scale deforestation for mechanized agriculture results in an ecological imbalance that affects the hydrological cycle, nutrient recycling, microclimatic, and biotic environments including soil micro-floral and faunal activity (Cunningham, 1963; Lal and Cummings, 1979; Lal, 1980). The result is soil compaction, water runoff, and accelerated soil erosion producing vast tracts of barren and unproductive lands where lushgreen tropical forest once prevailed.

And yet the food production must be increased immediately and substantially. The entire Amazon basin, covering approximately 4 million km^2 may be developed for food crops, pasture, and plantation crops within the next half century (Brunig, 1975). Though many environmentalists and conservationists have cautioned against this exploitation of the threatened resource (Richards, 1973), large-scale deforestation for mechanized agriculture is inevitable. However, the basic research infor-

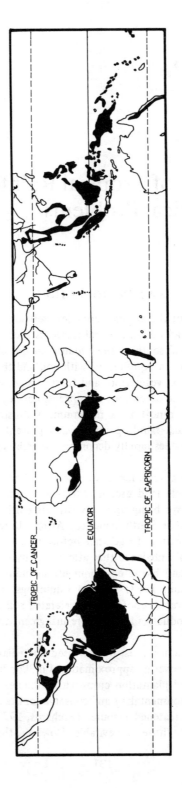

Figure 3.2 Tropical rain forest (Adapted from Richards, 1973)

mation that could provide guidelines for decision makers and planners to select suitable regions and appropriate methods of forest removal is not available. Most of the hydrological research in the tropics has been concerned with the general water balance studies. The effects of methods of deforestation and of soil management techniques on water runoff, on soil loss, and on productive potential of soil have not been quantified. This type of information is necessary prior to implementing the large scale deforestation schemes envisaged.

Though information relating the effects of deforestation on changes in the hydrological cycle and sediment transport is scanty in the tropics, a considerable body of information exists for the temperate latitudes. Hibbert (1967) and Pereira (1973) reviewed the effects of deforestation and change in land use on hydrological balance and water yield from catchments in the tropics and temperate latitudes. The increase in streamflow is generally proportional to the reduction in forest cover (Hibbert, 1967). The effects in East Africa (Pereira, 1962, 1965; McCulloch and Dagg, 1965; Pereira, 1973) indicated that where surface storage opportunities are small rational catchment management must aim at maintaining the maximum infiltration rate. For example they found that when converting some natural forest into a tea plantation in Kenya, clearing some 30 per cent of the forest preparatory to planting tea reduced the evapotranspiration for the area by only 11 per cent, but great care was taken to minimize soil erosion during the period of tea establishment. By the time the tea bushes gave an effectively complete canopy, the mean annual water consumption was virtually unchanged compared with natural forest (Blackie, 1972). On some newly-cleared land and before the development of canopy cover of the tea plantation, a 90-mm storm resulted in a maximum streamflow of 27 m^3 sec^{-1} km^{-2} from a cleared catchment compared with only 0.6 m^3 sec^{-1} km^{-2} from a forested control (Pereira, 1973).

Results obtained from hydrological investigations in tropical Asia have emphasized the importance of forest cover in proper management of soil and water resources (Kenworthy, 1969; Low and Goh, 1972; Pathak, 1974; Srivastava, 1974; Gupta *et al.*, 1975). Thijsse (1977) reported some problems of soil erosion resulting from deforestation in the humid tropics of Java. The protective effect of forest cover on water yield and erosion was demonstrated by an accidental fire in the Snowy Mountains of New South Wales in Australia (Brown, 1972). The flow pattern changed abruptly after the fires. There were changes in the shape of the flood hydrographs with the occurrence of pronounced sharp secondary peaks on the rising side of many flood hydrographs. The water yield and sediment load increased significantly in the first four years after fire. Recent investigations in the wet tropical coast of Queensland, with average annual rainfall of 4175 mm has indicated that average runoff even from a forested catchment may be as much as 63 per cent and most of it is attributed to storm runoff (Gilmour, 1977; Bonell *et al.*, 1979). Analyses of the physical properties of soil under forest cover indicated that the surface 10 cm of soil has saturated hydraulic conductivity of the order of 32 m day^{-1} (Gilmour and Bonell, 1977, 1979; Bonell and Gilmour, 1978). However, the saturated hydraulic conductivity decreases sharply with depth so that the 10–20 cm zone has a value of only 1.5 m day^{-1} and the 20–100 cm zone a value of only 0.3 m day^{-1}. The surface layer is quickly saturated and results in surface or sub-surface runoff.

There are few investigations reported from the tropics of South America. The research conducted by the coffee foundation in Colombia has demonstrated the protective effect of ground cover, similar to that reported from East Africa (CENI-CAFE, 1975).

Soil erosion under virgin forest conditions is generally negligible (Young, 1974). Brunig (1975) observed that soil erosion under virgin forest cover may be as low as 0.2 t ha^{-1} yr^{-1}. However, cultivation may increase the soil erosion to an alarming rate of 600 to 1200 t ha^{-1} yr^{-1}.

The objectives of this report are to describe the effects of methods of deforestation and of post-development soil management on runoff rate and amount, sediment density, and soil erosion in a tropical catchment in southwest Nigeria.

3.4.2. MATERIALS AND METHODS

These experiments were conducted at IITA during 1979. Different methods of land clearing and post-clearing management with field layout of the watershed have been described in another report in this volume by Couper *et al.* (Chapter 3.3) Each of the sub-plots was equipped with a 4.5-feet H-flume and a water stage recorder (Plate 5). Runoff samples were obtained by a combination of the Coshocton sampler, and multi-divisor, and storage tanks (Plate 6). Runoff amount, and total soil loss were computed for each storm during 1979. During this period maize was seeded in early May. In all no-till plots, cassava cuttings were planted in the standing maize towards the end of June. After harvesting ears, maize stover was manually slashed to facilitate cassava growth. In the conventionally-tilled plots, maize stover was ploughed in, and plots were ridged in early September. Contour ridges were made one metre apart, and cassava cuttings were planted on ridges at intervals of one metre. In plots with traditional clearing and farming, cassava had already been planted in May. After harvesting maize from these plots, maize stover, and bush re-growth was manually slashed to facilitate cassava growth.

3.4.3. RESULTS AND DISCUSSION

3.4.3.1. Sediment Density

Sediment concentration in water runoff was significantly affected by methods of deforestation and tillage systems (Table 3.9 below). Although there was measurable runoff from forested and traditionally managed catchments, leaf-litter, and canopy cover prevented soil detachment and resulted in a negligible sediment transport. Among completely cleared treatments, the sediment density in water runoff was

Plate 5 A 4.5 foot H-flume with a water level recorder (in box) and the Coshocton wheel sampler (foreground)

Plate 6 Multi-divisor and storage tanks

Facing page 134

Plate 7 High soil erosion in photograph cleared with tree pusher/root rake attachment

Plate 8 Gully erosion was observed even in the first season after clearing with the tree-pusher/root-rake attachment

Plate 9 Deposition of sediments at the flume entrance

the least from manually cleared and untilled catchments and the maximum from mechanically cleared and conventionally tilled catchments. The sediment density is related to the veolcity of water runoff and the surface soil conditions. The surface soil conditions, through their effects on infiltration rate and soil detachment, affect the velocity and sediment capacity of water runoff.

3.4.3.2. Water Runoff and Soil Erosion

Compared with no runoff from a forested watershed and a slight runoff from plots managed with traditional methods, there was significantly more runoff from other treatments (Table 3.9). Among the cultivated catchments, the amount of runoff was different from one another by several order of magnitude. Runoff was less from manual than mechanically cleared treatments, and from untilled than tilled catchments. The mean runoff from no-till catchments (average over the clearing methods) was 84.7 mm compared with 152.3 mm from tilled catchments. In contrast, the runoff from manually cleared catchments (average over the tillage methods) was 35 mm compared with 163 mm from mechanically cleared treatments. In addition, the peak runoff rates were also high from mechanically cleared tilled catchments.

Soil erosion was significantly affected by methods of land clearing and tillage systems (Table 3.9). The highest soil erosion was observed on catchments that were mechanically cleared with tree-pusher/root-rake attachments and were conventionally tilled (Plates 7 to 9). This high rate of erosion was observed in spite of the graded channel terraces that were constructed to minimize soil erosion. In general there was significantly less soil erosion on manually cleared than on mechanically cleared regions (2.5 t ha^{-1} yr^{-1} vs. 13.8 t ha^{-1} yr^{-1}), and from no-tillage than from conventionally tilled catchments (6.5 t ha^{-1} yr^{-1} vs. 12.1 t ha^{-1} yr^{-1}). Among mechanically cleared treatments, soil loss was less from the plots cleared by shear blade than those cleared by tree-pusher/root-rake attachments (3.8 t ha^{-1} yr^{-1} vs. 17.5 t ha^{-1} yr^{-1}). The soil erosion from plots cleared with the tree-pusher/root-rake and tilled catchments was underestimated because there was a considerable deposition of sediments at the flume entrance that could not be measured (Plate 9). The deposition was as much as 30 cm deep covering an area of about 100 m^2. Consequently, the soil loss per mm of runoff did not observe the same trend as that of measurable erosion. Nevertheless, the soil loss per unit quantity of water runoff was more from tree pusher than manually cleared (86.9 kg mm^{-1} vs. 54.7 kg mm^{-1}) and from tilled than untilled catchments with manual clearing (84.7 kg mm^{-1} vs. 25.8 kg mm^{-1}).

3.4.3.3. Hydrological Problems and Land Development

Bringing new land under cultivation is still one of the cheapest means of increasing food production in the tropics. In spite of the adverse environmental consequences,

Table 3.9. Effects of methods of deforestation and tillage systems on sediment density, water runoff, and soil erosion from maize–cassava rotation

Clearing treatment	Tillage system	Sediment density ($g\ l^{-1}$)	Water runoff ($mm\ yr^{-1}$)	Soil erosion ($t\ ha^{-1}\ yr^{-1}$)
Traditional clearing	Traditional seeding	0.0	2.6	0.01
Manual clearing	No-tillage	3.4	15.5	0.4
Manual clearing	Conventional tillage	8.6	54.3	4.6
Crawler tractor/shear blade	No-tillage	5.7	85.7	3.8
Crawler tractor/tree-pusher	No-tillage	5.6	153.1	15.4
Crawler tractor/tree-pusher	Conventional tillage	13.0	250.3	19.6

Sediment density reported here was from a rainstorm monitored on 31st May 1979.

vast areas of tropical forest will be developed for food production. Data presented indicates that other than the shifting-cultivation treatment, manually cleared plots produced less water runoff and soil loss than mechanically cleared treatments. However, manual clearing can be inefficient, time consuming, and uneconomical, and because of the labour shortage, manually cleared plots may not be ready for cultivation in time. Roots and stumps are not properly removed and pose hazards to the equipment for mechanized operations. Mechanical methods are, therefore, indispensable for large scale deforestation for agricultural purposes. To ensure sustained productivity, it is important to develop appropriate soil management systems that will minimize the adverse effects of deforestation by mechanical means.

It is evident from the data presented that tillage methods and appropriate management of soils and crops play an important role in soil and water conservation and in decreasing the rate of decline of soil quality. The ratio of soil loss to grain yield presented in Table 3.10 supports this argument. Those treatments that were mechanically cleared with front-mounted shear blade attachment and those that were manually cleared and tilled both lost about 2 kg of soil per kg of maize grains produced. The ratio of soil loss to grain yield was highest in mechanically cleared plots with the front-mounted tree-pusher and root-rake attachment. For the first season after clearing the no-tillage treatments were relatively ineffective because of lack of crop residue mulch on the soil surface. Subsequently with adequate crop residue mulch there should be a substantial decrease in water runoff and soil loss from untilled plots cleared with shear blade and tree-pusher attachments.

3.4.4. GENERAL DISCUSSION

Any method of deforestation can be detrimental, and more so with the indiscriminate use of heavy machinery. However some areas now under forest cover will have to be developed for food production. Wherever practical, priority should be given to the manual methods that cause least damage to the soil structure and suffer minimal losses due to water runoff and soil erosion. It may be difficult, however, to develop large areas of mature forest by the use of the manual labour. Enough manpower may not always be available in remote areas (such as Amazon or the Congo basin), or the manual operation may be expensive and time consuming. Under these conditions, there may be no choice but to resort to the mechanical means for forest removal.

Review of the currently available information indicates that there is a scope for improvements in the mechanical methods commonly employed, and their adverse effects can be appreciably decreased by careful management. The choice of appropriate attachments is important. Care must be exercised not to remove all roots and stumps and carry-off the top soil in windrows. Leaf litter and other biomass should be left on the soil surface as much as possible. In this connection, the importance of subsequent soil and crop management systems cannot be overemphasized. If weeds can be controlled by herbicides, mechanical seedbed preparation should be elimin-

Table 3.10. Effects of methods of deforestation and tillage systems on the ratio of soil loss to grain yield

Clearing treatment	Tillage system	Soil loss/maize grain yield (kg kg^{-1})	Relative soil loss/grain yield	Maize grain yield (t ha^{-1})
Traditional clearing	Traditional seeding	0.02	1	0.50
Manual clearing	No-tillage	0.25	13	1.56
Manual clearing	Conventional tillage	2.09	105	1.59
Crawler tractor/shear blade	No-tillage	1.92	96	1.98
Crawler tractor/tree-pusher	No-tillage	11.32	566	1.36
Crawler tractor/tree-pusher	Conventional tillage	11.20	560	1.75

ated. However, this is easier said than done. Appropriate herbicides for a given crop may not be available. Furthermore, the use of heavy clearing equipment may compact the soil that renders it unsuitable for cultivation of tropical root crops such as cassava (*Manihoc esculenta*) and yam (*Dioscorea rotundata*).

Crop covers (such as *Pueraria phaseolodies, Stylosnathes guianensis, Centrosema pubescens*, etc.) may play an important role in the management of soils in the humid tropics. If the forest is cleared at the end of the rainy season, appropriate crop covers may be planted immediately to provide a quick ground coverage. These covers may improve the soil structure that has been damaged by the use of heavy machinery. Subsequent seeding of rowcrops through chemically or mechanically suppressed sod may minimize the risk of soil erosion.

Adaptive research is needed to evaluate the performance of different land development techniques for soil and water conservation in diverse agro-ecological environments of the humid tropics. These experiments must be conducted on catchment basis, so that their effects on hydrological and ecological parameters can be quantified.

3.4.5. ACKNOWLEDGEMENT

Help received from Professor L. L. Harrold of Coshocton, Ohio, USA, in establishing the hydrological experiment and from some staff of the Farming Systems Program of IITA is gratefully acknowledged.

REFERENCES

Blackie, J. R., 1972, Hydrological effects of a change in a land use from rainforest to tea plantation in Kenya, Symposium Report and Experimental Basins, Wellington, *N.Z. Bull. Int. Ass. Sci. Hydrol. Publ.,* **97**, 312–329.

Bonell, M., and Gilmour, D. A., 1978, The development of overland flow in a tropical rainforest catchment, *J. Hydrol.,* **39**, 365–382.

Bonell, M., Gilmour, D. A., and Sinclair, D. F., 1979, A statistical method for modelling the fate of rainfall in a tropical rainforest catchment, *J. Hydrol.,* **42**, 251–267.

Brown, J. A. N., 1972, Hydrological effects of a bushfire in a catchment in south-eastern New South Wales, *J. Hydrol.,* **15**, 77–96.

Brunig, E. F., 1975, Tropical ecosystems: state and targets of research into the ecology of humid tropical ecosystem, *Plant Research and Development,* **1**, 22–38.

CENICAFE, 1975, Manual de conservation de sudes de ladera, *CENICAFE,* Chinchina, Caldas, Colombia, 267 pp.

Cunningham, R. K. 1963, The effect of clearing a tropical forest, *J. Soil Sci.,* **14**, 334–345.

Gilmour, D. A., 1977, Effect of rainforest logging and clearing on water yield and quality in a high rainfall zone of north-east Queensland, *Institute of Engineers of Australia Symposium on the Hydrology of Northern Australia,* Brisbane, Queensland 1977 Natl. Conf. Publ. No. 77/5, pp. 155–60.

Gilmour, D. A., and Bonell, M., 1977, Streamflow generation processes in a tropical rainforest catchment—preliminary assessment, *Institute of Engineers of Australia Symposium on the Hydrology of Northern Australia,* Brisbane, Queensland 1977 Natl. Conf. Publ. No. 77/5, pp. 178-179.

Gilmour, D. A., and Bonell, M., 1979, Runoff processes in tropical rainforests with special reference to a study in north-east Australia, In A. F. Pitty (Editor), *A Geographical Approach to Fluvial Processes,* Geobooks, Norwich (in press).

Gupta, S. K., Das, D. C., Tejwani, K. G., Srinivas, Chittaranjan, S., and Ram Babu, 1975, *Hydrological Investigations,* (Tejwani, K. G., Gupta, S. K., and Mathur, H. N., Editors), ICAR (1975), 17-615.

Hibbert, A. R., 1967, Forest treatment effects on water yield, *Proceedings of International Symposium on Forest Hydrology, Pennsylvania State University* (Pergamon Press), 537-543.

Kenworthy, J. B., 1969, Waterbalance in a tropical rainforest, A preliminary study in the Ulu Gombak forest reserve, *Malay. Nat. J.,* **22**, 129-135.

Lal, R., 1980, Management of soils for continuous production: controlling erosion and maintaining physical condition, In D. J. Greenland (ed.), *Characterization of soils in relation to their classification and management for crops production.* Oxford University Press, London, UK (in press).

Lal, R., and Cummings, D. J., 1979, Clearing a tropical forest: 1 Effects on soil and micro-climate, *Field Crops Res.,* **2(2)**, 91-107.

Low, K. S., and Goh, K. C., 1972, The water balance of five catchments in Selangor, West Malaysia, *J. Trop. Geography,* **35**, 60-66.

McCulloch, J. S. G., and Dagg, M., 1965, Hydrological aspects of protection forestry in East Africa, *East Afr. Agric. and For. J.,* **30**, 390-394.

Pathak, S., 1974, Role of forests in soil conservation with special reference to Ramganga watershed, *Soil Conservation Digest,* **2(1)**, 44-9.

Pereira, H. C. (ed.), 1962, Hydrological effects of changes in land use in some east-African catchment areas, *East Afr. Agric. and For. J.,* **27**, (Special Issue), 1-129.

Pereira, H. C., 1965, Land use and streamflow, *East Afr. Agric. and For. J.,* **30**, 395-397.

Pereira, H. S., 1973, *Land Use and Water Resources.* Cambridge University Press pp. 246.

Richards, P. W., 1973, The tropical rainforest, *Scientific American,* **229(6)**, 58-67.

Srivastava, S. N., 1974, Peak rate of runoff from watersheds of soils conservation research farm at Demetans—Hazaribagh, *J. Soil and Water Conservation, India,* 22-23, 28-31.

Thijsse, J. P., 1977, Soil erosion in the humid tropics, *Landbourwkundig Tijdshcrift,* **pt. 89(11)**, 408-411.

Young, R., 1974, The rate of slope retreat, *The Institute of British Geographers, Special Publication No. 7,* 65-78.

Tropical Agricultural Hydrology
Edited by R. Lal and E. W. Russell
© 1981, John Wiley & Sons Ltd.

3.5

Rainfall Redistribution and Microclimatic Changes over a Cleared Watershed

T. L. Lawson, R. Lal, and K. Oduro-Afriyie

3.5.1. INTRODUCTION

The distribution of heat and moisture over the earth's surface is the major determinant of climates both on the large and the small scale. Whereas large-scale transport of properties on the global scale by the general circulation of the atmosphere determines the climatic features thousands of miles downstream, climate on the meso- and microscale is more intimately tied to the local surface characteristics because of their profound and immediate impact on the local energy and moisture budgets.

Modification of the microclimate can therefore be readily brought about by changes in surface cover. These changes are often brought about quite inadvertently and may be of more than local significance; this is evidenced by the analysis of the causes of the expansion of the Rajasthan Desert in northwestern India by Bryson and his colleagues (1965, 1966). Numerous other examples of historical significance have been cited by Bryson and Murray (1977).

In general, modification of surface characteristics involving the clearing of forests for the purpose of crop or animal husbandary ignore all too often the need to preserve to the best possible extent the delicate balance between the environment and the climax vegetation to be removed. The consequence, not surprisingly, has in general been an increase in the 'harshness' of the environment, both aerial and edaphic, due mainly to the misuse of the cleared land (Goodland and Irwin, 1975).

There are however, admirable examples of the successful development of tropical forests into arable lands. The work of the Tennessee Valley Authority in the United States and the studies by Pereira (1973) and his co-workers in East Africa are but two clear illustrations in contrasting climatic situations. It is evident therefore that, to minimize risks, the initial environment must be understood and whatever changes need to be made are such as to hold their impact on water and energy balance to the minimum consistent with their overall goal. These imperatives constitute the underlying basis for the studies on the watershed management at IITA, an aspect of which is reported here. It is pertinent to point out that the results reported below are preliminary.

3.5.2. SITE CHARACTERISTICS

The study area lies to the west of a reservoir on the farm of The International Institute of Tropical Agriculture (IITA) at Ibadan, Nigeria. It is made up of two watersheds, the larger one of which was divided into 12 plots which were cleared by different methods (Lal, 1981), and the smaller was left in its initial state of secondary forest cover that includes such trees as *Ceiba pentandra, Trichilia mono- delpha*, and remnants of cacao (*Theobroma cacao*) and cola species (*Cola nitida, Cola acuminata*) plantations. It thus provides a control against which the various clearing treatments can be compared.

3.5.3. INSTRUMENTS AND OBSERVATIONS

Incoming radiation, air temperature and relative humidity, and soil temperature were measured both in the forest and in a cleared area near the edge of the forest. Measurements of wet and dry bulb temperatures, and periodic observations on soil temperature were also made on a few of the cleared plots, including those manually cleared and conventionally cultivated and those mechanically cleared with the shear blade and with the tree pusher followed by zero tillage. Rainfall was measured by a series of rain gauges in the cleared area (Nwa, 1975), throughfall was measured in the forest by a series of randomly sited plastic raingauges while stemflow was moni- tored by girdling representative trees of different diameters and canopy structure with a flow duct and collecting the water into large plastic bottles. Total runoff from the cleared watershed was gauged by means of two weirs (1:5 slope) that en- able separation of surface and subsurface flow. H-flumes were installed for the determination of runoff from individual plots. Two weirs were also installed in the uncleared area for monitoring surface and subsurface flow.

3.5.4. RESULT AND DISCUSSIONS

Basic climatological and hydrological observations over the two watersheds were initiated by Nwa in 1974 and 1975; (IITA, 1975; Nwa, 1975). The data reported here covers the first year following the clearing of the larger watershed.

3.5.4.1. Rainfall

As Nwa (1977) has deduced from an earlier study, spatial variability of rainfall at given incidents was high over the wateshed. The 1979 data show that in 30 per cent

Table 3.11. Rainfall and components over the forested watershed at the IITA catchment, 1979

Variables	May	June	July	Month August	September	Season
Number of storm sampled	5	6	7	5	7	30
Total rainfall[a] (P) mm	89.4	107.8	238.8	76.3	132.8	645.1
Throughfall (T) mm	68.4	85.0	170.0	50.8	98.8	473.0
Stemflow (S) mm	9.5	8.9	29.8	6.9	9.3	64.4
Interception (I) mm	11.5	13.9	39.0	18.6	24.7	107.7
T/P (%)	77	79	71	67	74	73
S/P (%)	11	8	12	9	7	10
I/P (%)	13	13	16	24	19	17

[a] Only some rainstorm events were sampled.

of all cases with rainfall greater than 5 mm, the difference between the highest and the lowest point rainfall exceeded 25 per cent of the areal mean value. These differences were particularly high during the months of prevailing thunderstorms, notably April and May. Even on the basis of the monthly totals for the individual raingauges, they amounted to 26 per cent and 31 per cent respectively for these two months.

The rains in 1979 were on the whole unusual in terms of monthly distribution and total amount received. The cumulative total at the end of October was 21 per cent above the corresponding long-term mean, with the months of March and June experiencing well below normal amounts, and April, May, July, August, and September well above. The period was particularly noteworthy because the expected August break in the rains did not occur.

The partitioning of the rainfall over the forested area, which was assumed to be that measured in the open outside the forest, into throughfall, stem flow, and interception is given in Table 3.11.

3.5.4.2. Throughfall

The proportion of the rain reaching the ground (more or less directly as throughfall) ranges from 79 per cent in June to a minimum of 67 per cent in August. It averages 73 per cent over the period sampled. This compares with values of 90 per cent and 86 per cent measured at two sites in the Banco forest in Ivory Coast, and 77 per cent observed in the Yapo forest not far from the preceding area (Bernhard-Reversat *et al.*, 1972). A corresponding seasonal value of 74 per cent has been given by Noirfalise (1956) for the Garamba forest in the Congo (now Zaire) while values deduced by the present authors from data quoted by Penman (1963) for middle latitude forests at Sauerland, Germany (Eidmann, 1959) show the same order of magnitude as evidenced by Table 3.12. The differences in values in these and other studies (Helvey and Patric, 1965; UNESCO/UNEP/FAO, 1978) result from a

Table 3.12. Throughfall (*T*), Stemflow (*S*) and Interception (*I*) for spruce and beech as percentage of rainfall (mm) (Adapted from Eidmann, 1959)

		Spruce			Beech		
Period	Rainfall	*T*	*S*	*I*	*T*	*S*	*I*
November–April	587	79	0.7	20	79	16.5	4
May–October	629	68	0.8	32	73	16.5	11
Year	1216	73	0.7	26	76	16.5	7.7

number of factors including the lack of standardized methods of measurement, and the dependence of observed values on such factors as canopy structure, diameter of trees (Rutter, 1963), and the intensity of rainfall and associated winds.

3.5.4.3. Stemflow

The mean value of 10 per cent in stemflow (Table 3.11) is an order of magnitude higher than the corresponding data reported by Malaise (1973) for a woodland in Zaire and by Nye (1961) for a semi-deciduous forest in Ghana. However, values as high as 18 per cent and 28 per cent have also been reported for tropical forests in Puerto Rico (Kline *et al.*, 1968) and Brazil (Freise, 1936) respectively. These apparent discrepancies are undoubtedly due to the same factors affecting measurements of throughfall.

3.5.4.4. Interception

Interception was calculated as the differences between the rainfall and the through-and-stemflow. It varied from 24 per cent in August to 13 per cent in May and June (Table 3.11). The seasonal average of 17 per cent compares well with figures generally quoted for other tropical areas; among these the 12–20 per cent obtained in Puerto Rico (Kline *et al.*, 1968), the 18–20 per cent average for Bamboo forest in Kenya (Pereira, 1952).

3.5.4.5. Runoff and Seepage

Under native forest vegetation, there is little or no water runoff. The 34-ha experimental site, before clearance, from 1974 to 1978 gave a runoff of less than 2.5 per cent of rainfall, and Figure 3.3 shows the hydrograph of this catchment for 1975. The maximum flow was 6 litres per second. The measurements of ground water table made in 1975 indicated that from July to October the ground water rose to 30 cm below the surface.

The runoff and seepage measurements made on the 34-ha cleared area are shown

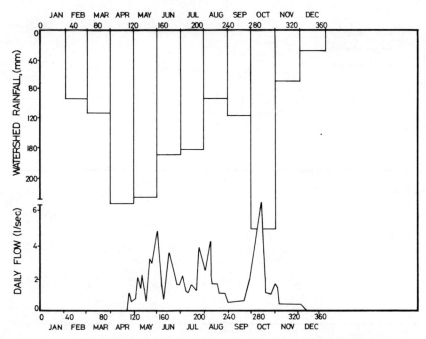

Figure 3.3 Hydrograph of uncleared watershed in 1975 (Data of Dr. E. Nwa)

in Figure 3.4. The effects of clearing this catchment in early 1979 on the surface and subsurface flow can be seen by comparing the figures of 1975 and 1979 shown in Figures 3.3 and 3.4 respectively. Compared with a negligible water runoff from the forested catchment, 23 per cent (339.4 mm out of 1445.3 mm of rainfall) of rain received was lost as runoff. Similarly, land clearing also resulted in a significant increase in the subsurface flow, both the duration and the quantity of which increased by deforestation. Data in Figure 3.4 indicate that tbout 1.5 per cent of the rainfall received was lost as subsurface flow. Though the runoff and subsurface flow observed at the IITA catchment in 1979 may be high due to an exceptionally wet year, the annual evapotranspiration from the cleared catchment (i.e. the difference between the rainfall and water yield) was 75 per cent of the rainfall received, compared with about 97 to 98 per cent of the rainfall collected from the forested catchment.

3.5.5. TEMPERATURE AND RELATIVE HUMIDITY

3.5.5.1. Air and Soil Temperatures

The air temperature regime in the forest and in the cleared and subsequently cropped area during the period of study is shown in Figure 3.5. As expected, the temper-

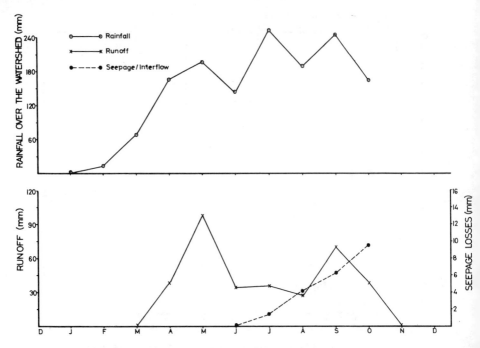

Figure 3.4 Rainfall, runoff, and seepage/interflow over the cleared watershed (IITA, 1979)

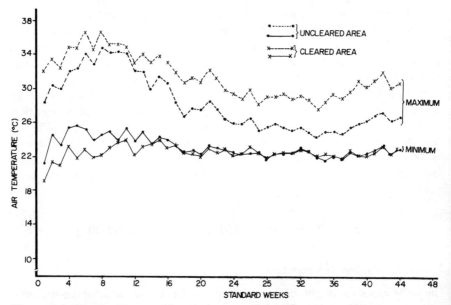

Figure 3.5 Comparative weekly maximum and minimum temperature in cleared and uncleared areas

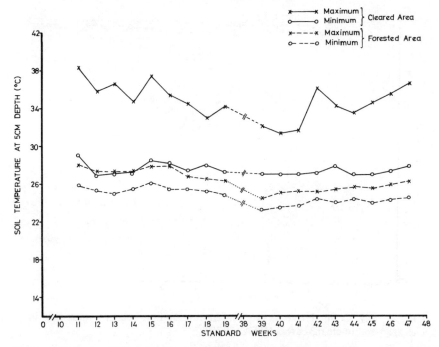

Figure 3.6 Comparative weekly maximum and minimum soil temperature at 5 cm depth in cleared and uncleared area (IITA, 1979)

ature variations are smaller under forest, with the maxima lower and the minima higher particularly in the dry season, than the corresponding values in the open. The higher maximum air temperatures over the cleared land are probably due to greater proportion of the incoming radiation that reaches the soil surface, and the lower minimum air temperature during the dry season, when the cloud cover during the night is much less than during the rains, probably reflects for the greater back radiation from the surface.

In parallel with the air temperatures, the soil temperature wave is also substantially damped by the forest, the reduction in maxima and minima being even more pronounced (Figure 3.6). Again the important role of available moisture in the rainy season in ensuring a preferential depletion of available energy and thus reducing differential soil heat flux is evident by the decrease in the temperature difference between the contrasting surfaces.

3.5.5.2. Relative Humidity

The generally more humid conditions in the forest in comparison with the cleared and cropped surfaces is shown by Figure 3.7. Differences in the maximum values are minor because of the near saturation conditions that prevail at night and in the early morning hours in this region regardless of time of year. The differences in

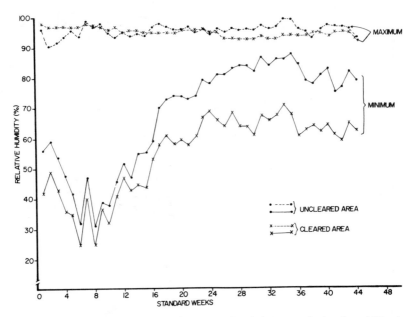

Figure 3.7 Comparative maximum and minimum relative humidity in cleared and uncleared areas (IITA, 1979)

minima are particularly pronounced during the rainy season and they partly reflect the warmer conditions over the cropped land. These results show a persistence in the trend established earlier (Lawson, 1978) and are in agreement with those reported from comparable ecological areas (Cachan and Duval, 1963; Hopkins, 1965; Richards, 1972).

3.5.5.3. Evaporation

Figure 3.8 shows the weekly mean pan evaporation measured in the cleared area close to the edge of the forest and corresponding values at the main climatic station on site. The rate of evaporation is seen to be lower at the edge of the forest than at the normally exposed climatic station. With clearing of the adjacent areas, the differences in annual evaporation is however reduced, which is not surprising because of the increased exposure of the pan on the watershed following clearing. It may be deduced that evaporation from the surface into the ambient air under forest conditions would be much lower. Based on the observed data in the cleared area the total evaporation over the period March through September amounts to 794 mm. Estimate of total pan evaporation for the cleared area for a complete water year, hence over the moisture cycle was derived from the multi-annual mean evaporation from the main climatic station on site and the ratio of the 1979 March through October evaporation for the cleared watershed to that of the main station. A value

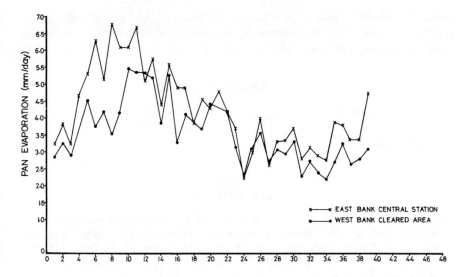

Figure 3.8 Comparative weekly mean pan evaporation

of 1409 mm was obtained. Using this and the value of 1083.5 mm or 75 per cent of the annual rainfall discussed earlier as the total evapotranspiration, E_T, over the cropped watershed, a ratio E_t/E_0 was found to be equal to 0.78. This compares closely with similar results from East Africa (Pereira, 1973).

3.5.6. SUMMARY AND CONCLUSION

The water budget and meso- and microclimatic parameters over a forested area and a cleared and cropped watershed were routinely monitored during the 1979 cropping season within the framework of a watershed management study at IITA, Ibadan.

Seventy-three per cent of the incoming rainfall was found to reach the forest floor as throughfall; the respective values for stemflow and interception were 10 per cent and 17 per cent. Surface runoff from the cleared area was 23 per cent of the rainfall, compared to the negligible pre-clearing value and the similarly negligible runoff from an adjacent forested watershed. Subsurface flow was about 2 per cent in the cleared area. Evapotranspiration (E_t) from the cropped watershed was estimated to be 75 per cent of the rainfall; it represented 78 per cent of the open water evaporation (E_0), i.e. $E_t/E_0 = 0.78$.

Soil and air temperature extremes were accentuated following clearing as expected; so were the relative humidity values. Clear differences were observed with treatments of the cropped areas, a zero-tillage treatment showing a more moderate soil temperature condition than a conventionally tilled plot, for instance.

It is reasonable to conclude that, although forest removal for crop or animal husbandry generally leads to an increased harshness of the environment with respect to water and energy budgets, a judicious selection of methods can considerably lessen the otherwise adverse impact.

3.5.7. ACKNOWLEDGEMENT

The authors gratefully acknowledge the contribution made by Dr. E. U. Nwa (formerly Soil and Water Engineer, IITA, now of Ahmadu Bello University) to this work in initially delineating the watershed and setting up some of the structures used in this study.

REFERENCES

Bernhard-Reversat, F., Juttel, C., Lemee, G., 1972, Quelques aspects de la periodicite ecologique et de l'activite vegetale saisonniere en foret ombrophile sempervirente de Cote-d'Ivoire, In Golly, F. B., and Golley, P. M. (eds.), *Tropical Ecology with Emphasis on Organic Production,* pp. 214–234, Athens, University of Georgia, 418 pp.

Bryson, R. A., 1966, Inadvertent climatic modification (A lecture presented to the Freshman Forum Class, September 1966), Mimeo, 18 pp., University of Wisconsin, Madison.

Bryson, R. A., and Baerreis, D. A., 1965, Possibilities of major climatic Modification and their Implications, *Northwest India, A case for study,* Paper presented at the Symposium on Economic and Social Aspects of Weather Modification, Boulder, Colorado, July 1965.

Bryson, R. A., and Murray, T. J., 1977, *Climates of Hunger,* The Univ. of Wisconsin Press, Madison, Wisconsin, 1977.

Cachan, P., and Duval, I., 1963, Variation Microclimatique Verticales et Saisonieres dans la Forêt Sempervirente de Basse Côte d'Ivoire, *An. Fa. Sci.,* Tome 8, Serie Sciences Animales, No. 1, 5–87.

Eidmann, F. E., 1959, Die Interception in Buchen—und Fichtenbestanden, *C. R. Ass. Int. Hydrologie Sci. Hannover Symp.,* **1**, 5–25.

Freise, F., 1936, Das Binneklimma Von Urwaldern im sub-tropischer Brasilien, *Petermanns Geogr. Mitteilungen,* **82**, 301–307.

Goodland, R. J. A., and Irwin, H. S., 1975, *Amazon Jungle: Green Hell to Red Desert?* Reprinted from *Landscape Planning,* Vol. 1, No. 2/3, 123–254.

Helvey, J. O., and Patric, J. H., 1965, Canopy and Litter Interception of Rainfall by Hardwoods of Eastern United States, *Water Resour. Res.,* **1**(2), 193–206.

Hopkins, B., 1965, Vegetation of the Olokemeji Forest Reserve, Nigeria, III The Microclimates with special Reference to their Seasonal Changes, *Jour. Ecol.,* **53**, 125–138.

Kline, J. R., Jordan, C. F., and Drewry, G., 1968, Tritium movement in soil of a tropical rain forest (Puerto Rico), *Science,* **160**, 550–557.

Lal, R., 1981, Deforestation of tropical rainforest and hydrological problems. In R. Lal and E. W. Russell (eds) *Tropical Agricultural Hydrology.* J. Wiley & Sons, Chichester, U.K. 131–140.

Lawson, T. L., 1978, Inadvertent modification of climate on the mesoscale: A case study, *Paper presented at the Pre-WAMEX (West African Monsoon Experiment) Symposium,* November 1–3, 1978, Univ. of Ibadan, Nigeria (Proceedings in press).

Malaise, F., 1973, Contribution a l'etude de l'ecosysteme foret claire (miombo) Note 8. *Le projet miombo Ann. Univ. Abidjan, E.,* **6(2)**, 227–250.

Noirfalise, A., 1956, *Exploration du Parc National de la Garamba Fase, 6. Le Milieu Climatique,* Inst. Pares Nat. Congo Belge (Zaire), Brussels, 1956.

Nwa, E. U., 1975, *Annual Report,* IITA, Ibadan, Nigeria.

Nwa. E. U., 1977, Variability and error in rainfall over a small tropical watershed. *J. Hydrology,* **34,** 161–169.

Nye, P. H., 1961, Organic matter and nutrient cycles under moist tropical forest, *Plant and Soil,* **13,** 333–346.

Penman, H. L., 1963, *Vegetation and Hydrology Tech. Comm. No. 53,* Commonwealth Bureau of Soils Harpenden, Commonwealth Agricultural Bureaux, Farnham Royal, Bucks, England.

Pereira, H. C., 1952, Interception of rainfall by cypress plantations, *E. Afr. Agric. J.,* **18,** 73–76.

Pereira, H. C., 1973, *Land Use and Water Resources in Temperate and Tropical Climates,* Cambridge University Press, Cambridge.

Richards, P. W., 1972, *The Tropical Rain Forest,* Cambridge University Press, Cambridge.

Rutter, A. J., 1963, Studies in the water relations of *Pinus sylvestris* in plantation conditions, 1 Measurements of rainfall and interception, *J. Ecol.,* **51,** 191–203.

UNESCO/UNEP/FAO, 1978, *Natural Resources Research XIV Tropical Forest Ecosystems,* UNESCO, Paris, 1978.

Tropical Agricultural Hydrology
Edited by R. Lal and E. W. Russell
© 1981, John Wiley & Sons Ltd.

3.6

Nutrient Losses in Water Runoff from Agricultural Catchments

B. T. KANG AND R. LAL

Nutrient losses through leaching, and in water runoff and eroded sediments contribute to the depletion of soil fertility of cropped land. Although the amount of nutrient loss in solution phase may be small, research information on the magnitude of nutrient losses in runoff water from tropical catchments is scanty. Nitrogen is one of the nutrients that can be leached with seepage water or carried off in surface runoff. Kowal (1972) reported an average loss ranging from 7 to 19 kg ha^{-1} yr^{-1} for some soils in northern Nigeria. Lal (1976) observed that the magnitude of N loss in runoff is affected by cropping systems and mode of crop residue management. Losses of NO_3-N in the runoff water during one season in 1973 were 3.3 kg ha^{-1} from bare fallow, 1.5 kg ha^{-1} from conventionally tilled maize, 0.1 kg ha^{-1} from no-till maize, and 0.04 kg ha^{-1} from mulched maize plots. These observations were made on small plots of 100 m^2. The losses were greater by many orders of magnitude in the eroded soil, and in the seepage water.

A relatively small quantity of P in water runoff can be a major source of eutrophication and pollution of surface waters. There exists a little information concerning the dynamics of P movement from tropical catchments. Lal (1976) also reported that P losses measured from runoff plots in tropical regions may be as low as 0 to 2.0 kg ha^{-1} season^{-1}. Movement of P from agricultural lands was primarily with the soil solids (Lal, 1976). In addition to N and P, losses of K, Ca, Mg, and Na are also of agricultural significance. Turvey (1973) investigated the mean concentrations of K, Ca, N, Mg, and Na in a tropical rain forest catchment in Papua, New Guinea. His values and those from other studies on small tropical streams as summarized by himself showed that in general concentrations of these nutrients in natural catchments are low. Potassium, Ca, Mg, and Na concentrations ranges were respectively: 0.43 to 1.55; 0.28 to 1.33; 0.20 to 2.68 and 1.61 to 6.05 p.p.m.

Barnett et al. (1972) in their study of Puerto Rican soils observed K concentrations from 0.01 to 2.29 p.p.m. in runoff from fertilized plots. Losses of some of these nutrients in tropical West Africa can be substantial. Kowal (1972) reported an average annual loss of Ca, Mg, and Na in runoff water and eroded soil ranging from 14 to 30 kg ha^{-1}. Lal (1976) also reported large runoff losses of K, Ca, and Mg from runoff plots. K losses ranged from 0.1 to 13.4 kg ha^{-1} season^{-1}. Ca from 0 to 4.2 kg ha^{-1} season^{-1}.

From the above limited review, it is clear that our knowledge on nutrient losses with runoff water from agricultural land in the tropics is far from adequate. Data on losses of sulphur and micronutrients are not available. This report describes the effects of clearing methods and post soil management practices on nutrient losses with runoff.

3.6.1. MATERIALS AND METHODS

These observations were made at the watershed experiment of IITA described in this volume by Couper *et al.* (Chapter 3.3). The soils of the experimental site belong to Egbeda and Iwo (oxic paleustalf) associations (Moormann *et al.*, 1975). The site was of moderate fertility and was over 15 years of decidious forest at time of clearing. The average chemical composition of the surface soils at time of planting from the various clearing treatments is shown in Table 3.13. Use of the crawler tractor/tree pusher caused the most disturbance of the surface soils and resulted in lower nutrient levels than the other treatments.

Trees and underbush growth in traditional and manually cleared catchments were burnt *in situ* about 3 to 4 weeks after felling (Plate 10). Unburnt trees, particularly the trunks of oil palm (*Flaeis guianeensis Jacq*) were removed to plot boundaries.

Runoff samples were collected below the H-flume with a Coshocton wheel sampler. The sample collected by the wheel was partitioned by the multi-slot divisor system and only $^{1}/_{25}$ of the sample was collected in the storage tank. Runoff water in the storage tank was thoroughly stirred prior to taking two one-litre samples in a plastic container.

A uniform crop rotation of maize–cassava was adopted in all treatments. In the first year, the crop did not receive any fertilizer. Runoff samples were analysed from the maize part of the rotation only.

3.6.1.1. Runoff Water Analyses

Runoff samples were centrifuged to remove suspended sediments. The samples were then treated with toluene and refrigerated prior to analysis. Subsamples were taken for measurements of pH and conductivity. NO_3-N and NH_4-N were determined by distillation (Bremner and Keeney, 1966), PO_4-P was determined colorimetrically after colour development using NH_4-molybdate (Murphy and Riley, 1962). K, Ca, and Na were determined using flame photometer and Mg, Cu, Fe, Mn, and Zn determined with atomic absorption spectrophotometer.

Plate 10 Ash and unburnt trees left on the surface of a manually cleared treatment

Table 3.13. Some properties of surface soil (0–15 cm) from various clearing treatments before cropping in 1979

Clearing treatments	Tillage systems	pH–H$_2$O	Organic C (%)	1 N Am-acetate Exchangeable				Extractable-P Bray-1 p.p.m.
				Ca	Mg (me 100g^{-1})	K	Na	
1. Manual clearing	Conventional tillage	6.8	1.61	11.24	1.65	0.59	0.07	19.5
2. Manual clearing	No-tillage	6.8	1.54	10.20	1.71	0.70	0.08	12.9
3. Crawler tractor/tree-pusher	Conventional tillage	6.5	1.56	10.53	1.70	0.48	0.06	9.7
4. Crawler tractor/tree-pusher	No-tillage	6.3	1.43	8.36	1.48	0.32	0.06	4.2
5. Crawler tractor/shear-blade	No-tillage	6.2	1.72	12.86	1.99	0.58	0.07	10.2
6. Traditional clearing	Traditional seeding	6.6	1.55	9.81	1.68	0.60	0.07	14.0

3.6.2. RESULTS AND DISCUSSION

There was a considerable variability in nutrient levels of runoff water between duplicate catchments at the same treatment. This variability is attributed to initial differences in soil fertility and also to differences in ash levels on the soil surface, particularly in the vicinity of the H-flume. Results in subsequent years will be expected to differ from these results obtained just after clearing.

3.6.2.1. Nutrient Composition of Runoff Water

To illustrate the effects of land clearing and soil management treatments on the quality of the runoff water, the pH, conductivity, and nutrient concentrations observed with the lowest and highest runoff values over the month of July, are shown in Table 3.14. The pH and conductivity tend to be lower with high discharge. Runoff from the traditionally cleared plots was very low and occurred mainly from the edge of the plots. Its pH and electrical conductivity were high due to the small quantity of water which dissolved the soluble substances in the ash from the burnt vegetation.

The other treatments do not show any effect on the NO_3-N and NH_4-N concentrations. With the manual clearing, the tree pusher, and the bulldozer clearings, the NO_3-N levels are generally low, and agree with the value < 2 p.p.m. reported by Taylor *et al.* (1971) for farmlands in Ohio. On the average the runoff water contains more NO_3-N than NH_4-N.

The PO_4-P concentrations in the runoff water with manual and traditional clearings are low. The values of < 0.01 p.p.m. P observed is equivalent to the value of 0.015 p.p.m. P reported in stream water from Ohio by Taylor *et al.* (1971). Ryden *et al.* (1973) also reported lower concentrations of dissolved inorganic P in stream water from Washington with 0.007 p.p.m. P. With the tree pusher and bulldozer clearing, PO_4-P concentrations are substantially higher, indicating more loss of P in the runoff water. The PO_4-P concentration also appears to decrease with increasing runoff.

In general, the K and Na concentrations in the runoff water show an increase with increasing discharge. Turvey (1973) in a study on water quality from a forested catchment area in Papua New Guinea also reported that K contributes significantly greater proportion of the dissolved solids at high flows and a smaller proportion at low flows. In other words, high flow removed more K per hectare of land contributing to the flow. The Ca and Mg concentrations are related, both nutrients decreased in concentrations with increasing rate of runoff. High K, Ca, Mg, and Na concentrations in runoff with manual no-tillage and traditional clearings may be attributed to reasons mentioned earlier. In general the concentrations of K, Ca, Mg, and Na in the runoff water irrespective of treatments corresponds to figures reported by Lal (1976) from small plots at IITA, and are substantially higher than the concentrations observed in small tropical streams (Turvey, 1973). The high concentrations of these cations in the present study is the result of the clearing method (burning) and due to solubles from the ash.

Table 3.14. pH, conductivity and nutrient concentrations of runoff water associated with lowest and highest runoff values in July, 1979, for different land clearing and soil management treatments

Clearing treatments	Tillage systems	Runoff mm	pH	mmho 25°C	NH$_4$-N	NO$_3$-N	PO$_4$-P	K	Ca	Mg (p.p.m.)	Na	Cu	Fe	Mn	Zn
1. Manual clearing	No-tillage	0.06	7.1	0.230	0.18	0.53	0.04	12.0	37.0	4.5	6.6	<0.1	23.0	1.4	<0.10
		2.74	6.8	0.053	0.18	0.53	<0.01	24.0	2.9	1.1	1.6	<0.1	7.4	<0.1	<0.10
2. Manual clearing	Conventional tillage	0.001	7.3	0.120	0.18	0.53	<0.01	10.0	20.0	2.7	6.8	<0.1	1.6	<0.1	0.27
		17.2	7.2	0.100	0.53	1.40	<0.01	13.0	10.2	2.0	28.0	<0.1	<0.1	<0.1	<0.10
3. Crawler tractor/shear-blade	No-tillage	0.31	6.7	0.070	0.53	1.40	0.18	10.0	9.0	1.9	5.6	<0.1	30.0	0.7	0.15
		11.70	6.6	0.053	0.53	0.88	0.11	12.0	4.9	1.4	22.0	<0.1	10.1	<0.1	<0.10
4. Crawler tractor/tree-pusher	No-tillage	0.37	6.5	0.044	0.18	0.70	0.14	10.0	6.0	1.0	6.2	<0.1	1.1	0.2	<0.10
		19.27	7.1	0.058	0.18	0.35	0.06	9.6	2.8	1.3	1.4	<0.1	9.2	0.1	0.32
5. Crawler tractor/tree-pusher	Conventional tillage	0.001	6.9	0.074	0.35	1.40	0.14	10.0	10.0	1.5	6.1	<0.1	11.5	0.2	<0.10
		33.10	6.5	0.051	0.53	0.35	0.07	9.6	6.2	1.1	26.0	<0.1	4.5	<0.1	<0.10
6. Traditional clearing	Traditional seeding	0.001	7.5	0.300	0.18	7.35	<0.01	14.0	35.0	9.3	5.8	<0.1	0.9	<0.1	0.72
		1.36	7.0	0.280	5.60	0.53	<0.01	16.0	28.0	3.0	32.0	<0.1	<0.1	<0.1	<0.10

This data is an average of two replications, except treatment 6 that is reported for one replicate only.

The micronutrient concentrations in the runoff water are generally low. No distinct treatment effect is seen. Fe shows the largest and highest variation in concentration, while Cu shows the lowest concentration.

Changes in solute concentrations over the months July, August, and September are shown in Table 3.15. The NH_4-N and NO_3-N showed large changes though no distinct pattern can be observed. The concentrations of K, Ca, Mg, and Na on the other hand showed a decrease from July to September in almost all the treatments. This may be due to the fact (1) that the soluble nutrients from ash and soil surface have been removed during early runoff and that the soil is approaching a new equilibrium condition, and (2) that there is better crop cover if the nutrient losses in runoff water are reduced.

3.6.2.2. Nutrient Removal in Runoff Water

Data on nutrient removal in the runoff water is shown in Table 3.16. With mechanical clearing substantial runoff losses took place, and resulted in higher amounts of plant nutrients being removed with the runoff water. Tree pusher–conventional tillage > Tree pusher no-tillage > Bulldozer-shear blade > Manual clearing–conventional tillage > traditional clearing > Manual clearing–no-tillage. The losses of nitrogen in the runoff water is relatively small during the three months from July to September.

Massey *et al.* (1953) also reported small annual losses of NO_3-N from 0.19 to 1.13 kg ha^{-1} yr^{-1} in a well-maintained rotation in Wisconsin. The small amount of PO_4-P losses in the runoff water was expected, considering the low solubility of phosphorus. Since much of the P is retained strongly by soil particles, a large portion of the P loss with runoff is associated with sediments (Ryden *et al.*, 1973).

Losses of cations with mechanical clearing and conventional tillage are rather substantial. Substantial losses of K, Mg, and Zn which are in short supply and needed for crop growth may result in soil fertility decline even a few years after land development.

3.6.3. SUMMARY AND CONCLUSIONS

The results reported here were obtained in the 3 months following land clearance when the soil structure is generally at its best, so when runoff and nutrient losses are likely to be low. Or results show that for those treatments with low runoff losses, nutrient losses are also low, but for treatments giveing appreciable runoff, nutrient losses increase with runoff, except for N and P which were low in all treatments. The treatments giving the largest runoff and nutrient losses were those involving mechanical clearing with tree-pusher blades and root-rakes, and seed beds prepared by mechanical methods gave larger losses than those in which no tillage was used. In detail, the concentration of Ca, Mg, K, and Na tended to be highest in the first month of the experiment, when the rainfall was highest, and to be lower in

Table 3.15. Average concentrations of N, K, Ca, Mg, and Na of runoff water during the months July, August, and September for different land clearing and soil management treatments

Clearing treatments	Tillage systems		Total runoff mm	NH$_4$-N	NO$_3$-N	K	Ca	Mg	Na
						p.p.m.			
1. Manual clearing	No-tillage	July	5.05	0.44	0.55	13.2	13.1	2.6	14.4
		August	0.01	0.27	0.88	5.5	10.6	1.0	1.7
		September	0.26	2.52	2.29	4.9	10.9	2.3	8.2
2. Manual clearing	Conventional tillage	July	14.72	0.40	1.19	13.4	9.9	2.1	18.8
		August	0.46	0.18	0.35	5.1	7.2	3.4	7.8
		September	8.85	0.30	0.58	5.0	1.9	1.0	2.3
3. Crawler tractor/ shear-blade	No tillage	July	17.38	0.41	1.07	12.7	8.2	2.3	13.8
		August	0.41	0.75	0.63	10.8	7.0	2.1	2.9
		September	13.45	1.20	1.49	11.7	4.6	1.5	3.3
4. Crawler tractor/ tree-pusher	No-tillage	July	31.84	1.05	1.13	14.1	10.3	2.6	11.5
		August	6.19	0.76	1.29	10.3	6.5	2.0	4.8
		September	28.71	0.50	0.67	6.2	2.2	1.2	3.2
5. Crawler tractor/ tree-pusher	Conventional tillage	July	55.00	0.47	2.17	12.7	7.9	1.6	16.7
		August	7.82	0.40	0.48	6.3	8.5	4.2	4.9
		September	31.81	0.41	0.62	6.4	3.6	1.3	3.4

Rainfall: July 239 mm, August 76 mm, and September 133 mm.

Table 3.16. Runoff and nutrient loss in runoff water for different land clearing and soil management treatments for the months July, August, and September

Clearing treatment	Tillage systems	Runoff mm	NH_4-N	NO_3-N	PO_4-P	K	Ca	Mg $kg\,ha^{-1}$	Na	Cu^{3+}	Fe	Mn	Zn	Total
1. Manual clearing	No-tillage	5.3	0.03	0.04	tr	1.0	0.4	0.09	1.1	tr	0.2	0.05	0.11	2.8
2. Manual clearing	Conventional tillage	24.1	0.1	0.23	tr	2.4	1.6	0.41	4.3	tr	0.8	0.07	0.09	9.9
3. Crawler tractor/ shear-blade	No-tillage	31.2	0.4	0.7	0.01	3.7	1.4	0.52	3.3	tr	2.1	0.26	0.17	12.6
4. Crawler tractor/ tree-pusher	No-tillage	66.7	0.6	0.5	0.08	6.7	4.0	1.36	7.0	tr	3.1	0.17	0.18	23.7
5. Crawler tractor/ tree-pusher	Conventional tillage	94.6	0.4	1.7	0.07	10.0	7.9	1.30	15.2	tr	4.3	0.35	0.58	41.9
6. Traditional clearing	Traditional seeding	12.8	0.1	0.1	tr	1.1	2.1	0.46	1.1	tr	0.1	0.04	0.09	5.3

tr = trace < 0.01

the subsequent two months, the first of which the rainfall and runoff were low and the second in which both were higher. There was no clear or consistent relation between the amount of runoff and the concentration of these nutrients in the runoff, though the concentration of Ca and Mg tended to decrease more rapidly with time than the K.

This experiment is a long-term experiment so future results will show the effects of continued cultivation and possibly fertilizer application.

3.6.4. ACKNOWLEDGEMENTS

The assistances of Dr. Poonam Rao and Mr. W. A. Alawode with the analysis of the runoff water samples is gratefully acknowledged.

REFERENCES

Bremner, J. M., and Keeney, D. R., 1966, Determination of isotopic ratio or different forms of nitrogen in soils 3: Exchangeable ammonium, nitrite and nitrate by extraction distillation methods, *Soil Sci. Soc. Am. Proc.,* **30**, 577.

Kowal, J., 1972, The hydrology of a small catchment basin of Samaru, Nigeria. IV. Assessment of soil erosion under varied land management and vegetation cover. *Niger. Agr. J.,* **7**, 134–147.

Lal, R., 1976, Soil erosion problems on an Alfisol in western Nigeria and their control, IITA monograph no. 1, IITA, Ibadan, Nigeria.

Massey, H. F., Jackson, M. L., and Hays, O. E., 1953, Fertility erosion of two Wisconsin soils, *Agron. J.,* **45**, 543–547.

Moorman, F. R., Lal, R., and Juo, A. S. R., 1975, The soils of IITA, *IITA Technical Bulletin no. 3.,* Ibadan, Nigeria.

Murphy, J., and Riley, J. P., 1962, A modified simple solution method for the determination of phosphate in natural water, *Annal. Chim. Acta,* **27**, 31–36.

Ryden, J. C., Syers, J. K., and Harris, R. H., 1973, Phosphorus in runoff and streams, *Adv. Agron.,* **25**, 1–45.

Taylor, A. W., Edwards, W. M., and Simpson, E. C., 1971, Nutrients in stream draining woodland and farmland near Coshocton, Ohio, *Water Resources Res.,* **7**, 81–89.

Turvey, N. D., 1973, Water quality in a tropical rainforested catchment, *J. Hydrol.,* **27**, 111–125.

Tropical Agricultural Hydrology
Edited by R. Lal and E. W. Russell
© 1981, John Wiley & Sons Ltd.

3.7

Results of the East African Catchment Experiments 1958–1974

K. A. Edwards and J. R. Blackie

3.7.1. INTRODUCTION

The expansion of agricultural activity in East Africa following the Second World War led to an increased pressure on the small percentage of the land area considered to have a high potential for agriculture. A good proportion of this land consisted of forested highlands conserved by the Forest Departments to protect water supplies. The timber yield of these indigenous forests was low and economic pressures demanded that the land they occupied should be utilized more efficiently. At the same time, a growing body of scientific evidence was indicating that forests used more water than shorter vegetation (Law, 1956). While not denying the important role of forests in soil conservation, this challenged the widely held view that forests were essential to maintaining streamflow (Nicholson, 1936). There was a clear need to evaluate, by means of controlled experiments, the hydrological effects of land use changes in forested areas which incorporated good soil conservation practices.

These questions were discussed at a Planning Committee Meeting at EAAFRO,* Muguga, in June 1956, during which it was decided that a series of experiments be initiated to investigate the hydrological effects of land use change and that they should be conducted with the strictly practical purpose of providing Government Ministers and their advisers with technical data on the consequences of land use policy decisions. In practice, however, the experiments which were established have yielded numerous results of direct scientific value (Russell, 1962) and continue to supply some of the highest quality hydrological data in Kenya for research purposes (TAMS, 1979).

The preliminary results of the experiments were published in 1962 (Pereira *et al.*,

---

*East African Agriculture and Forestry Research Organization—now the Agricultural Research Department of the Kenya Agricultural Research Institute.

1962) and the final report, marking the formal conclusion to the Institute of Hydrology's commitment to the programme, in 1979 (Blackie, Edwards, and Clarke, 1979). Numerous papers on individual aspects of the experiments have been published and are included in the list of references in the final report. In addition, a data summary containing daily values of rainfall, streamflow, and evaporation has been published (Edwards *et al.*, 1976). A non-technical summary of the results has also been prepared by the initiator of the original experiments, Sir Charles Pereira, for the use of officials of the respective governments who are charged with land use planning policy (Pereira, 1979).

This paper is intended as a condensation of the technical results in order to bring to the notice of a wider audience the existence of the hydrological analyses and data summaries contained in the above reports. For further reference and details of the associated research projects which lend support to the conclusions outlined here, the Special Issue of the *East African Agricultural and Forestry Journal* (Blackie *et al.*, 1979) should be consulted.

3.7.2. DESCRIPTION OF THE EXPERIMENTS

Four distinct areas in East Africa which were experiencing land-use problems were chosen as sites for the experiments (Figure 3.9). Two were in Kenya and dealt with the replacement of indigenous forest, first, by plantation tea at *Kericho* and, secondly, by exotic conifers at *Kimakia*. A third experiment was at *Mbeya* in southern Tanzania, where an opportunity arose to compare the streamflow and sediment yield in two catchments of similar physical characteristics except for land use; one being under indigenous forest cover and the other cultivated without soil conservation. The fourth experiment dealt with the serious problem of degradation of grass savanna to thorn scrubland—so common in the rangelands of East Africa. At *Atumatak* in north-east Uganda, the soil moisture regime and the streamflow of two adjacent catchments were monitored whilst bush clearing; grass recovery and subsequent controlled grazing took place on one and uncontrolled grazing continued on the other. This experiment, unlike the others, was designed to measure the beneficial effects of a controlled land management policy rather than the possibly deleterious effects of land use changes.

All the experiments were started in 1957 and 1958. Kericho and Kimakia still continue as representative basins under the Kenya Ministry of Water Development, but the detailed monitoring of the effects of the land use change was terminated in June 1974. Mbeya closed in 1969 following increased difficulties in managing the experiment from Muguga, some 1600 km away. Atumatak, the Ugandan experiment, suffered throughout its life from the difficult communications with Muguga but was finally closed in December 1970 following a renewal of stock theft in the region which left 23 people dead, the observers terrorized and all the government cattle stolen.

From 1967 onwards, the Overseas Development Administration of the British

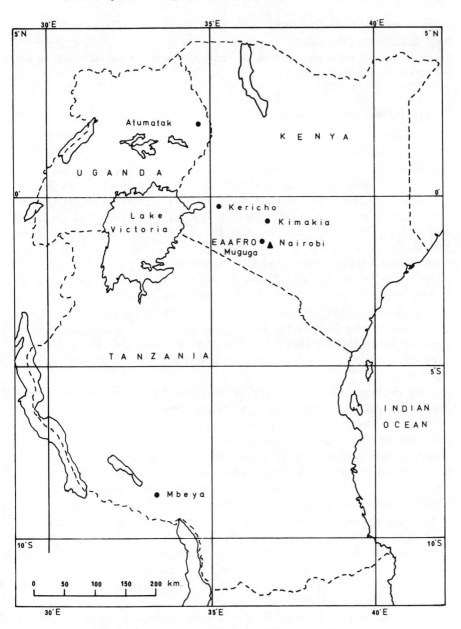

Figure 3.9 Location of the experimental catchments

Government assisted with the management of the experiments and with the task of processing the huge volume of data which had accumulated during the course of the studies. At the same time, the assembly of the data on magnetic tapes was undertaken to allow the development of mathematical models as well as making the data bank available for other users.

The following paragraphs outline the essential details of the individual experiments and the practical objectives in each case.

3.7.2.1. The Kericho Experiments

The successful introduction of tea as a plantation crop in the high rainfall areas of East Africa led to a demand for more forest to be opened up. There was some opposition to any further expansion of the plantations, however, because of the risk to water supplies if forest excision were allowed to continue. Little was known, at the time, of either the water use of tea or the consequences of removing the forest with its protective canopy and deep litter layer. It was decided, therefore, to conduct a controlled experiment near the centre of tea production in Kenya. Two similar catchments on the edge of the South West Mau Forest near Kericho were instrumented to monitor rainfall, potential evaporation, streamflow, soil moisture, and sediment yield before, during, and after a controlled change in land use from indigenous forest to tea plantation in one of them.

One catchment, the Sambret Valley, was excised from the forest and leased to Brooke-Bond Liebig (Kenya) Ltd. to be cleared and planted in accordance with an experimental programme specified by EAAFRO. The other catchment, the Lagan tributary of the Saosa river, remained as a control under broad-leaved montane rain forest.

Both catchments lie less than 0.5° south of the Equator at altitudes between 2000 m and 2800 m. At the beginning of the experiment both were under unbroken forest cover. The soils are deep (6 m), stone free and physically uniform, being derived *in situ* from the phonolitic Tertiary lavas. Rainfall is over 2000 mm per annum, with only three months of the year when it is less than potential open water evaporation (EO). The climatic environment of the two catchments is summarized in Table 3.17 which shows the mean monthly climatological data from the adjacent Tea Research Institute of East Africa.

Basic instrumentation was completed by the end of 1957 and forest clearing in the Sambret catchment began in 1959. By June 1960, the first 120 ha of the new estate were completed and, by 1964, 380 ha of the 702 ha catchment had become tea plantations in accordance with the programme. Because of the presence of the substantial amount of bamboo forest in the remainder of the catchment, a sub-catchment of 186 ha which included the bulk of the bamboo and mixed-bamboo forese was also gauged. The rainfall network in the catchment consisted of 21 daily read gauges including 3 recording gauges. Soil moisture was sampled monthly to 3 m at sites representing the upper, middle, and lower parts of the catchments. From 1968 onwards, access tubes were installed at 14 sites and the neutron probe was used for soil moisture determination.

In the control catchment (544 ha), six gauges were used to sample rainfall and soil moisture access tubes were also situated at the same six sites.

Table 3.17. Mean climatological data (1958–1974) Kericho (Tea Research Institute) altitude 2073 m, latitude 0° 21′ S, longitude 35° 20′ E

	Rainfall	Temperature			Humidity	Wind	Radiation	Sunshine
	Monthly total	Max	Min	Mean	Saturation deficit	Mean speed at 2 m	Gunn Bellani radiometer	
	mm	°C	°C	°C	mb	km hr^{-1}	MJ m^{-2}	hr
Jan	92.6	23.9	9.0	16.5	7.6	5.9	24.0	8.1
Feb	104.8	24.1	9.1	16.6	7.5	5.7	23.8	7.8
Mar	171.5	24.1	9.5	16.8	7.0	5.7	23.4	7.5
Apr	264.4	22.8	10.0	16.4	4.7	4.5	18.9	5.9
May	282.8	21.9	9.8	15.8	3.6	4.6	18.0	6.0
Jun	209.8	21.3	9.1	15.2	3.9	5.3	18.7	6.5
Jul	197.0	20.5	9.2	14.9	3.8	5.4	17.2	5.6
Aug	213.2	20.8	9.1	15.0	4.1	5.7	17.7	5.7
Sep	181.8	21.9	8.6	15.3	4.6	5.8	19.1	6.1
Oct	172.3	22.3	9.1	15.7	5.0	5.6	18.6	6.0
Nov	151.0	22.3	9.7	16.0	5.3	5.4	18.6	5.8
Dec	98.2	23.0	9.1	16.1	6.4	5.7	21.9	7.3

Apart from the major monitoring programme, a number of subsidiary projects have been carried out during the course of the experiment. During the establishment of the tea estate, storm runoff from a 12 ha administrative area was compared with a similar area of forest (Dagg and Pratt, 1962). A hydraulic lysimeter study of the water use of tea was conducted at the Tea Research Institute (Dagg, 1970; Wang'ati and Blackie, 1971). An erosion plot experiment was undertaken by the staff of the Tea Research Institute to measure the effects of different conservation techniques on soil and water loss from young tea (Othieno, 1979). In 1975, attempts were made to calculate the direct water loss from a mature tea crop by use of the eddy correlation technique (Callander and Woodhead, 1979) and by use of the Zero Flux Plane method and the Hydraulic Conductivity-Potential Gradient method (Cooper, 1979). All these projects were concerned with providing background information on the main objective, i.e. the hydrological effects of a change in land use from indigenous forest to a well-managed tea estate.

3.7.2.2. The Kimakia Experiments

In the late 1950s, the expansion of the city of Nairobi led to concern about safeguarding its water supply. The city depends heavily on the perennial streams which originate on the forested slopes of the Aberdare Mountains in Central Kenya. These streams also provide critical dry season supplies to the coffee industry and to densely populated rural areas. At the same time, the Forest Department was eager to exploit the high potential of the Aberdares for economic softwood timber production and other forms of land use, such as high density sheep grazing, were being considered. To determine whether such changes in land use would affect total streamflow yield or its seasonal distribution, catchment studies were proposed to compare the water yields from indigenous bamboo forest, softwood plantations, and sheep pasture.

The three catchments studied lie at the southern end of the Aberdares, 0.5° south of the Equator and at a mean altitude of 2440 m. The central catchment is a 65 ha tract of bamboo forest (*Arundinaria alpina*) with scattered evergreen forest species such as *Podocarpus milanjianus*. Nine raingauges, including two recording gauges, were installed in the catchment either in clearings or at canopy level on tree-mounted platforms and on telescopic masts. The experimental catchment (36 ha) was cleared of bamboo in 1956, except for a protective strip on the steep river banks, and planted with pine seedlings (*Pinus patula*) in April 1957. Following the usual Kenyan system of plantation development, vegetables and maize were grown among the pine seedlings until 1960, when the closing pine canopy inhibited interplanting. A further 37 ha valley was instrumented in 1966, partially cleared, planted to Kikuyu grass (*Pennisetum clandestinum*) and subsequently used for a high density sheep grazing trial supervised by the Ministry of Agriculture.

The three catchments are situated on Miocene basalts and agglomerates which give rise to deep, porous soils with high available water capacities (765 mm in the top 3.2 m of soil). Soil moisture was sampled at three sites in each catchment until the introduction of the neutron soil moisture meter made it possible to sample at six sites in the control and the pine catchments and three sites in the grass catchment.

A meteorological enclosure at the Forest Station serves all catchments and a summary of climatological data is given in Table 3.18. Mean annual rainfall is 2235 mm; of the same order as Kericho but with a more pronounced bi-modal distribution.

All catchments are equipped with water level recorders operating on compound sharp-crested weirs. Suspended sediment samples were taken at the weirs during the establishment of the pine plantation and during the last year of the experiment.

3.7.2.3. The Mbeya Experiment

The protection of forest to conserve water supplies has for many years been the cornerstone of forest policy. Despite pressure to release these areas of high agricultural potential, Forest Departments have opposed any proposals to allow further excision until it can be shown that serious and irreversible damage to soils and water supplies will not ensue. At the same time, they have encouraged quantitative studies on the effects of land use change such as the Mbeya experiment in southern Tanzania.

With the gazetting of the Forest Reserve in the Mbeya Range, cultivation of certain areas within the boundary ceased and the local Wasafwa people accepted monetary compensation and moved to an adjacent valley outside the Reserve. An opportunity arose, therefore, to measure the effects of cultivation on sediment and water yield and to compare the hydrological regime of a cultivated and a forested catchment.

With the assistance of the Forest Department and the Department of Water Development and Irrigation, EAAFRO instrumented the catchments and began measurements in 1957. The two catchments chosen are situated on volcanic ash which overlies weathered gneiss of the Pre-Cambrian Basement Complex. The combination of these two parent materials gives rise to a very porous but structurally stable soil. The forested catchment is 16.3 ha in area with very steep valley sides at an average slope of $30°$ near the weir. The cultivated catchment is as steep as the forested control and 20.2 ha in area, of which about 50 per cent is cultivated in any given season. Both catchments have compound V-notch and rectangular sharp-crested weirs with water level recorders. The cultivated catchment has a sediment trap upstream of the weir and for a time a stormflow sediment sampler was also in operation (Pereira and Hosegood, 1962). Rainfall was measured by six gauges in the forested catchment and seven in the cultivated catchment. Six raingauges were also installed in a regenerating catchment adjacent to the forested control. This catchment was later abandoned as a hydrological experiment but the rainfall network was continued. Of the 19 gauges, five were Dines tilting-syphon recording gauges.

Soil moisture monitoring in the two catchments was accomplished by gravimetric sampling and the installation of electrical resistance blocks.

Table 3.18. Mean climatological data (1958–1974) Kimakia Forest Station altitude 2438 m, latitude 0° 48′ S, longitude 36° 45′ E

	Rainfall	Temperature			Humidity	Wind	Radiation	Sunshine
		Max	Min	Mean	Saturation deficit	Mean speed at 2 m	Gunn Bellani radiometer	
	Monthly total							
	mm	°C	°C	°C	mb	km hr^{-1}	MJ m^{-2}	hr
Jan	90.4	20.1	6.9	13.5	3.7	9.5	26.4	8.3
Feb	118.8	20.7	7.3	14.0	4.5	9.6	26.9	8.5
Mar	195.1	20.4	8.5	14.4	4.5	10.6	26.2	8.2
Apr	435.5	19.0	9.5	14.3	3.1	9.5	22.3	6.9
May	363.9	17.8	9.1	13.4	2.2	7.6	18.5	5.3
June	126.9	16.5	7.5	12.0	2.0	6.6	17.2	4.9
Jul	86.5	14.8	7.2	11.0	1.4	5.5	13.2	3.1
Aug	85.3	15.0	7.1	11.1	1.4	5.9	9.6	3.3
Sep	78.0	17.5	6.8	12.2	2.6	7.9	21.3	6.0
Oct	226.0	18.4	8.3	13.4	3.4	9.7	23.3	7.1
Nov	306.9	18.2	8.8	13.5	2.8	10.7	21.6	6.4
Dec	121.5	19.1	7.1	13.1	3.1	8.5	24.6	7.7

A meteorological site was positioned between the two catchments and a summary of climatological data is shown in Table 3.19. It can be seen that the mean annual rainfall of 1733 mm has a pronounced unimodal distribution with a long dry season, in contrast to the two Kenyan catchments.

3.7.2.4. The Atumatak Experiments

Large areas of potentially productive rangeland in East Africa suffer from the effects of overgrazing. The grasslands degrade into bushland and dry thicket and the removal of the protective vegetation, combined with trampling of the surface by animal hooves, leads to a rapid removal of soil, a lowering of the infiltration rate, and flash flooding. In regions of high rainfall erosivity and high soil erodibility, the cycle is self-propagating and may lead to the widespread removal of soil by sheet and gully erosion.

If the cycle can be interrupted in its early stages, however, a comparatively modest expenditure on bush clearing and fencing, together with some system of controlled grazing, can bring about a remarkable recovery of degraded grasslands. The social and economic problems of encouraging pastoralists to limit their herds are formidable and it was thought that a demonstration of the beneficial results of controlled grazing would be an effective means of introducing modern concepts of range management. By choosing two adjacent catchments, one to act as a control and remain under traditional grazing practices and the other to have a system of inexpensive treatment and management imposed, it was hoped that the effect on runoff and on soil moisture regimes could also be demonstrated.

The catchments chosen lie at the head of the Atumatak valley near Moroto in northeastern Uganda. Together they form a well-defined 811 ha catchment draining north into the Omanimani River. The soils are derived from the Basement Complex peneplain and are in an advanced state of erosion. Soil profiles are truncated with stone mantles or erosion pavements exposed throughout the catchments. The climate of the area is summarized in Table 3.20 which shows the mean annual rainfall to be 753 mm. Potential evaporation is of the order of 2000 mm yr^{-1} but, with high wind speeds and large saturation deficits at certain times of the year, there is considerable heat advection. A dense network of 18 daily and 5 recording raingauges was installed in the two catchments. Soil moisture tension was recorded at 20 sites by means of electrical resistance blocks and a meteorological site was established. Because of the large range in discharge and the silt-laden nature of the streamflow, compound rectangular-throated flumes were used to measure discharge. Unfortunately, silt deposition during the recession of the flow was a constant problem. Although remedial measures were attempted during the course of the experiment, they were never completely successful and a considerable amount of flow data was lost.

From 1957 to 1961, the runoff patterns under typical local grazing intensities were determined and surveys of the soil and vegetation were made. In 1961 and

Table 3.19. Mean climatological data (1958–1969) Mbeya range altitude 2428 m, latitude 8° 50′ S, longitude 33° 28′ E

	Rainfall	Temperature			Humidity	Wind	Radiation	Sunshine
	Monthly total	Max	Min	Mean	Saturation deficit	Mean speed at 2 m	Gunn Bellani radiometer	
	mm	°C	°C	°C	mb	km hr^{-1}	MJ m^{-2}	hr
Jan	315.3	18.0	10.4	14.2	2.3	5.6	17.9	4.0
Feb	323.8	18.2	10.3	14.2	2.2	5.8	18.1	3.8
Mar	399.0	18.1	10.5	14.3	2.1	5.4	17.6	3.8
Apr	244.0	18.1	10.3	14.2	2.3	7.2	17.8	5.3
May	28.1	17.9	8.2	13.1	3.1	6.7	20.8	7.6
Jun	2.0	16.0	6.2	11.1	3.8	7.4	20.8	8.1
July	0.1	17.2	6.0	11.6	4.7	8.7	23.8	9.2
Aug	0.1	18.4	7.1	12.7	5.8	9.1	24.6	9.1
Sep	7.9	20.0	9.0	14.5	7.1	9.6	24.4	8.4
Oct	25.2	21.1	10.3	15.7	8.1	8.6	23.9	7.9
Nov	100.4	20.2	10.8	15.5	5.7	6.6	20.3	5.5
Dec	287.5	18.3	10.7	14.5	3.1	5.1	17.5	4.0

Table 3.20. Mean climatological data (1958–1974) Atumatak altitude 1524 m, latitude 2° 14′ N, longitude 34° 39′ E

	Rainfall	Temperature			Humidity	Wind	Radiation	Sunshine[a]
	Monthly total	Max	Min	Mean	Saturation deficit	Mean speed at 2 m	Gunn Bellani radiometer	
	mm	°C	°C	°C	mb	km hr^{-1}	MJ m^{-2}	hr
Jan	10.4	29.9	15.0	22.5	13.1	8.8	27.0	9.8
Feb	22.2	30.3	15.4	22.8	12.7	8.0	26.6	9.2
Mar	56.7	29.7	15.8	22.8	11.6	8.0	26.2	8.7
Apr	121.7	28.2	16.1	22.2	9.0	6.5	23.5	7.9
May	105.8	27.4	15.1	21.3	6.8	4.1	23.3	8.2
Jun	54.3	27.6	14.2	20.9	7.5	4.0	23.2	8.6
Jul	110.1	26.2	14.6	20.4	6.6	3.8	21.7	7.2
Aug	93.3	26.9	14.4	20.7	7.3	4.4	23.6	8.0
Sep	59.6	28.3	13.9	21.1	8.6	4.8	26.0	8.9
Oct	49.8	28.6	15.8	22.2	10.5	7.3	25.5	8.7
Nov	49.3	28.4	16.0	22.2	10.8	8.7	24.8	8.5
Dec	20.0	28.9	15.0	21.9	11.5	9.0	25.8	9.5

[a] Sunshine records incomplete (1971–1974)

1962, one catchment was cleared of bush and fenced to exclude cattle while the other remained under the traditional grazing system. By 1964, a grass cover had re-established itself in the cleared catchment and, having demonstrated that this minimum treatment produced good rehabilitation, the next stage was to introduce grazing on a controlled basis to determine the optimum grazing densities. Grazing started in 1965 but, because of the many difficulties experienced (including theft of fencing wire, illicit grazing, and theft of stock), the proposed rotational grazing scheme was never fully implemented.

Towards the end of the experiment, the problems of maintaining and operating instrument networks became overwhelming and the continuity and quality of the catchment data deteriorated. So little is known about the hydrology of these semi-arid areas, however, that an attempt has been made at the interpretation of the experimental data.

3.7.3. METHOD OF ANALYSIS OF EXPERIMENTAL DATA

Previous hydrological experiments of this nature required lengthy calibration periods or repeated trials to establish statistically valid results (Bates and Henry, 1928; Wicht, 1966). With the advent of better methods of estimating the evaporation from natural surfaces (Penman, 1948) and a general improvement in methods of measuring the major components of the hydrological cycle, it became possible to calculate the water balances of individual catchments with reasonable precision, to check the results with the energy balance and to produce comparative estimates of the water use of different types of vegetation. The 'paired catchment' method became the basis for the EAAFRO experiments and implicit in this method is the use of the water balance equation to evaluate the unknown components. This equation can be written in general form as:

$$AE = R - Q - \Delta S - \Delta G$$

where R, Q are precipitation and streamflow, AE is actual evapotranspiration and ΔS, ΔG are changes in soil moisture and groundwater storage, respectively, over whatever period is specified.

Because of the slow transfer of infiltrated water from soil moisture to groundwater and finally to base flow, the calculation of AE (actual evapotranspiration) by difference becomes most precise when the ΔS and ΔG terms are evaluated in the dry season. At this time, ΔS can be measured with great precision and ΔG can be estimated from base-flow recession curves with some confidence (Blackie, 1972). From this emerges the concept of a 'water year', a time interval of approximately one calendar year running from one dry season to the following equivalent dry season, as a period over which consecutive estimates of water use can be made with precision.

Thus the water balance equation can be used to calculate actual water use of the paired catchments over a series of water years. Similar values of water use should be obtained during the calibration period, but if the imposed land use change coincides with a major climatic fluctuation, as in the case of the Kimakia and Kericho experiments, the resulting variations in hydrological regime from very dry to very wet years may confuse the comparison of the values of AE derived before and after the change. If the values are consistent with theoretical estimates of evaporation of intercepted water and of transpiration, it lends confidence to the assumption that the measured changes in streamflow are entirely attributable to the land use change. If not, there will be reason to believe that the differences in streamflow are due to other factors such as the failure of the recession curve method to estimate total groundwater storage change or the presence of systematic errors in the data.

Generally speaking, the water balance equation was used initially to detect systematic errors and then when it was considered that all major errors had been corrected, the streamflow regimes were accepted as being characteristic of the respective vegetation types. To interpret these regimes, use has been made of evidence from the studies of the physical processes of evaporation and transpiration and from the application of simple conceptual models to simulate the streamflow response to changes in physical parameter values such as interception storage, the rate of evaporation of intercepted water, and crop albedo. It is not possible to enter into a detailed description of the results of either the process investigations or the mathematical modelling in this condensed report and the results presented are confined to estimates of water use of the different types of vegetation derived from the water balance and to differences in the seasonal patterns of streamflow.

It has become clear during the analysis of the 130 catchment years of data that the precise measurement of rainfall and streamflow are of paramount importance in studies of this kind. Fortunately the rainfall networks were designed with great care (McCulloch, 1962) and the precision of the areal rainfall estimates is high. The measurement of streamflow, on the other hand, has given cause for concern in some of the catchments and retrospective corrections to streamflow based on reassessments of the rating curves account for the differences between previously published values and those presented here. Throughout the analysis, the Penman method has been used to estimate open water evaporation (EO) for each pair of catchments and the ratio of actual water use (AE) to EO is used as the index of comparative water use.

3.7.4. RESULTS OF THE CATCHMENT EXPERIMENTS

Water balance results are presented for the Kericho, Kimakia, and Mbeya experiments. Atumatak flow data was discontinuous and did not allow water balance calculations to be made.

3.7.4.1. Annual Water Balance

Table 3.21 lists the mean water year AE/EO ratios for the years following the establishment of each new land use. These show that none of the land use changes

Table 3.21. Comparison of mean water year AE/EO ratios for the Kericho, Kimakia, and Mbeya catchments for the periods after each new land use was fully established

Location	Catchment	Dominant vegetation	Period	Mean rainfall (mm)	SEE	AE/EO	SEE
Kericho	Lagan	Montane rainforest		2219	±149	0.93	±0.032
	Sambret sub-catchment	Bamboo	1967–73	2026	±140	0.86	±0.022
	Sambret	Tea		2011	±139	0.84	±0.030
Kimakia	C	Bamboo		2143	±158	0.76	±0.012
	A	Pines	1967–73	1997	±151	0.76	±0.020
	M	33% Grass		2062	±137	0.70^a	±0.023
Mbeya	C	Montane rainforest	1958–68	1924	±143	0.93	±0.065
	A	Cultivated crops		1658	±120	0.64	±0.025

[a] R–Q only

resulted in an increase in long-term water use. Thus, total water available to the downstream user has not been adversely affected. As indicated by the standard errors in Table 3.21, some year to year variability in the AE/EO ratios was observed. A detailed examination of the year to year variations (Tables 3.22–3.28) showed that most if not all of the variations could be accounted for by the enhancement of AE values in wet years, when evaporation of intercepted water became a more significant factor, and by the reduction of AE values in dry years when large soil moisture deficits brought about a reduction in transpiration.

Simulation of the evaporative process using a mathematical model indicated that the rate of evaporation of intercepted water from each tall vegetation type was considerably in excess of EO. Wet season water use can be expected to exceed the mean value for the year, therefore, and the annual mean itself will fluctuate from year to year with variation in rainfall amount and frequency. Similar simulation of the control exerted by soil moisture deficit on rates of transpiration showed that some of the residual variability in AE/EO rates could be explained for the forest and bamboo catchments at Kericho, but not for the bamboo and pines at Kimakia. This may merely be a reflection of the greater frequency of severe drought conditions at Kericho than at Kimakia. In the case of tea, both the soil sampling results and the heat flux studies indicated a significant decrease in transpiration rates with increasing soil moisture deficit.

The apparent year to year variations in AE/EO for the Mbeya catchments (Tables 3.27 and 3.28) are almost certainly inaccurate indications of the true variability. A stone mantle made it impossible to sample soil moisture reliably below a

Table 3.22. Water use estimates, AE, for the control catchment (indigenous forest) at Kericho

Water year Ref. No.	Starting date	Rain mm	Flow mm	ΔS mm	ΔG mm	AE mm	EO mm	AE/EO
0	201260							
1		2727	1040	−89	−4	1780	1791	0.99
	200262							
2		2080	916	+114	+4	1046	1066	0.98
	211162							
3		1843	582	+46	+4	1211	1199	1.01
	200963							
4		2450	1013	−55	+6	1486	1691	0.88
	211164							
5		1507	269	+69	−3	1172	1481	0.79
	201165							
6		1560	622	−172	+3	1107	1321	0.84
	211066							
7		2367	851	+23	−10	1503	1981	0.76
	190168							
8		2230	1014	+118	+16	1082	1118	0.97
	161168							
9		1581	478	−6	−3	1112	1166	0.95
	030969							
10		2840	1129	+2	−6	1716	1806	0.95
	021270							
11		2199	944	+16	−12	1251	1383	0.90
	041171							
12		2388	775	+10	+15	1589	1698	0.94
	281272							
13		1926	614	−12	0	1324	1275	1.04
	021173							
Total 1–13		27698	10247	+64	+10	17379	18976	0.92

depth of a metre, whereas qualitative methods (gypsum block profiles) indicated that considerable deficits developed to at least 20 m under the cultivated catchment and to well below 3 m under the forest, where roots have been traced to 8 m depth. Thus the ΔS values are unreliable. Using a conceptual model (Blackie *et al.*, 1979) with parameters derived from the Kericho forest catchment an attempt has been made to estimate annual water use of the forested catchment (Table 3.27) and hence the probable error in the storage changes.

The AE/EO values given in Table 3.21 can be used to predict annual water losses from similar vegetation in similar rainfall and climatic regimes. From the preceding discussion, it is clear that the dependence of water use on the intensity and frequency of rainfall and on total water availability within the profile must be taken

Table 3.23. Water use estimates, AE for the experimental catchment (tea) at Kericho

Water year Ref. No.	Starting date	Rain mm	Flow mm	ΔS mm	ΔG mm	AE mm	EO mm	AE/EO
	161059	2538	1059	−34	0	1513	1683	0.90
0								
	131260	2547	1113	+39	+24	1371	1759	0.78
1								
	100262	2260	1212	+4	−27	1071	1214	0.88
2								
	111262	1908	857	+21	−3	1033	1209	0.85
3								
	111063	2419	1075	−17	+27	1334	1555	0.86
4								
	101164	1591	306	−32	−27	1344	1596	0.84
5								
	101265	1718	571	+12	+27	1108	1205	0.92
6								
	101066	2457	1012	−9	−17	1471	1978	0.74
7								
	100168	1932	837	+36	−7	1066	1163	0.92
8								
	151168	1388	324	−25	+3	1086	1282	0.85
9								
	200969	2383	984	−12	−7	1418	1699	0.83
10								
	031270	1993	912	+18	+7	1056	1366	0.77
11								
	021171	2163	794	−3	−3	1375	1705	0.81
12								
	271272	1767	615	−66	0	1218	1272	0.96
13								
	311073							
Total 0–13		29064	11671	−68	−3	17464	20686	0.84

into account if the results are to be extrapolated to markedly different climatic regimes.

3.7.4.2. Seasonal Distribution of Streamflow

The assessment of the extent to which land use changes have affected the seasonal distribution of streamflow is made particularly difficult because factors other than land use can enter into 'before and after' or ' between catchments' comparisons, making a simple analysis impossible. For instance, the change from drier-than-average conditions in the late 1950s to the unusually wet years of 1961 and 1962 coincided

Table 3.24. Water use estimates, *AE* for the bamboo sub-catchment at Kericho

Water year Ref. No.	Starting date	Rain mm	Flow mm	ΔS mm	ΔG mm	AE mm	EO mm	AE/EO
0		—	—	—	—	—		
1		—	—	—	—	—		
	100262							
2		2280	1240	+2	−31	1069	1214	0.88
	111262							
3		1975	931	−32	+9	1067	1209	0.88
	111063							
4		2468	1087	+43	+27	1311	1555	0.84
	101164							
5		1547	274	−5	−56	1334	1596	0.84
	101265							
6		1708	545	−28	+56	1135	1205	0.94
	101066							
7		2508	998	−15	−32	1557	1978	0.79
	100168							
8		1844	863	+70	+5	906	1017	0.89
	071068							
9		1504	335	−16	−10	1195	1429	0.84
	300969							
10		2496	1011	−55	+15	1525	1699	0.90
	031270							
11		1945	807	+2	−5	1141	1366	0.84
	021171							
12		2085	674	+23	−5	1393	1705	0.82
	271272							
13		1798	622	−64	0	1240	1272	0.97
	011073							
Totals 2–13		24158	9387	−75	−27	14873	17245	0.86

most unfortunately with the major period of land use change in the catchments. Furthermore, it has been seen that a most important factor affecting the water use of forest is the frequency of canopy wetting and, hence, seasonal and between catchment variations in rainfall input can have a strong influence on the pattern of streamflow. In the analysis of seasonal streamflow data, it became apparent also that the relationship of the recession curve to inherent geometric, soil and aquifer characteristics varied even between adjacent paired catchments.

Rather than attempt to oversimplify these relationships, with the risk of drawing unwarranted conclusions, statements on the effects of the land use changes on seasonal streamflow distribution have been confined to comments on their effects on the most important processes contributing to streamflow. These are surface run-off, infiltration, and groundwater recharge.

Table 3.25. Water use estimates, AE for the control catchment (bamboo) at Kimakia

Water year Ref. No.	Starting date	Rain mm	Flow mm	ΔS mm	ΔG mm	AE mm	EO mm	AE/EO
	260258							
58		2322	1381	−97	−52	1090	1497	0.73
	240259							
59		1858	792	+49	+24	993	1404	0.71
	280160							
60		1966	895	−21	−16	1108	1583	0.70
	260161							
61		3456	2160	−57	+20	1333	1707	0.78
	280262							
62		2431	1246	+159	+47	979	1260	0.78
	290163							
63		2656	1533	−28	−28	1179	1483	0.80
	300164							
64		2758	1516	−83	−36	1361	1496	0.91
	100365							
65		2219	1043	+66	+19	1091	1256	0.87
	270166							
66		2253	1208	−130	−26	1202	1430	0.84
	160167							
67		2192	956	+70	+30	1136	1540	0.74
	100168							
68		2645	1502	−51	−7	1202	1467	0.82
	210169							
69		1875	534	+62	−10	1289	1690	0.76
	280270							
70		2020	1008	−55	−13	1080	1361	0.79
	110271							
71		2036	803	+111	+43	1079	1441	0.75
	260172							
72		2705	1332	+18	+38	1317	1722	0.76
	270273							
73		1527	706	−162	−77	1060	1477	0.72
	020274							
Totals 58–73		36919	18615	−149	−44	18497	23814	0.78
Totals 67–73		15000	6841	−7	+4	8162	10698	0.763

3.7.4.2.1. *Surface Runoff and Infiltration*

All the catchments at Kericho, Kimakia, and Mbeya are on volcanic soils with high infiltration rates. Surface runoff constitutes only a small percentage of total stream-flow (two to three per cent) when the indigenous forest cover is present and this

Table 3.26. Water use estimates, *AE* for the experimental catchment (pines) at Kimakia

Water year Ref. No.	Starting date	Rain mm	Flow mm	ΔS mm	ΔG mm	AE mm	EO mm	AE/EO
	040258							
58		2416	1519	−35	+4	928	1496	0.62
	060259							
59		1813	874	+29	−6	916	1548	0.59
	050260							
60		1842	936	−34	−26	966	1619	0.60
	080261							
61		3319	2096	−59	+6	1276	1706	0.75
	140362							
62		2536	1219	+117	+20	1180	1414	0.83
	140363							
63		2716	1505	+8	+12	1191	1439	0.83
	110364							
64		2297	1551	−122	−39	907	1362	0.67
	200365							
65		2255	1027	+86	+4	1138	1356	0.84
	240266							
66		1989	1116	−109	−14	996	1368	0.73
	310167							
67		2029	844	+98	+23	1064	1539	0.69
	240168							
68		2524	1390	−32	+10	1156	1379	0.84
	210169							
69		1763	454	+47	−50	1312	1690	0.78
	280270							
70		1882	829	−60	+11	1102	1361	0.81
	110271							
71		1816	619	+145	+14	1038	1441	0.72
	260172							
72		2520	1084	+22	+62	1352	1722	0.79
	270273							
73		1443	604	−192	−64	1095	1477	0.74
	020274							
Totals 58–73		35160	17667	−91	−33	17617	23917	0.74
Totals 67–73		13977	5824	+4	+6	8119	10609	0.765

proportion did not change materially except during the transition stages of the land use changes (Pereira *et al.*, 1962). At Kimakia this result is not so surprising since the pine plantation rapidly developed both a protective canopy and a deep litter layer. At Kericho, it was attributable to the efficiency of the soil and water conservation measures used in establishing the tea estate. Othieno (1979) demonstrated that the same soils can give higher rates of runoff when less effective measures are

Table 3.27. Water balance (mm) for Mbeya catchment C (forested)

Water year	Rain	Flow	Apparent AE	Simulated AE^a	EO	Apparent AE/EO	Simulated AE/EO	Estimated error in $\Delta S + \Delta G$
1958–59	1421	214	1189	1522	1722	0.69	0.88	−333
1959–60	2043	564	1470	1418	1526	0.96	0.93	+52
1960–61	1332	330	1070	1564	1773	0.60	0.88	−494
1961–62	2753	842	1784	1330	1406	1.27	0.95	+454
1962–63	1878	534	1405	1335	1453	0.97	0.92	+70
1963–64	2199	652	1516	1380	1481	1.02	0.93	+136
1964–65	1512	446	1105	1322	1482	0.75	0.89	−217
1965–66	2013	564	1437	1284	1380	1.04	0.93	+153
1966–67	1681	453	1228	1314	1435	0.86	0.92	−86
1967–68	2404	814	1610	1364	1441	1.12	0.95	+246
Mean	1924	541	1381	1383	1510	0.92	0.92	−2
	±143	±62	±73	±29	±42	±0.06	±0.01	±90

^{a} Modelled using parameter values for canopy storage, interception evaporation, and transpiration for the forested area derived from the Kericho forested catchment and mean water use of 0.65 EO for the grass area.

Table 3.28. Water balance, Mbeya catchment A (cultivated)

Period	R	Q	ΔS^a	ΔG	AE	EO	AE/EO
20.10.58–10.10.59	1320	329	+4	0	987	1625	0.61
10.10.59–10.10.60	1718	578	+6	+90	1044	1527	0.68
10.10.60–10.10.61	1190	391	+18	−109	890	1687	0.53
10.10.61–10.10.62	2248	1112	−27	+163	1000	1462	0.68
10.10.62–10.10.63	1548	628	+14	−72	978	1488	0.66
10.10.63–13.10.64	1884	854	−3	+15	1018	1449	0.70
13.10.64–10.8.65	1369	418	+25	−15	941	1165	0.81
10.8.65–10.10.66	1485	548	−22	−21	980	1691	0.58
10.10.66–10.10.67	1570	485	+18	−22	1089	1368	0.80
10.10.67–10.10.68	2240	1326	+24	+114	776	1379	0.56
Mean 1958–68	1657	667	6	14	970	1484	0.65
	±116	±104	±6	±27	±28	±51	±0.03

^{a}Measured to 1 m depth only and known to underestimate total storage change.

adopted. At Mbeya, the very small increase in surface runoff when forest gives way to smallholder cultivation incorporating no effective measures to prevent erosion is a surprising result and atypical of experience elsewhere in East Africa. This is attributed to the remarkable wet season stability of the ash-derived soils, combined with the low erosivity of rainfall in this part of Tanzania (Moore, 1978).

At Atumatak, records from the soil moisture tension blocks demonstrated a recovery in infiltration rates (Table 3.29) following the re-establishment of grass cover in the cleared catchment. Starting from a slightly lower frequency of penetration to consecutive depths, the cleared catchment (B) ended up with a higher frequency of penetration at most sites down to a depth of 60 cm. A corollary to the recovery in infiltration was the reduction in storm runoff from the improved catchment. Table 3.30 shows how the runoff coefficient (Q/R per cent) increased during the clearing phase and then decreased significantly when the grass began to recover. Mean runoff coefficients can be deceptive since the actual values vary widely with antecedent soil moisture conditions and rainfall intensity. Table 3.31 shows the same data as a frequency distribution for various class intervals of runoff percentages and reinforces the striking and rapid recovery following the modest management programme.

The extent of recovery at Atumatak was very largely a function of the degree to which soil cover had been removed prior to rehabilitation. Whereas the least eroded soils recovered rapidly, steeper or less protected slopes still showed signs of a change in grass succession 12 years after clearing. The speed of recovery reflects the importance of maintaining infiltration rates on pasture land. As the colonizing grasses decreased surface runoff and increased infiltration, more moisture became available in the root range to sustain growth and facilitate propagation.

Table 3.29. Frequency of rainfall penetration at Atumatak

	Year	Mean number of readings		Depth in cm 15	30	45	60	90	120	180	240
	1959	A	31	38	38	10	6	13	0	0	0
		B	31	38	18	4	8	0	0	0	0
Pre-clearing	1960	A	48	47	43	19	11	21	14	0	0
		B	50	44	35	7	21	0	0	2	0
	1961	A	48	66	59	52	39	30	12	17	17
		B	49	62	48	32	38	12	12	4	4
Mean for pre-clearing phase 1959-1961		A		50	46	27	18	21	9	6	6
		B		48	34	14	22	4	2	2	1
Values for clearing phase 1962-1963	1962	A		Very few readings							
		B									
	1963	A	44	48	37	31	38	13	18	52	32
		B	43	50	50	38	45	35	23	14	9
	1964	A	48	22	15	9	13	1	3	6	3
		B	47	32	34	18	12	4	2	0	0
	1965	A	31	19	19	12	9	0	0	0	13
		B	31	27	18	13	17	0	0	0	0
	1966 1967 1968			Very few readings							
Post-clearing	1969	A	46	8	4	4	3	0	0	0	0
		B	43	19	19	11	15	1	0	2	0
	1970	A	50	20	19	16	14	1	2	0	8
		B	53	44	42	37	41	1	0	2	0
Mean for post-clearing phase 1964-1970		A		17	14	10	10	0	1	2	6
		B		30	26	19	21	22	1	1	0

B is the cleared catchment and *A* the control.
Figures indicate the mean percentage frequency with which available moisture was indicated by the resistance blocks.

3.7.4.2.2. Groundwater Recharge

Recharge of the catchment aquifers is governed by the amount of rainfall which infiltrates into the soil moisture store and by the rate at which this store is depleted by transpiration. For instance, changes in the interception characteristics of the vegetation will affect the amount of water available for infiltration and a higher transpiration rate will lead to a larger proportion of the infiltrated moisture being

Table 3.30. Depths of runoff and rainfall (mm) for a sample of storms (*n*) at Autumatak

	n	Catchment *A* (control, untreated)			Catchment *B* (bush-cleared)		
		Q	*R*	*Q/R%*	*Q*	*R*	*Q/R%*
Bush-clearing 1959–1961	59	70.73	678.8	10.4	76.02	785.9	9.7
During clearing 1962–1963	53	65.40	838.3	7.8	135.44	868.0	15.6
After clearing 1964–1968	61	160.09	1109.1	14.4	74.86	1092.8	6.8

Table 3.31. Number of occurrences of *Q/R%* values fitting in percentage classes at Atumatak

		Percentage						
		0–4.9	5.0–9.9	10.0–14.9	15.0–19.9	20.0–29.9	30.0–39.9	40 and over
1959–1961	*A*	32	11	3	8	3	2	5
	B	30	11	9	5	3	3	3
1962–1963	*A*	34	6	3	4	4	1	0
	B	22	9	3	3	7	4	4
1964–1968	*A*	21	8	13	5	8	5	1
	B	40	9	2	5	2	2	1

A and *B* are respectively the control and treated catchments

required to replenish the soil moisture store before groundwater recharge may begin.

Direct measurements of interception storage, soil moisture movement, and rates of transpiration were carried out only on the tea catchment at Kericho. In the other catchments, these processes were simulated in the mathematical modelling. At Kericho the process and modelling studies implied some change in groundwater recharge and hence in seasonal baseflow distribution due to lower interception losses and higher transpiration rates from the tea. In the particular environment at Kericho, however, these effects were small enough to be obscured by between-catchment differences in rainfall distribution and aquifer characteristics. In the change from bamboo to pines at Kimakia no significant differences in these processes were detected by the modelling; it may be concluded that there was no significant modification to baseflow. Thus whilst the catchment studies gave the practical answer of no change in seasonal streamflow distribution at Kericho and Kimakia, the data were sufficiently detailed to support modelling of the processes; this, in turn, identified the need for further investigation of interception and transpiration before global predictions can be made.

At Mbeya the change from evergreen forest to cultivation resulted in a marked decrease in interception and in dry season transpiration in the cultivated catchment, giving an overall increase in baseflow; observed baseflow levels were, on average, twice as high in the cultivated catchment. In the Atumatak study the marked increase in infiltration resulted in a substantial improvement in the grass cover, the enhanced transpiration of which did not permit any significant groundwater recharge and hence any significant base flow.

3.7.4.3. Sediment Yield

In the Kimakia and Kericho experiments steady-flow suspended sediment data were collected and, apart from some inconsequential increases during the transition phases, no significant differences in sediment yield from the different catchments could be detected. In all catchments, the sediment concentrations were low and more recent measurements at Kimakia, made with an automatic sediment sampler, confirmed the low sediment yields even in a catchment which had only recently (1973) been partly cleared and converted to smallholder cultivation.

At Kericho, once the tea crop had become established, the only appreciable sediment movement was a direct result of runoff from the estate roads.

At Mbeya, more detailed measurements were made. A stormflow sediment sampler was installed at the weir in the cultivated catchment and a bed-load trap was constructed just upstream. The results suggest that by far the largest percentage of total sediment load was contributed by stormflow (60 per cent). Bed load constituted some 36 per cent and the steady-flow suspended sediment was about four per cent. The total load was estimated as 9 t ha^{-1} yr^{-1} in the cultivated catchment compared with practically nil in the forested catchment. This quantity is surprisingly small considering the steepness of the cultivated land ($c.$ 30°) and the absence of any conservation measures, apart from ineffective bunds of maize stalks.

3.7.5. CONCLUSIONS

The conclusion of these land use studies may be summarized as follows.

(a) Replacement of rain forest by tea estate at Kericho resulted in an overall reduction in water use, combined with no significant increase in surface runoff or of sediment loss. This result will apply on similar soils experiencing similar rainfall distribution provided only that equally efficient soil conservation measures are adopted. Whilst, at Kericho, no significant change in seasonal flows was observed, process and modelling studies revealed differences in interception and transpiration which, in other environments, might alter the seasonal distribution of streamflow.

(b) Replacement of bamboo forest by pine softwood plantations at Kimakia initially decreased the water use; once the pine canopy had closed, no significant differences in water or sediment yield could be detected. The modelling suggested that this result could be applicable in a wide range of climatic conditions, provided only that the soils are equally stable. The effects of the felling phase remain to be investigated.

(c) At Mbeya, the replacement of evergreen forest by smallholder cultivation on very steep slopes resulted in a large increase in water yield. Because of the remarkably stable, porous nature of the ash-derived soils, only marginal increases in surface runoff and sediment loss were recorded but the dry season baseflow was doubled. Whilst a similar increase in water yield can be expected following this land use change in other unimodal rainfall areas, maintenance of seasonal flow patterns and of water quality is critically dependent on soil type.

(d) Bush clearing followed by several years of cattle exclusion resulted in a remarkable grass recolonization of the severely overgrazed rangeland at Atumatak. This recolonization increased infiltration rates and drastically reduced the peak flows. Subsequent controlled grazing did not affect the hydrological stability of the improved regime.

The scientific and practical results outlined above, together with the pointers to further areas of necessary research, prove the value of well designed and executed catchment studies. Apart from producing answers to important local practical questions, they also provide the quality of data necessary to test theories or models of how the processes controlling the hydrological cycle work and interact.

Such models are necessary for the global prediction of the effects of land use changes. The individual components may be constructed on the basis of the results of process studies but, to achieve acceptability, the complete model must be shown to work on long runs of high quality catchment data. These data must achieve an accuracy, particularly in areal estimates of input and output far exceeding that normally expected from a national hydrometeorological network.

3.7.6. ACKNOWLEDGEMENTS

The success of the experiments, which continued while very great changes were taking place in the structure of government in the three countries involved, is a tribute to the foresight of the instigator of the experiments, Sir Charles Pereira, and to the willingness of the various government departments as well as the UK Ministry of Overseas Development to sponsor and to staff the project. The quality of the data bank which is now available from the Institute of Hydrology and the Kenya Agricultural Research Institute, is also a tribute to the diligence of the observers, some of whom worked for seventeen years in the same catchment.

REFERENCES

Bates, C. G., and Henry, A. J., 1928, Forest and streamflow experiments at Wagon Wheel Gap, Colorado, *Monthly Weather Rev. Suppl.,* **30**, 79 pp.

Blackie, J. R., 1972, The application of a conceptual model to some East African Catchments, *Unpubl. M. Sc. Thesis,* Imperial College, London.

Blackie, J. R., Edwards, K. A., and Clarke, R. T., 1979, Hydrological Research in East Africa, *E. Afr. Agric. For. J.,* Special Issue, **45**, in press.

Callander, B. A., and Woodhead, T., 1979, Eddy correlation measurements of convective heat flux and estimation of evaporative heat flux over growing tea, *E. Afr. Agric. For. J.,* Special Issue, **43**, in press.

Cooper, J. D., 1979, Water use of a tea estate from soil moisture measurements, *E. Afr. Agric. For. J.,* Special Issue, **43**, in press.

Dagg. M., 1970, A study of the water use of tea in East Africa using a hydraulic lysimeter, *Agric. Meteorol.,* 303–320.

Dagg, M., and Pratt, M. A. C., 1962, Relation of stormflow to incident rainfall. *E. Afr. Agric. For. J.,* Special Issue, **27**, 31–34.

Edwards, K. A., Blackie, J. R., Cooper, S. M., Roberts, G., and Waweru, E. S., 1976, Summary of hydrological data from the EAAFRO experimental catchments, *EAAFRO EAC Printer,* Nairobi, 301 pp.

Law, F., 1956, The effect of afforestation upon the yield of water catchment areas, *Proc. Brit. Ass. Adv. Sci.,* Sheffield, 489–494.

Moore, T. R., 1978, An initial assessment of rainfall erosivity in East Africa, *Tech Commun. 11,* University of Nairobi Fac. Agricul., Dept. Soil Sci. Mimeo, 40 pp.

McCulloch, J. S. G., 1962, Measurements of rainfall and evaporation, *E. Afr. Agric. For. J.,* Special Issue, **27**, 27–30, 64–67, 88–92, 115–117.

Nicholson, J. W., 1936, The influence of forests on climate and water supply in Kenya, *E. Afr. Agric. J.,* **2**, 48–53, 164–170, 226–240.

Othieno, C. O., 1979, An assessment of soil erosion on a field of tea under different soil management procedures, *E. Afr. Agric. For. J.,* Special Issue, **45**, in press.

Penman, H. L., 1948, Natural evaporation over open water, bare soil, and grass, *Proc. Roy. Soc. (A),* **193**, 120–145.

Pereira, H. C., 1979, Effects of land use changes on water resources—a summary of 25 years of studies on East African catchment areas, *ARD(KARI) Govt. Printer,* Nairobi (in press).

Pereira, H. C., *et al.,* 1962, Hydrological effects of changes in land use in some East African catchment areas, *E. Afr. Agric. For. J.,* Special Issue, 27, 131 pp.

Pereira, H. C., and Hosegood, P. H., 1962, Suspended sediment and bed-load sampling in the Mbeya Range catchments, *E. Afr. Agric. For. J.,* Special Issue, **27**, 123–125.

Russell, E. W., 1962, Foreword in: Hydrological effects of changes in land use in some East African catchment areas, *E. Afr. Agric. For. J.,* Special Issue, **27**, 1–2.

TAMS, 1979, *National Master Water Plan, Stage 1, Vol.1,* Water Resources and Demands, Tippetts, Abbett, McCarthy, Stratton, Nairobi.

Wang'ati, F. J., and Blackie, J. R., 1971, A comparison between lysimeter and catchment water balance estimates of evaporation from a tea crop, in *Water and the Tea Plant,* ed. M. K. V. and S. Carr, Tea Research Inst. of E. Africa, Kericho, Kenya, 9–20.

Wicht, C. L., 1966, Trends in forest hydrological research, *S. Afr. For. J.,* **57**, 17–25.

Tropical Agricultural Hydrology
Edited by R. Lal and E. W. Russell
© 1981, John Wiley & Sons Ltd.

3.8

Recent Studies on Soil Erosion, Sediment Transport, and Reservoir Sedimentation in Semi-arid Central Tanzania

L. STROMQUIST

3.8.1. INTRODUCTION

Tanzania is situated between latitudes $1°$ and $12°$ S. It covers an area of 884,000 square kilometres. The estimated population is 16 million out of which more than 93 per cent live in rural areas. Broadly speaking Tanzania has two types of environments with a high soil erosion hazard: the humid mountains with steep cultivated slopes and the semi-arid plains (cf. Figure 3.10 and 3.11). The semi-arid region is defined by the Tanzania government (1977) as the area included within the 800 mm yearly rainfall isohyet, thus covering most of the central provinces and the new national capital at Dodoma. Approximately 3 million people live within this area, from time to time affected by water shortage and local human starvation.

The aim of this paper is to give a brief presentation of some recent studies of soil erosion made within the country, with an emphasis on those from the dry lands. An excellent survey of the past and present erorion rates and processes was presented in fifteen papers published as *Studies of Soil Erosion and Sedimentation in Tanzania* (Rapp, Berry, and Temple, eds., 1972a). The study was made within the Dar-es-Salaam–Uppsala Soil Erosion Research project (DUSER) jointly managed by the universitities in Dar-es-Salaam and Uppsala. These papers form an integrated geographical study of erosion and sedimentation in some tropical environments under heavy human influence. The main approach to the project was geomorphological and hydrological as surface runoff, erosion and sedimentation were documented in a number of catchment areas. The studies of the watersheds were supplemented by earlier data from various erosion plots, etc. Hence the DUSER-project included types of environments sensitive to severe soil erosion: the semi arid savannas and the humid mountains. The project was carried out between 1968 and 1972. In two

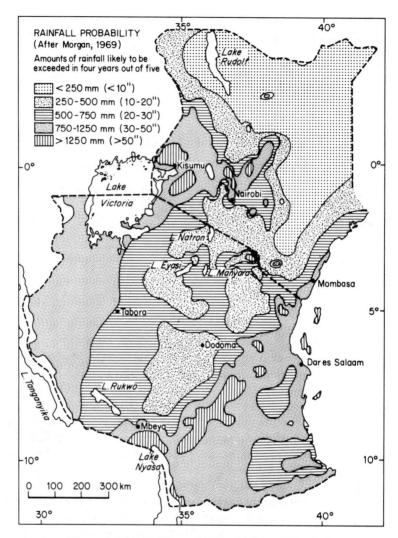

Figure 3.10 Rainfall probability in Kenya–Tanzania

of the studied areas further investigations have been made: by Christiansson (1978) in the Dodoma area and by Lundgren (1975; 1978) in the Ulugure mountains (cf. Figure 3.11).

One of the largest development schemes in semi arid Tanzania, is the Kidatu hydroelectric power plant in the Great Ruaha river and the river storage reservoirs at Kidatu and Mtera, a site 1975 km upstream from the plant. The main environmental impact in the river basin will be caused by the Mtera storage reservoir, which after the impoundment in 1980 will cover some 600 km² of land close to the con-

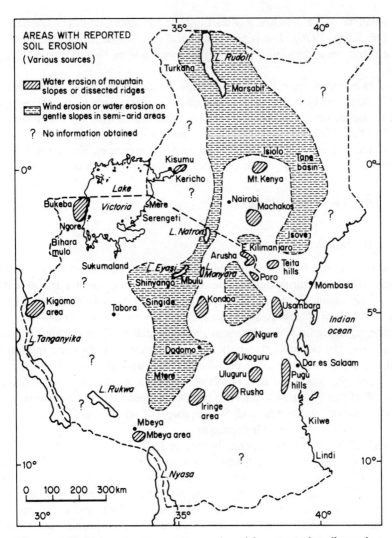

Figure 3.11 Areas in Kenya–Tanzania with reported soil erosion (From Lundgren, ed., 1975)

fluence of Great Ruaha, Little Ruaha, and Kisigo rivers. Mtera is situated about 120 km to the south of the Dodoma area. An ecological study of the area within the power development scheme (Johansson, ed., 1976). These studies included surveys based on LANDSAT-1 on the extent of soil erosion in the catchment area (65,000 km²) upstream from the reservoir (Stromquist, 1976) and in the reservoir region (cf. Stromquist and Johansson, 1978).

Studies of soil erosion and land management planning have also been made in conjunction with various regional development schemes, water master plans, etc. A good example is the Mwanza Region Integrated Planning Project in western Tanzania (cf. Rapp, 1976). The geomorphological part of the study focused on the changed land use and erosion pattern caused by the formation of the new 'ujamaa' village.

Two case-studies from the semi-arid area are described below. The first is the DUSER studies at Dodoma using mostly air-photo interpretation, field surveys, and reservoir sedimentation as means of estimating and monitoring the soil erosion. The second is the SWECO-project (Swedish Consulting Group) in the Mtera area using satelite image interpretation, dendrochronological measures, fluvial-tile transport and vegetative patterns to quantify and describe the actual erosion.

3.8.2. TWO CASE STUDIES—THE DODOMA AND MTERA AREAS

3.8.2.1. The Dodoma Area

The Dodoma region is characterized by inselberg plains developed in a Precambrian bedrock with late faulting and fracturing. The altitude is about 1,100 m and the relative relief about 200 to 300 m. The dominating climatic features are a long dry season and a short intense rainy season. The dry season lasts for about 7–8 months, generally beginning in April. The mean annual rainfall is 573 mm during an average 54 rain days. The potential evaporation is 2,123 mm yr^{-1}. The natural vegetation is woodland, which remains only in small parts. Man and his domestic animals have transformed the woodland into farmland, bushland, or thicket. The bushland is composed of shrubs 3 to 4 m high and occasional trees, and has a poorly developed ground cover (cf. Johansson and Stromquist, 1978a). The soils form a catena sequence and the rivers are all intermittent.

The DUSER research project on the rate of soil erosion and reservoir sedimentation was carried out during 1968–72. The water sheds and reservoirs of Ikowa, Matambula, Msalatu and Imagi near Dodoma were surveyed. The catchment areas are under intense land use and subject to severe soil erosion due to over cultivation, overgrazing, and a high rate of extraction of firewood. The rate of sedimentation in each reservoir was determined by repeated surveys of cross profiles. Inventories of erosion features within the catchments were based on air-photo interpretaton and field checks. Christiansson (1978), who is a continuator of the studies in the area, also use LANDSAT-satelite images for monitoring the extent of eroded land.

Within the drainage basins (cf. Rapp *et al.*, 1972a, b; Christiansson, 1978) gullying is the most striking form of erosion as gullies appear in distinct zones on the upper pediment slopes. However the gully erosion is of less quantitative importance than splash and wash erosion.

The investigated reservoirs have very high rates of sedimentation (Table 3.32). Two will be filled with sediments only about 25 to 30 years after construction, which also indicates the severe erosion within the drainage basins. An example of a catchment area inventory map is shown on Figure 3.12 (From Rapp *et al.*, 1972).

Table 3.32. Reservoir data, sedimentation and soil denudation rate for five catchments in semi-arid areas of Tanzania

Catchment	Catchment area (km2)	Relief ratio (m km⁻¹)	Annual sediment yield (m³ km⁻²)	Soil denudation rate (mm yr⁻¹)	Reservoir completed (sediment survey) year	Capacity m³	Percentage of original volume	Annual loss of capacity through sedimentation[a] (%)	Expected total life of reservoir years
Ikowa	640	730/50	1957-74 191	0.1-0.36	1957	3,807,000	100.0	1957-74 2.8	30-40
			1957-60 362		(1960)	3,110,000	81.6	1957-69 6.13	
			1960-63 193		(1963)	2,740,000	71.9	1960-63 3.23	
			1963-69 111		(1969)	2,315,000	60.8	1963-69 1.85	
			1969-74 99		(1974)	2,000,000	52.5	1969-74 1.66	
Matumbulu	(18.1) effective 15.0	257/4.4	1962-74 581	0.44-0.63	1962 (−60)[b]	333,000	100.0	1962-74 2.6	35-45
			1962-71 626		(1971)	248,500	74.6	1962-71 2.8	
			1971-74 445		(1974)	228,500[c]	68.6	1971-74 2.0	
Msalatu	8.7	183/4.1	1944-74 556	0.44-0.62	1944 (theor)	421,000[c]	100.0	1944-74 1.15	80-90
			1944-50 623		(1950)	388,500	92.3	1944-50 1.3	
			1950-60 443		(1960)	358,000[d]	85.0	1950-60 0.9	
			1960-71 622		(1971)	298,000[c]	70.8	1960-71 1.3	
			1971-74 536		(1974) (theor)	284,000[c]	67.4	1971-74 1.1	
Imagi	2.2	122/1.6	1934-71 610	0.52-0.70	1934 (−29)[e]	171,500	100.0	1934-71 0.8	120-130
			1934-50 521		(1950)	152,000	88.6	1934-50 0.67	
			1950-60 659		(1960)	146,500[d]	85.4	1950-60 0.85	
			1960-71 703		(1971)	129,500	75.5	1960-71 0.90	
Kisongo[f]	0.3	225/5.7	1960-71 481	0.45-0.64	1960	121,000	100.0	1960-71 3.7	25-30[g]
			1960-69 447		(1969)	83,600	69.1	1960-69 3.3	
			1969-71 640		(1971)	71,700	59.3	1969-71 4.7	

[a] Changes in capacity due to raised spillway and excavations have not been accounted for in this column.
[b] Katumbulu reservoir was completed in 1960. In February 1961 a part of the embankment was washed away. The embankment was repaired in 1962.
[c] The spillway of Msalatu reservoir was raised by 2 feet in 1950 and by 1 foot in 1972. The spillway of Imagi reservoir was raised by 4 feet in 1932-33 and by 2 feet in 1972.
[d] Both Msalatu and Imagi reservoirs have been subject to sediment excavations. Imagi in 1952 (9000 m³) and Msalatu in 1953 (8000 m³).
[e] Imagi reservoir was completed in 1929. However, the first proper survey of the total volume of the reservoir was not undertaken until 1934.
[f] Data from Rapp, Murray-Rust, and Christiansson (1972b).
[g] In early 1974 the embankment gave way. By the end of 1974 it had not yet been repaired.

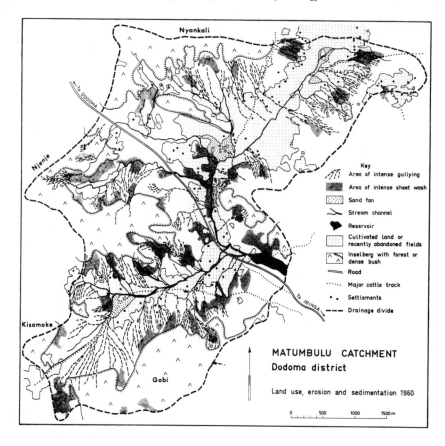

Figure 3.12 Map of land use, erosion, and sedimentation, Matumbulu catchment, Tanzania. Based on air photographs from 1960 and field checking during 1969–1971. Note the zones of erosion and deposition: gullied upper pediments with intense sheet wash, cultivated lower pediments, stream channels with three sand fans and reservoir with heavy sedimentation. Map by C. Christiansson (After Rapp, Murray-Rust, Christiansson, and Berry 1972)

According to a recent paper by Christiansson (1978) the annual sediment yield corresponding the reservoir sedimentation is about 200 to 600 m^3 km^{-2} (or 300 to 1,000 t km^{-2}). He and Rapp and Hellden (1979) recommend a reduction in stock numbers and rangeland burning as a useful means of increasing biological production, soil water infiltration, and life span of the reservoirs. However, the formation of a reservoir in a grazing area is likely to lead to an increase in stock numbers due to an improved water supply, which will lead to increased grazing pressure in the area and an acceleration in the rate of erosion and reservoir sedimentation. This illustrates the need for integrated planning of the rural development in the country.

3.8.2.2. The Mtera Area

The Mtera reservoir will be situated in the central part of the east African rift valley. The geomorphology is characterized by steep rift valley scarps and series of tilted and inclined plateaux below which extensive pediments, wide flood plains, and large areas of 'mbuga' (playa) clays are dominating elements of the landscape (Figure 3.13) The reservoir will cover most of the flood-plains and 'mbuga' deposits, hence its shorelines will develop on the gentle, in parts severely eroded, pediments.

The yearly rainfall at Mtera is very low (450 mm yr^{-1}) and concentrated in a rainy season between November and April. Most of the rain falls in December and January. January is the month which shows the least variation in number of rain days and the highest daily rainfall intensitites, hence making it the month with most erosive rainfall (cf. Stromquist, 1976; Johansson and Stromquist, 1977a; Stromquist and Johansson, 1978). The potential evaporation is 3,261 mm yr^{-1}. The vegetation can be divided into three major types: woodland, bushland, and grassland. The vegetation as we see it today is a result of an interaction of both natural and cultural processes. The very large number of domestic grazing animals, mainly cattle and goats, have severely depleted and locally exterminated the perennial grasses and palatable herbs leaving only short-lived annual grasses and unpalatable shrub with large areas of bare ground. The soil (and vegetation) types of the Mtera basin often form a marked catena sequence. Figure 3.13 (from Stromquist 1976) illustrates the geomorphological variation in the area interpreted from LANDSAT-1 images and Table 3.33 summarizes the relation between geomorphology, vegetation, and soils.

The reports by Johansson (1976) and Johansson and Stromquist (1977b, p. 19) draw attention to the need for monitoring and control of the land use of the basin after the impoundment: 'After the creation of the Mtera reservoir a rational land use policy has to consider how best to obtain an optimal and continuous yield from both farming and grazing without causing increased soil erosion and subsequent sedimentation in the reservoir. Agriculture will include both the use of traditional methods on the land above the highest shoreline as well as new methods on the land which is flooded annually.'

At present (cf. Figure 3.13, symbol 4) the southern pediment close to the perennial Great Ruaha River is severely affected by soil erosion and a similar evolution can be expected on the other pediments after the impoundment.

The ecology study made by SWECO (cf. Johansson, 1976) includes studies of geomorphology, vegetation, forestry, limnology, disease ecology, sociology, and planning. The geomorpholical studies have included:

(a) A land systems map with a description of the catchment area upstream from the reservoir (based on LANDSAT-1 satellite images, aerial survey, and limited field checks), stressing the extent of soil erosion (Stromquist, 1976).

(b) A study of land use, soils and conservation potential of the reservoir region

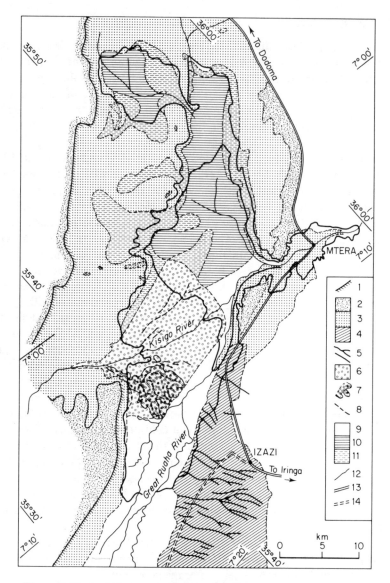

Figure 3.13. Landforms, erosion, and sedimentation in Mtera area

based on conventional air-photo interpretation (cf. Johansson and Stromquist, 1977b).

(c) A study of soil erosion and fluvial transport (cf. Johnasson and Stromquist,
1978a).

(d) An investigation of the expected environmental changes below the Mtera
reservoir (cf. Johansson and Stromquist, 1978b).

Table 3.33. The correlation between soils and vegetation of the land facets in the Mtera basin

Land facet	Soils	Vegetation
(1) Rift escarpment and tectonic Hills	Red, coarse grained acidic soils	Brachystegia woodland (miombo) (1) Acacia bushland (2)
(3) Pediments	Brown-grey sandy silts (upper parts), hardpan soils in certain areas, silty sands (upper parts)	Open bushland of mixed Acacia species (*A. tortilis* and *A. drepanolobium*)
(4) Seasonal streams	Well-sorted sand	Gallery forest on the bluffs
(5) Alluvial fan	Well-sorted sand, silt and clay	Riparian forest and ground-water bushland
(6) Floodplains	Well-sorted, sandy-to-clay grey soils	Riparian forest and ground-water bushlands
(7) Mbugas	Alkaline clay	Grasslands intersected by Acacia bushlands (*A. seyal* and *A. Stuhlmannii*)

The LANDSAT-1 satellite images have more recently been evaluated for mapping the geomorphology and vegetation in semi-arid Tanzania (cf. Johansson and Stromquist, 1978c).

Experiments were carried out to examine the feasibility of using growth rings of trees for calculating the annual soil erosion (Johansson and Stromquist, 1978a; Stromquist and Johansson, 1978). A variety of trees and shrubs were used initially for testing the method; it is suggested, however, that future studies should deal with only a few species within each area studied. *Acacia drepanolobium* which is a short shrubby tree and *Combretum hereroense*, a shrub, proved to be very useful because of both they have distinct growth rings and because of their ability to protect the ground from splash erosion. The rate of erosion was calculated from the difference in level between the ground protected by the canopy of the shrubs and that of the surrounding area. This method estimated the yearly soil loss as about 2,900 t km^{-2} or nearly three times as much as in the Dodoma area (cf. Christiansson, 1978; and Table 3.32). Combining the air-photo and satellite image interpretation with the soil loss estimated by dendrochronology the yearly transport into the reservoir from the surrounding pediments will be 0.8-1 million t yr^{-1}.

A study of the suspended sediment transport by the three main rivers was made during the 1976-1977 rainy season. The water sampling programme was made flexible as regards sampling intervals, which were shorter during the flood periods than during the recession stages. This procedure was adopted because of the discharge regime which is characterized by flash floods responsible for the bulk of sediments (cf. Table 3.34). Previous studies of sediment transport in the Great Ruaha River only used a weekly sampling programme, which often may have excluded the flashy peak floods and high sediment concentrations. This new procedure gave estimates for the total suspended sediment load four times greater than those obtained by the previous study. This clearly illustrates the need for a sediment sampling programme that takes into account the wide variation in magnitude and frequency of the erosion processes within the catchment area.

To simplify a future control programme of soil erosion, erosion rods (pegs) have been installed at different sites, each site representing a different situation in terms of soil vegetation and present land use. These results will be combined with a renewed air-photo–satellite image interpretation programme to quantify the amount and extent of future soil erosion around the reservoir.

3.8.3. CONCLUSIONS

Summarizing the main results from the recent Tanzanian studies on soil erosion, the most important result is the illustration of the necessity to observe the environmental reactions of exploitation and to draw rational conclusions for a better land use and planning from these observations. 'Every land development scheme should to some extent also be a research project' (Rapp, 1972), as land degradation is a common phenomenon but there is no common solution of the problem.

Table 3.34. Discharge and sediment concentration at Mawande, Little Ruaha River, Tanzania, 21 December 1976

Time (hr):	6 00	9 00	12 00	12 30	13 45	14 25	15 00	18 25
Discharge ($m^3 sec^{-1}$)	9.5	10.0	12.5	15.0	20.0	23.0	26.0	12.0
Concentration ($mg\ l^{-1}$)	39	7640	5784	6110	3290	3491	2143	1707
Sediment load ($t\ hr^{-1}$)	4	275	258	330	237	289	200	74

The studies have to be tailored to the local physical conditions and man. That has been illustrated by SWECO studies at Mtera as well as by the DUSER project from various parts of the country.

The two studies from semi-arid Tanzania clearly illustrate the importance of remote sensing methods but also the need for a good basic knowledge of the landscape and its process activity.

The different methods to quantify soil erosion described in this paper also illustrate the need of an adaptation of field techniques to the local environment and processes.

REFERENCES

Christiansson, C., 1978, Relations Between Heavy Grazing, Cultivation, Soil Erosion and sedimentation in the Semiarid Parts of Central Tanzania, *Proceedings of the first Int. Rangel. and Congress,* 1978, Denver.

Johansson, D. (ed.), 1976, Great Ruaha Power Project, Tanzania, *Ecological Studies of the Mtera Basin,* Swedish Consulting Group (SWECO), Stockholm.

Johansson, D., and Stromquist, L., 1977a, Great Ruaha Power Project, Tanzania, *Studies of Soil Erosion, Vegetation, and Fluvial Transport of Mtera Reservoir Region, Tanzania,* SWECO, Stockholm.

Johansson, D., and Stromquist, L., 1977b, Great Ruaha Power Project, Tanzania, *Land Use and Conservation Potential of the Proposed Mtera Reservoir Region, Tanzania,* SWECO, Stockholm.

Johansson, D., and Stromquist, L., 1978a, Termite mounds near Dodoma, Central Tanzania, *Svensk, Geografisk. Arsbok,* Lund, Also as Med. Upps. Univ. Geogr. Instn., Ser A, No. 272.

Johansson, D., and Stromquist, L., 1978b, Great Ruaha Power Project, Tanzania, *Expected Environmental Changes below the Mtera Reservoir,* SWECO, Stockholm.

Johansson, D., and Stromquist, L., 1978c, Interpretation of geomorphology and vegetation of LANDSAT satellite images from semiarid central Tanzania, *Norsk Geogr. Tidsskr.,* 1978:4, Oslo.

Lundgren, V. (ed.), 1975, Landuse in Kenya and Tanzania, *The Physical Background and Present Situation and an Analysis of the need for its Rational Planning,* Royal College of Forestry, Stockholm.

Lundgren, L., 1978, Studies of soil and vegetation development on fresh land-slide scars in the Mgera Valley, Western Uluguru mts., Tanzania, *Geogr. Annlr.,* **Vol 60A,** Stockholm.

Rapp, A., 1972, Conclusions from the DUSER soil erosion project in Tanzania, in Rapp *et al.* (eds.), 1972.

Rapp, A., 1976, An assessment of soil and water conservation needs in the Mwanza region, Tanzania, *Lunds Univ. Naturgeogr. Instn. Rapporter och Notiser*, 31, Lund.

Rapp, A., Berry, L., and Temple, P. (eds.), 1972a, Studies of soil erosion and sedimentation in Tanzania. *Geogr. Annlr.*, **Vol 54A**, Stockholm.

Rapp, A., Murray-Rust, D. H., Christiansson, C., and Berry, L., 1972b, Soil erosion and sedimentation in four catchments near Dodoma central Tanzania, in Rapp *et al.* (eds.), 1972.

Rapp, A., and Hellden, U., 1979, Research on environmental monitoring methods for land use planning in African drylands, *Lunds Univ. Natgeogr. Instn. Rapporter och Notiser*, 42.

Stromquist, L., 1976, Land systems of the Great Ruaha drainage basin upsteam the Mtera dam site, in Johansson, D. (ed) 1976, also as *Med. Upps. Univ. Geogr. Instn.*, Ser. A, No. 268.

Stromquist, L., and Johansson, D., 1978, Studies of soil erosion and sediment transport in the Mtera Reservoir Region, Central Tanzania, Zeitschrift fur Geomorphology, *Suppl. Bd. 29*, Stuttgart, also as *Med. Upps. Univ. Geogr. Instn.*, Ser. A, No. 275.

Tanzanian Government, 1977, The threat of desertification in central Tanzania. A technical paper for the UN conference on desertification, *Mimeographed Report*, Dar-es-Salaam.

Tropical Agricultural Hydrology
Edited by R. Lal and E. W. Russell
© 1981, John Wiley & Sons Ltd.

3.9

Sediment Transport and River Basin Management in Nigeria

L. OYEBANDE

3.9.1. STATUS OF RIVER BASIN DEVELOPMENT

River basin management involves soil and water with the plant and animal communities which together form a complete organism and produce a number of products and services. The objective of such management often is to maximize these yields and at the same time minimize the adverse effects of the productive activities.

According to Thorne (1963) river basin management often includes a number of complex measures such as administration, data collection and evaluation, planning and designing, construction and implementation of projects, fiscal management, and enforcement of laws and research. Often, too there are competing demands in basin management activities. For instance a dam built upstream of an aggrading floodplain will reduce flooding problem, but may also induce ecological changes in the floodplain which is perhaps being adapted to grazing and fadama farming. A case in point is the Yobe basin which is discussed below in detail. It is also known that trees transpire large quantities of water and also intercept the rainfall, thus reducing the quantity of water reaching the soil. Many studies even show that large stream flow increases result from deforestation, but soil stability may be reduced concomitantly, resulting in accelerated erosion and sediment transport as well as deposition elsewhere.

In Nigeria, efforts within the river basins have been directed mainly towards water supply projects and irrigation development undertaken by individual States of the Federation. No deliberate efforts however have been made for integrated basin development, and often activities within the same river basin have not been properly coordinated. This lack of consciousness of the integrating influence of a river basin undermined the importance attached to comprehensive hydrological data collection. In recent years, however, several types of planning studies have been undertaken. These include feasibility studies of individual river basins, sub-basins, or sub-regions. The studies provide some crude economic and hydrologic analyses and projections relating to the development of water and associated land resources.

One of the most far-reaching steps towards effective river basin management in Nigeria has been the setting up in 1976 of eleven semi-autonomous River Basin Development Authorities (RBDA). A Coordinating Committee was also established to ensure integrated and even development of the different basins (Figure 3.14). The RBDA were given wide powers to exercise the following functions to which the present civilian government is seeking to add other roles such as integrated rural development.

(i)   To undertake comprehensive development of groundwater resources for multipurpose use.

(ii) To undertake schemes for the control of floods and erosion for watershed management.

(iii) To construct and maintain dams, dykes, polders, wells, boreholes, irrigation, and drainage systems, etc.

(iv) To develop irrigation schemes for the production of crops and livestock and to lease the area concerned with the approval of the Commissioner.

(v)   To provide water from reservoirs, wells, and boreholes, under the control of the authority concerned for urban and rural water supply schemes on request by the state government and when directed to do so by the Commissioner.

(vi) The control of pollution in rivers and lakes in the authority's area in accordance with nationally laid down standards.

(vii) To resettle persons affected by the works and schemes specified in (iii) and (iv) above or under special resettlement schemes.

Actually three such basin authorities had been in existence since 1973, but there was little to show in the way of basin management during those three years. The activities and powers of the new basin authorities can be expected to have profound effect on the status of hydrology in general, and basin management in particular, throughout the country.

The present paper studies some implications of the reciprocal relationship between river basin planning and development on one hand, and the production and transportation of sediment by the major river systems on the other; and it examines in some detail the pattern of sediment transport by these rivers in order to provide a better understanding of the major processes and factors at work in the Nigerian environment. The report also examines ways of checking the adverse effects of sedimentation and erosion through improved watershed management.

3.9.2. EROSION AND SEDIMENT TRANSPORT

3.9.2.1 Some General Considerations

Sediment originates in erosion of the drainage basin, with the major part of it coming from the upland areas where streams are torrential and dissect the land

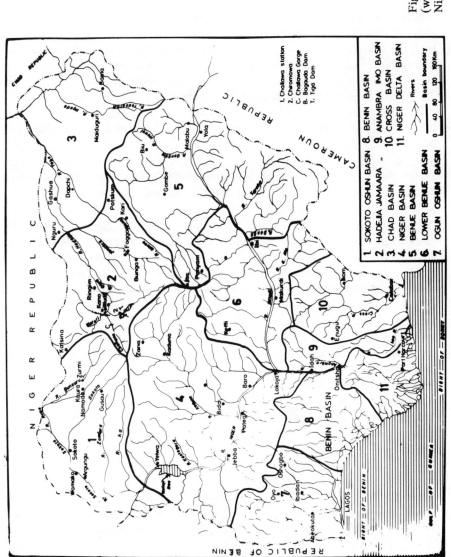

Figure 3.14 Hydrological (water resources) regions of Nigeria

effectively. Geologically, every hill and mountain is destined to be reduced to a peneplain and its material deposited elsewhere. The rate and duration of such reduction is the point of significance.

Sheet erosion which affects much larger areas has been found to be no less significant then gully erosion in Nigeria. The latter often produces more spectacular and sometimes monumental features, as in the Awka-Orlu escarpment of the Anambra-Imo basin and in the Challawa watershed in Upper Hadejia in the Hadejia-Jamaare basin (Figure 3.14). These two areas together with the Jos Plateau are notorious for erosion, but are less thought of as sources of sediment production. A primary factor of the accelerated erosion in these areas is the nature of the sandy ferralitic or lateritic soils which are highly vulnerable to erosion, particularly gullying once the protective forest cover has been removed. In the Awka–Orlu district, the dissection of the lateritic horizons exposes beds of unconsolidated pebbly, silty sands which erode very easily (Floyd, 1965; Ofomata, 1965). Man's interference with the ecological balance through overfarming and overgrazing by the dense population is also an important, if secondary, factor causing erosion.

Rainfall intensity usually features prominently in equations used to predict rate of sheet erosion. Both the Musgrave and the so-called Universal Soil Loss Equation (Chow, 1964), include the maximum 30-minute rainfall. The latter also incorporates the storm energy in foot tonnes per acre-inches. Wishmeier and Smith (1958) have shown that the yield of sediment is proportional to the product of the maximum 30-minute intensity experienced during the storm and the kinetic energy of the rainfall. Thus a large proportion of the sediment produced may be attributed to the medium–high intensities which occur more frequently. Table 3.35 shows the range of the 2-year 30-minute rainfall for the different watersheds. The values vary from 64–66 mm in Sokoto-Rima; 72–82 mm in the Niger; Cross and Benue river systems. On the Jos Plateau, Challawa, and Anambra basins the amount is 65, 65, and 85 mm respectively. But the higher mean annual amount and more frequent falls in the Anambra basin increase the effectiveness of rainfall as factor of erosion in the absence of protective cover.

Not all eroded materials in a watershed reach the river system. For instance, in flat areas with low drainage density, eroded material may not be transported to down stream points in the watershed. On the other hand, where the interaction of climate and geology is such that produces very dense drainage net, sediment transport becomes more efficient, and the sediment–delivery ratio (proportion of erosion material removed down stream) increases. In general this proportion of gross erosion resulting sediment yield may vary from 5-100 per cent (Chow, 1964, pp. 17-12).

The load of a river is carried in three main ways—bedload, including saltation, suspended load, and solution load. The separation of bedload and suspended load is somewhat arbitrary, and Bagnold (1954) defines bedload as that whose weight is carried on the solid bed. The bedload is determinant for the behaviour of the river bed as regards scouring and deposition. He (Bagnold) also defines suspended load as

Table 3.35. Some hydrologic and sediment characteristics of selected river basins

Basin authority	River	Station	Basin area (10³ km²)	Annual mean basin rainfall (mm)	2-yr 30 minute rainfall (mm)	Annual discharge volume 10⁹ m³	Sediment load per area tonnes/km²	Soil loss from basin (10⁶ tonnes)	Average sediment concentration (p.p.m.)
Sokoto Rima	Rima	Wamako	35.37	778	66	1.60	155	5.48	3422
	Sokoto	Gidan Doka	12.59	852	68	0.78	392	4.94	3074
	Gagere	Kaura Namoda	5.67	956	68	1.57	292	1.65	1052
	Bunsuru	Zurmi	5.90	818	66	1.37	438	2.58	1886
Hadejia-Jamaare	Hadejia	Wudil	17.4	918	64	2.01	355	6.17	3072
	Kano	Chiromawa	6.98	1000	64	1.21	219	1.53	1266
	Challawa	Challawa	6.89	841	65	0.69	739	5.09	7380
	Watari	Gwarzo Rd.	1.45	852	66	0.12	483	0.71	5841
	Jamaari	Bunga	7.98	1001	64	2.08	459	3.66	1760
	Misau	Kari	5.60	950	65	0.69	9	0.05	71
Chad	Yobe	Gashua	62.4	765	64	1.56	141	0.41	262
Niger	Niger	Koji	1080.9	1380	74	154.9	19	20.8	134
	Niger	Baro	730.4	1100	76	72.22	12.5	8.76	63
	Kaduna	Kaduna	18.42	1213	64	5.3	52.33	0.96[c]	182
Upper Benue	Gongola	Bare	55.5	1028	71	6.4	77	4.35[b]	670
	Taraba	Gassol	21.3	1630	78	12.0	80	1.70[b]	140
	Donga	Nyankwola	19.8	1620	79	13.0	56	1.11[b]	85
	Benue	Yola	107.0	1055	68	23.0	56[a]	5.99[b]	260
Lower Benue	Benue	Makurdi	304.3	1525	72	100.5	43	13.2	132
	Katsina Ala	Sevav	22.0	1491	81	23.0	64	1.41[b]	60
Cross	Cross	Ikom	16.9	3320	82	33.96	72	1.23	36

[a] Based contributing area which is half of the total basin area and represents discharge of a tapering stream whose streamflow is only about 78 per cent of the flow at Wudil
[b] Estimates by NEDECCO (1959)
[c] Estimated from partial data provided by Kanuna State Water Board (1973) and NEDECCO (1959).

Table 3.36. Volume of different types of sediment transported (after NEDECCO, 1959)

River/station	Period	Bed load	Suspended load		
			Sand	Silt and Clay	Total
Upper Niger (Baro)	1915–57	0.31 (6.5)	0.55 (11)	4 (82.5)	4.9
Lower Niger (Shintaku)	1915–57	0.88 (5)	1.3 (8)	15 (87)	17
Lower Niger (Onitsha)	1915–57	1.0 (5.5)	1.7 (9)	16 (88.5)	19
Upper Benue (Yola)	1939–57	0.20 (6)	0.17 (5)	3.0 (89)	3.4
Lower Benue (Makurdi)	1932–57	0.61 (5)	0.97 (8)	10 (87)	12

Assume sediment specific weight of 2 tonnes m^{-3}. Figures in brackets (per cent of total sediment)

having its immersed weight carried by the fluid and thus finally by the interstitial fluid between the bed grains. The greater part of the suspended load is carried off to the sea or deposited on the floodplains. The amount of suspended load is usually much greater than that of bedload. Investigations in Nigeria by NEDECCO (1959) indicate that the proportion of bedload varies from 5.0 per cent in the Lower Benue and Lower Niger to 6.5 per cent in the upper reaches of the two rivers (Table 3.36). It was also found that the suspended load was made up largely of silt with little fine sand which is responsible for the building up of the alluvial valley, and important for the rate of reservoir sedimentation.

3.9.2.2. Sediment Load of Nigeria's River Systems

Table 3.37 shows the average grain sizes which account for the indicated characteristic percentage figures (50, 60, and 90), of the bed, saltation, and suspended sediment samples. Taking the median size we observe that in the Upper Niger only 10 per cent of the bedload is greater than 0.25 mm and 50 per cent less than 0.54 mm, while for the suspended load, grains less than 10 per cent have sizes larger than 0.37 mm and 50 per cent less than 0.2 mm.

Sediment analysis made for a number of rivers between 1966 and 1968 and reported by Colson (1969) provide useful information on particle sizes of suspended sediment in the Chad and Hadejia-Jamaare basins. The Hadejia basin contains the Hadejia river, formed by the confluence of the Challawa and Kano rivers just upstream of Wudil, and the Watari river is a major tributary of the Challawa. For the Challawa river at Challawa 14 per cent of the sediment is the particles finer than 0.002 mm for moderate flows, but this falls to 5–10 per cent for high flow. The bulk of the sediment has particles in the range 0.05–0.1 mm.

In the Watari Watershed, of all the sediment samples corresponding to discharges of 0.15 to 175 m^3 sec, not more than 4 per cent are larger than 0.2 mm and the

Table 3.37. The size distribution of particles in the transported sediments of some Nigerian rivers
(d_{50} and d_{90} give the upper limit of the equivalent diameter of particles, in nm, constituting the finest 50 and 90 per cent of the sediment by weight)

River/station	Bed load		Saltation load		Suspended load	
	d_{50}	d_{90}	d_{50}	d_{90}	d_{50}	d_{90}
Benue						
Yola	0.66	2.3	0.32	0.71	0.21	0.42
Ibi	0.85	3.0	0.41	1.5	0.18	0.49
Makurdi	1.04	4.3	0.63	1.7	0.22	0.69
Upper Niger						
Pategi	0.54	1.3	0.28	0.56	0.18	0.37
Baro	0.52	1.6	0.23	0.55	0.21	0.40
Lower Niger						
Shintaku	0.78	3.3	0.23	0.75	0.15	0.31
Onitsha	0.52	1.8	0.20	0.43	0.13	0.25
Gongola						
Bilachi	0.81	3.7	0.36	1.1	0.16	0.39
Faro						
Kossel	0.75	2.8	0.55	1.9	0.16	0.34
Taraba						
Sendirdi	0.82	3.1	0.68	2.3	0.21	0.46
Donga						
Nyankwola	0.88	2.2	0.42	0.8	0.13	0.23
K'Ala						
Confluence	0.85	2.7	0.50	1.7	0.27	0.77

Bedload samples taken using the Bedload Transport Meter Arnhem and Saltation and sediment samples using Delft bottles DF_1 and DF_2 respectively.
Source: 3000 samples taken between 1955–58 by NEDECCO (1959, p. 462).

bulk lies within the 0.002 and 0.05 mm size range, but during floods greater than 170 m^3 sec^{-1}, some 60–80 per cent of the sediment exceeds 0.05 mm in size with little difference in the proportion coarser than 0.2 mm. Not more than 9 per cent of the sediment from the Misau River have sizes greater than 0.2 mm during high flows exceeding 250·m^3 sec^{-1} from June to October. For the Hadejia at Wudil, not more than 3 per cent of the particles are coarser than 0.2 mm, while 80-95 per cent are finer than 0.1 mm.

The fairly large particle size of the bulk of the sediment from the Challawa river tends to indicate the relative importance of gully erosion which derives its material from the granitic hills in the Challawa headwaters. Areas that are predominantly affected by sheet erosion produce more fine-grained sediment since pre-channel flow seldom exceeds 0.9 m sec^{-1}, and in such cases with further sorting and abrasion up to 95 per cent of the sediment may have grain-size diameter less than 0.05 mm.

The sediment transport problems are exceedingly complex and their solution often proceeds with the help of hydraulics and fluid mechanics. For example sediment transporting capacity can be evaluated by the application of the relevant methods of hydraulics (Bogardi, 1970). Most sediment load formulas are in turn based on transporting capacity, which is the largest quantity of sediment that a water course can carry. The actual load is usually much less. Sediment transport studies can however go to the extreme of using purely theoretical methods, employing the accepted laws of hydraulics and fluid mechanics. In the present study sediment load is computed from measured sediment and discharge made between 1955 and 1977 in the various river systems by a number of agencies. Motor Columbus *et al.* (1978) measured suspended solid load on the Niger at Koji, Benue at Makurdi, and Cross River at Ikom during July 1977 to February 1978. Cross-section sampling yielded samples at five positions, and at each vertical position, at five different depths by Neyrptic turbisondo. A bank sample was taken at the sixth position 2 m from the bank at 0.3 m depth.

In the Hadejia–Jamaare, Chad, and Sokoto–Rima basins suspended sediment load samples were collected with depth-integrating samplers. The measured sediment data were used to develop sediment transport rating curves. Fortunately, too, rating curves have been reported by MRT (1978) for several rivers in the Hadejia and Sokoto–Rima basins. The rating curves obtained are shown in Figure 3.15 (a) to (f). Rating equations of the form

$$Q_s = kq^n$$

where Q_s is the sediment load in t day^{-1} and q the river discharge in m^3 sec^{-1} and k is the intercept when q is the unity on the logarithmic plot, and n is the slope of the line. The annual sediment load determined by careful application of the rating curves to continuous record of discharge for 13 rivers at specified stations is shown in Table 3.38. The quantities vary from 49 thousand tonnes in Misau, to 5 million for the Challawa and 21 million tonnes for the Niger at Koji near Onitsha.

The size of drainage area exercises great influence on the sediment yield from a watershed. Chow (1964) using data from about 1100 measurements in the USA illustrates the decline in sediment production rate as the area of the watershed increases. Watersheds of less than 26 km^2 in area on the average produce more than 7 times as much sediment per unit area as those exceeding 2600 km^2.

Table 3.39 is derived from such a few number of measurements that no generally applicable conclusions can be drawn from it. Nevertheless, the data appear to indicate that watersheds whose size is less than 10,000 km^2 can yield sediment per unit area more than 8 times as much as those whose size exceeds 100,000 km^2. The explanation is perhaps not far to seek. The wide flood plains, an ideal environment for deposition, which exist in the Lower Niger and Benue, and even in the Yobe River do not exist in the steeper channels of smaller headwaters tributaries of Watari, Challawa, and Bunsuru Rivers just to mention a few. In the following sections we take a closer look at the pattern of sediment transport in some of the River Basin Authority areas.

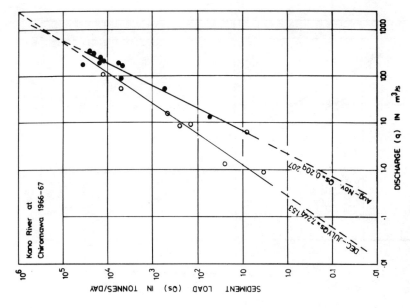

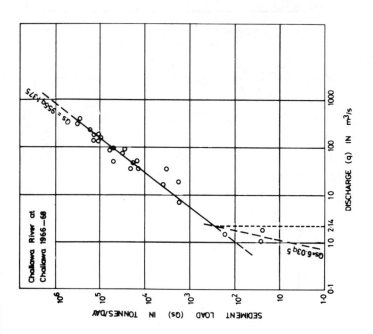

Figure 3.15(a) Sediment transport rating

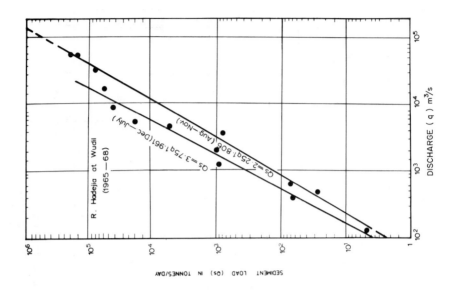

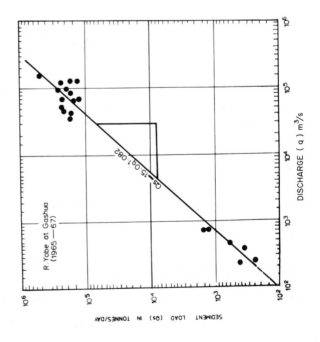

Figure 3.15(b) Sediment transport rating

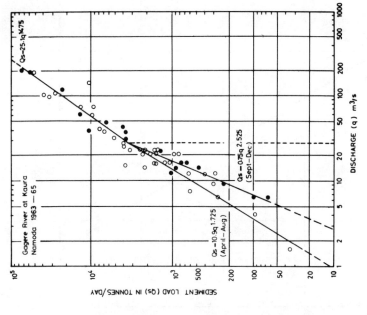

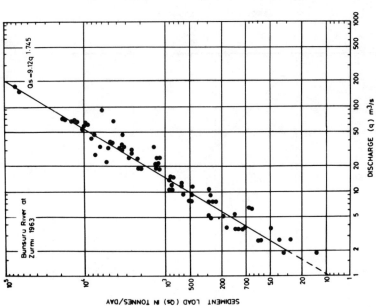

Figure 3.15(c) Sediment transport rating

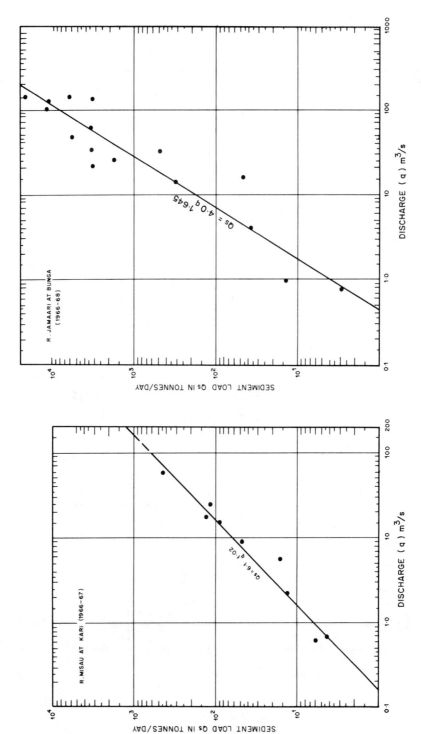

Figure 3.15(d) Sediment transport rating

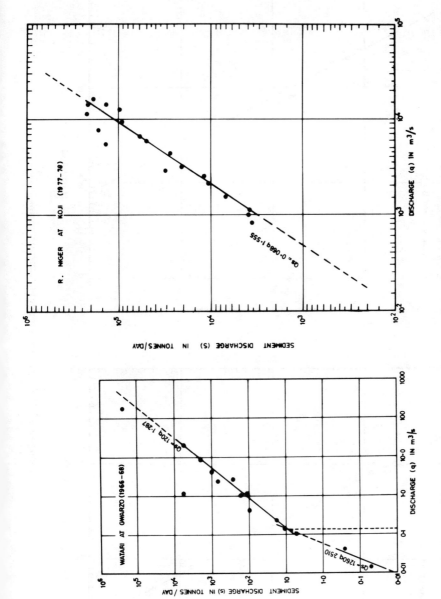

Figure 3.15(e) Sediment transport rating (left) and suspended sediment transport rating (right)

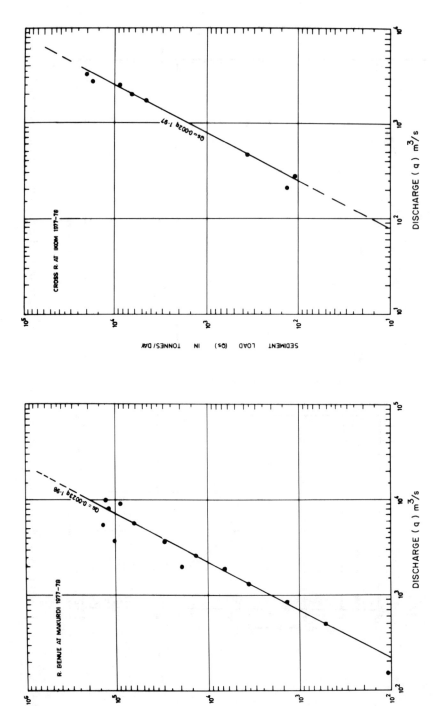

Figure 3.15(f) Suspended sediment transport rating

Table 3.38. Sediment load of some Nigerian Rivers in thousand tonnes

River	Station	J	F	M	A	M	J	J	A	S	O	N	D	Year	Period
Niger	Koji	276	237	197.2	164.7	172.3	411.8	1114	2945	6579	6919	1463	421	20,800	1963–77
Rima	Wamako	0.5	0.2	0.1	0.1	7.1	148.9	6091	22528	22750	176.3	4.4	1.0	5,476	1962–73
Sokoto	Gidan Doka	0	0	0	0	6.3	130.3	4820	20152	22148	92.2	0.6	0	4,941	1962–73
Gagere	Kaura Namoda	0	0	0	0	24.4	124.5	4187	6437	3984	42.8	0.2	0	1,653	1962–77
Bunsuru	Zurmi	0	0	0	0	28.8	81.8	3377	15999	5246	11.0	0	0	2,584	1962–77
Hadejia	Wudil	0	0	0	2.5	3.0	34.7	1647	2224	1794	127	3.4	0.1	6,175	1964–73
Kano	Chiromawa	0	0.1	0	4.2	15.7	55.8	2025	7214	5233	8.6	0.1	0.2	1,532	1964–71
Challawa	Challawa	0	0	0	0.6	125.0	4175	10441	22697	11811	53.1	1.3	0	5,092	1964–70
Watari	Gwarzo	0	0	0	0	3.8	53.8	1755	3590	1093	3.4	0.1	0	705	1965–70
Jamaare	Bunga	0	0	0	0.3	14.1	54.8	264	1944	1354	20	0.8	0.1	3,661	1964–74
Misau	Kari	0	0	0	0	0	0.1	2.7	23.3	20.4	1.8	0.4	0	49.2	1964–72
Yobe	Gashua	3.4	0.8	0.4	0.3	0.1	8.0	28.0	63.5	88.0	119.5	64.9	0.1	409	1964–72
Benue	Makurdi	19.8	7.8	6.0	7.8	29.4	172.2	579	1940	4917	4887	562	53.0	13,235	1963–77
Cross	Ikom	1.9	1.3	2.9	4.5	12.2	57.8	2163	2801	3784	2320	345	5.1	1,227	1962–77

Monthly sediment transport less than 100 tonnes is assumed to be zero.

Table 3.39. Sediment production rates for groups of selected drainage areas in Nigeria

Size of watershed (km^2)	Annual sediment (t km^{-2})	No. of watersheds
<10,000	377	7
10,000–50,000	119	7
50,000–100,000	77	1[a]
>100,000	44	4

[a] Yobe at Gashua falls within the range, but is not included on account of its pronounced peculiarities.

3.9.2.2.1. Hadejia–Jamaare and Chad Basins

The Yobe River system has its upper and middle segments in the Hadejia–Jamaare but its lower valley also represents the major source of surface water in the Chad basin. The peak discharge at Gashua occurs more than a month later than at Wudil, which is less than 320 km upstream. The upper reaches of the Kano, Jamaare, Challawa, and Watari, particularly the latter two with their high sediment carrying capacities continue to build up gradually silt and fine sand in the floodplain and channels of the sluggish middle and lower reaches of the Hadejia downstream of Ringim. Aggradation with consequent further flattening of the slopes has already caused serious flooding problem at Ringim, Hadejia, Nguru, and Gashua (Colson, 1969). The eventual solution to the problem would probably necessitate expensive dredging of the channels or building of protective levels around the affected towns.

However the construction of Bagauda and Tiga dams in the Kano River and Karaye in the Challawa watershed are expected to mitigate the flooding problem. Also under construction are dams at Watari on the Watari River, Challawa Gorge in the Challawa River (with storage capacity of 969×10^6 m^3), and several others in the Hadejia and Jamaari river systems. Challawa Gorge Dam is expected to trap 95 per cent of the total suspended load of the River Challawa, that is, some 4.84 million t yr^{-1}. This represents more than 78 per cent of the total suspended load of the Hadejia at Wudil during the pre-1974 period. Thus the combined effect of Tiga Dam and the proposed Challawa Gorge Dam may completely eliminate the present aggradation and flood problem downstream of Wudil. The dams at present store more than half of the total flow in the basin and also trap substantial amounts of the sediment load.

3.9.2.2.2. Sokoto–Rima and Niger Basins

The Niger enters Nigeria at Malenville after draining nearly 440,000 km^2 of land and carries in little sediment load even during the 'black' flood (December–March).

It has lost much of its sediment transport in the swamps of the Interior Niger Delta near Timbuctu. Within Nigeria, the Upper Niger receives several intermittent tributaries which are sediment-laden during the floods. The Sokoto–Rima system is the first major tributary of the Niger in Nigeria, and a major source of sediment load during the 'white' flood (July–October). The rivers carry sediment load of 155 to 438 t km^{-2} annually and have average concentration of 1052 to 3424 mg l^{-1} (Table 3.35). The high sediment load results in high turbidity which made the stream water unattractive for urban water supply development particularly in the headwater areas. The sediment load is derived mainly from the crystalline basement rocks in the eastern half of the basin where rainfall is highest and infiltration least, with runoff coefficient of 10–15 per cent at Zurmi, Kaura Namoda, and Gusau on the Bunsuru, Gagere and Sokoto rivers, respectively.

In the Upper Niger around Baro, Pategi, and at Jebba, average silt concentration was 60–110 mg l^{-1} and maximum and minimum observed in the pre-Kainji period at the present site of the dam were 250 and 27 mg l^{-1}. At Baro the total sediment load was about 9 million t yr^{-1} (Table 3.36). But in 1969 the average concentration of suspended load was only 40–86 mg l^{-1} downstream of the dam during the white flood. During the same period, the concentration in the open lake was 128 to 178 mg l^{1-} (Imevbore, 1975). The dam reservoir with full capacity of 15 km^3 and a surface area of 1250 km^2 had started to trap much of the sediment load and had considerably lowered the turbidity of the water of the Niger downstream.

The annual suspended sediment load carried during 1915–1957 by the lower Niger (Onitsha) was about 35 million tonnes (Table 3.36). During 1963–1977 a total of 20.8 million tonnes was transported past Koji, a station 115 km upstream from Onitsha (Table 3.34). The difference represents a decrease of 40 per cent. During the Sahelian drought of 1971–73 annual sediment load was 48 to 58 per cent of the normal. On the whole the drought accounts for nearly 10 per cent of the observed decrease.

The Kainji dam retains much of the 8 million tonnes of the suspended sediment which enters it. The amount trapped in the reservoir is perhaps upward of 6 million t yr^{-1}. It also appears that with the completion of the Lokoja dam with a planned capacity of 68 km^3 the upbuilding of silt in the Lower Niger and the concomitant flooding will be substantially reduced. This would in turn facilitate reclamation of large areas of otherwise rich lands now lost to farming as a result of poor drainage.

3.9.2.2.3. Benue Basins

There are two River Basin Development Authority areas in the drainage area of the Benue River. They are the Upper and Lower basins (Figure 3.14). Table 3.36 indicates that much suspended sediment is produced between Yola and Makurdi, a net load of some 15 million t yr^{-1}. The Gongola River alone carried some 4.5 million tonnes; the Doga 2.5 million tonnes, the Katsina Ala 1.5 million tonnes, and the Taraba 2 million tonnes. The total suspended load transported by Benue at Makurdi during 1932-1957 was about 22 million tonnes. However as Table 3.38 indicates the load was only 13.2 million tonnes during 1963-1977, a decrease of 40 per cent, just as in the case of the Lower Niger. This reduction in sediment transport by the Niger's principal tributary would be expected to account for the bulk of the remaining deficit of 5-6 million tonnes of sediment in the Lower Niger during 1963-1977.

Most of the Jos Plateau lies within the Benue's drainage area. The headwaters of Rivers Gongola, Ankwe, Wase, and Mada dissect the Plateau severely and derive much sediment load from this area of widespread tin mining spoil. Sheet erosion is quite considerable on the Plateau surface, but the Jos Plateau is also notorious for gully erosion. The gullies there are smaller but more numerous than those of the Awka-Orlu area (Dorman, 1978). Estimates of the extent of gullying on the Jos Plateau by Jones (1975) by means of air photographs gave the total length of the gullies as 7240 km. This figure represents a total soil loss of 100,000 tonnes. An average rate of headwater extension of the gullies of more than 16 m yr^{-1} was also observed.

3.9.2.2.4. The Cross River Basin

The total suspended load and its average concentration for the Cross River at Ikom appears to be very moderate in comparison with other measurements. This tends to give the impression that the basin is free from severe erosion. This is far from the true situation. Ikom is very close to the border with the Cameroons Republic and the obvious conclusion is that like the Upper Niger the river carries little sediment load into Nigeria. Much more information is needed on the middle and lower reaches of the Cross River before firm conclusions can be drawn. This is in view of the tremendous work of erosion and carrying of sediment load which intense rain storms and the headwaters of the Aboine river, a major tributary of the Cross, are known to be doing on the Enugu-Udi-Awgu-Okigwi escarpment on the western margins of the basin. It is also known that the Ikot Ekpene-Itu-Uyo triangle and eastwards across the main river itself experience equally insidious, if less pronounced, sheet and gully erosion as the Awka-Orlu escarpment further west (Floyd, 1965). The severe erosion and the resultant soil loss and deterioration are, as mentioned earlier due to over-farming and destructive cultivation methods which expose the susceptible soil structure to the irresistible energy unleashed by the intense and frequent falls of rain.

3.9.3. DISCHARGE MAGNITUDE AND SEDIMENT TRANSPORT

Rare, very severe rainstorms, cyclones, and floods are known to cause important alterations in the courses of alluvial channels and cause very severe gullying of hillslopes. Leopold *et al.* (1964) rightly pointed out that the cumulative effect of floods (and rainfall intensities) which recur once a year or perhaps once in every three to five years are less easily recognized. Much has been written for instance

about the dominant discharge and its effect upon the characteristics of a river channel in general, and the sediment load in particular. Some recognize the bankful stage
discharge as the most significant in this regard, while some others claim formative
agent of the river (King, 1966). NEDECCO (1959) made useful distinction by
asserting that

> 'No single steady discharge can be envisaged that would shape a river-bed with
> all the characteristics similar to those that result from a superposition of the
> influences of the varying discharges'

Rather, for each single channel characteristic it will be possible to define a charac-
istic discharge that is dominant for the origin and nature of that quality. Nevertheless Prus-Chacinski has argued that the mean annual discharge is more likely to be
the dominant discharge, in that it is responsible for the shape of the channel, because
it occurs for a great proportion of the time, 20-25 per cent. Much of the reported
case studies have indicated however, that the dominant discharge for several channel
characters is nearer to the rare bankful stage discharge and occurs with a frequency
of once in one or two years. The reason is that they are capable of doing more work
in moving bed material than the rare, but very large flows, or small mean annual discharge. Yet it appears that much of the work of carrying solution load is achieved
during mean flow, as the concentration tends to be lowered during flood periods.

Table 3.40 shows the frequency of the different magnitudes of discharge in some
of Nigeria's river systems. For Rivers Yobe, Niger, and Benue large flows with frequency of less than 2.5 per cent transport only 10-12 per cent of the total suspended load. But for the Hadejia and its sediment-laden tributary, Challawa, the
same magnitudes of flow carry 30-40 per cent of the load. We observed earlier the
bulk of the particles in the Challawa sediment are reasonably large. This particle-
size distribution and the flashy nature of the river flow may explain the above
pattern. At the lower end of the duration curve data, we find that flows smaller
than the mean (that is, greater than the 25 per cent frequency) transport less than
10 per cent of the total suspended sediment load. This implies that discharge
levels occurring on one out of four days (that is more than 7,000 m^3 sec^{-1} at Koji,
55 m^3 sec^{-1} at Gashua, 2.7 m^3 sec^{-1} at Challawa, 4000 m^3 sec^{-1} at Makurdi, and 113
m^3 sec^{-1} at Wudil transport more than 90 per cent of the load.

Furthermore, certain shorter discharge intervals also appear to be particularly
important in the sediment transport of some rivers. For example, the discharge
range of 10,000-20,000 m^3 sec^{-1} alone transport 70 per cent of the load in the
Lower Niger. For the Lower Benue the interval is 7-10,000 m^3 sec^{-1} and accounts
for some 40 per cent of the load. 40 per cent of the suspended load is also carried
in the Challawa river by discharge levels between 56 and 170 m^3 sec^{-1}.

3.9.4. IMPLICATIONS FOR RIVER BASIN MANAGEMENT

Some implications of erosion and sedimentation for river basin planning and development have been mentioned above in passing. However when one examines closely

Table 3.40. Flow characteristics and sediment loads of selected rivers

Discharge Interval (m³ sec⁻¹)	Flow volume (%)	Cumulative flow volume (%)	Flow duration (%)	Cumulative flow duration (%)	Annual sediment load (10^3 tonnes)	Sediment load (%)	Cumulative sediment load (%)
			(a) R. Yobe at Gashua	($\bar{Q}$ = 49.4 m³ sec⁻¹)			
55–104.9	21.7	86.9	14.0	34.6	74	18.0	76.0
105–178.0	44.6	65.2	16.2	20.6	16	39.3	58.0
178.1–198.0	7.4	20.6	2.0	4.4	27	6.6	18.7
198.1–283.0	6.5	13.2	1.4	2.4	25	6.1	12.1
283.1–396.2	6.7	6.7	1.0	1.0	25	6.0	6.0
			(b) R. Niger at Koji	($\bar{Q}$ = 4910.2 m³ sec⁻¹)			
1000–3000	16.2	100.0	47.5	100.0	374	1.8	100
3001–7000[a]	16.6	83.8	22.5	52.5	2870	13.8	98.2
7001–10000	11.2	67.2	8.3	30.0	666	3.2	84.4
10001–15000	20.1	56.0	10.1	21.7	5408	26.0	81.2
15001–20000	10.7	35.9	3.3	11.6	2974	14.3	55.2
17001–20000	19.1	25.2	6.5	8.3	6324	30.4	40.9
20000–23000	6.1	6.1	1.8	1.8	2184	10.5	10.5

(c) R. Challawa at Challawa ($\bar{Q} = 21.9$ m³ sec⁻¹)

Discharge interval							
56.6–113.2	25.4	84.6	6.9	12.5	1085	21.3	80.0
113.3–169.8	18.7	59.2	3.1	5.6	987	19.0	58.7
169.9–226.4	9.4	40.5	1.1	2.5	545	10.7	39.7
226.5–283.0	7.9	30.6	0.7	1.4	377	7.4	29.0
283.1–339.6	2.3	22.7	0.2	0.7	158	3.1	21.6
339.7–566.0	15.5	20.4	0.3	0.5	489	9.6	18.5
> 566.0	4.9	4.9	0.2	0.2	453	8.9	8.9

(d) R. Benue at Makurdi ($\bar{Q} = 3186$ m³ sec⁻¹)

Discharge interval							
4001–7000	18.7	75.5	9.9	26.9	1946	14.7	92.2
7001–10000	34.5	56.8	11.8	17.0	5479	41.4	77.5
10000–15000	22.3	22.3	5.2	5.2	4906	36.1	36.1

(e) R. Hadejia at Wudil ($\bar{Q} = 63.9$ m³ sec⁻¹)

Discharge interval							
56.6–113.2[a]	9.0	93.3	6.5	23.4	183	3.0	97.6
113.3–169.8	11.0	84.3	4.8	16.9	303	4.9	94.6
169.9–283.0	17.7	73.3	4.7	12.1	793	12.8	89.7
283.1–424.8	25.1	55.6	4.3	7.4	1617	26.2	76.9
424.9–566.0	13.2	30.5	1.6	3.1	1115	18.1	50.7
566.1–849.0	14.8	17.3	1.3	1.5	1640	26.6	32.6
> 849.0	2.5	2.5	0.2	0.2	369	6.0	6.0

[a] The Discharge interval contains the mean (Q).

the functions of the River Basin Development Authorities (RBDA) in Nigeria as listed in Section 3.9.1, the magnitude of the problem and the possible conflicts become more striking.

3.9.4.1. Water Quality

In the attempt to provide urban and rural water supply, power and water for irrigation and to control floods, the RBDA's in collaboration with the appropriate State Water boards and Corporations build dams and impoundments in a number of rivers. The high turbidity of some rivers have not made them attractive for such development, especially where there are alternatives. And where there is no choice, expensive treatment and filtration have to be undertaken. In the Sokoto-Rima basin we have observed that the Bunsuru, Gagere, and Upper Sokoto are usually sediment-laden. Similarly, the Challawa and Watari Rivers in the Hadejia–Jamaare basin have muddy appearance and were not chosen for a long time for water projects. However the Kano and Misau Rivers have excellent quality, with relatively low sediment loads. The Kano River has been dammed and more than 95 per cent of its flow is controlled by Bagauda and Tiga dams completed in 1970 and 1974 respectively. More dams have been designed for sites at Kafin Chiri, Tudun Wada, and Ruwan Kanya, all in the Kano watershed. However, some dams have been planned for the Challawa and Watari basins too. The Challawa Gorge Dam for instance is under construction, and will have a storage capacity of 969 million m^3.

3.9.4.2. Basin Degradation, Soil Loss, and Deterioration

The effect of gully and sheet erosion in producing material for transport has been discussed. Table 3.38 shows the magnitude of basin degradation or soil loss that may be expected to occur in the basin upstream of the specified stations. It ranges from less than 9 t km^{-2} in Misau and 19 t km^{-2} in Upper Niger Basin to 740 t km^{-2} in the Challawa watershed. Of course, erosion is not uniform over the entire basin, may be highly localized and is rather sporadic in time. The figures however give indications of the danger of soil loss and the changes that can occur in the moisture regime as a result of such lowering. There is certainly an urgent need for soil conservation measures in the Challawa, Bunsuru, and Watari watershed. The recent afforestation programmes launched for these Sahelian basins is therefore welcome as necessary though not a sufficient measure. For even where the whole soil is not washed away, much of its natural fertility may be lost already and much effort and expense would be needed to make such soils productive once again. The badly gullied lands of the Upper Challawa and Mamu River (Anambra basin) are good examples of badland topography produced by erosion. Not only is the land polluted and robbed of its aesthetic value, but often the numerous gully heads threaten to cut roads and other communication lines in a number of areas, such as around Enugu; and many thousands of hectares of land are badly scarred in these watersheds.

Rehabilitation of such badlands though expensive is a matter of necessity as productive land area is diminishing fast in such areas. By 1950 the Control Unit of the Soil Conservation had carried out work in the Anambra basin on some 134 gullies, built 805 check dams, some 40 km of contour ridges and 55 km of pathways which were protected against severe erosion with the aid of 4336 sumps to detain overland flow. Large areas were planted with trees and villagers were instructed on how to check erosion. During the first five years a sum of about N50,000 was spent on the restoration and preventive words (Udo, 1970).

3.9.4.3. Valley Aggradation and Reservoir Sedimentation

There are two sides to the problem of sediment transport. While soil loss occurs in the sediment producing areas, often undesirable accummulations are experienced in the alluvial valley below. The examples of the Hadejia and Yobe basins have been discussed. Also in the Cross River basin the lower valleys of the tributaries of the Aboine River, particularly the Nyaba reveals aggraded and braided streambed, with a confused pattern of sedimentation similar to the Lower Hadejia. In the latter example, sand dunes further complicate the drainage pattern, but the effect is more or less the same. The accummulations of sediment often prevents the tributary streams from joining the main river and small lakes have resulted from their damming. The Lower Hadejia is however fortunate to have accummulation of fertile silty soils which have proved useful environment for grazing and for floodplain (Fadama) farming, while sterile sands dominate the valleys accummulations in Mamu watershed.

When a dam is built across a sediment-laden stream the storage reservoir often traps much of the load. Most of the dams built in Nigeria are large in comparison to the river they control and store most of the streamflow reaching that section, so that upward of 90 per cent of the sediment load is retained over the years. Thus in planning a reservoir considerations must be given to the probable rate of sedimentation in order to determine whether the useful life of the proposed reservoir will be sufficient to warrant its construction. The percentage of the inflowing sediment which is retained in a reservoir is referred to as trap efficiency of the reservoir and is a function of the ratio of the reservoir capacity to total flow (Linsley and Franzini, 1972). Table 3.41 shows the rates of sedimentation expected if the rivers were dammed at the specified sections, assuming a trap efficiency of 95 per cent.

For the dam being constructed at Challawa Gorge (near Challawa) and in the Bunsuru at Dutsi-ma the sedimentation in 50 years represents a loss of 20 per cent of the total storage. Similarly a dam across the Benue near Yola with a capacity of about 1000 million m^3 would be reduced by 21 per cent of its capacity in a quarter of a century.

Table 3.41. Reservoir sedimentation data[a]

River	Station	Annual storage loss 10^6 m^3	50-year storage loss 10^6 m^3
Bunsuru	Zurmi	1.31	65.3
Rima	Wamako	2.77	138.4
Sokoto	Gidan Doka	2.50	124.8
Gagere	Kaura Namoda	0.84	41.8
Watari	Gwarzo	0.36	17.8
Challawa	Challawa	2.57	128.7
Kano	Chiromawa	0.77	38.7
Hadejia	Wudil	3.12	156.0
Jamaare	Bunga	1.85	92.5
Misau	Kari	0.02	1.2
Niger	Koji	10.51	525.5
Benue	Makurdi	6.69	334.4
Cross	Ikom	0.62	31.0

[a] Assume a sediment specific weight of 2000 kg or 2 t m^{-3}, 95 per cent Trap efficiency, and add 6 per cent bedload.
Source: Table 3.38.

3.9.5. CONCLUSION

We have observed that the accomplishment of some of the river basin management objectives in Nigeria may produce conflicts. Such conflicts can be profitably resolved through proper planning, coordinated, and scientific management of the basin as an organic whole, using adequate and relevant statistical data. Sediment transport and the erosion which yields the material represent important phenomena in basin planning and development activities in the country and must be accorded their proper place through emphatic research support.

A number of the world's rivers are large sediment transporters. The Colorado (USA), transports nearly 350 million tonnes of silt annually into Lake Mead; the King Ho (China) and Rio Puerco (USA) have sediment concentrations of 145 to 150 g Γ^1. Others such as the Ganges (India), Nile (Egypt), and the Amazon (Brazil) have average concentrations of 1 to 2 g Γ^1. On the other hand the average concentration of Nigeria's comparable rivers is only of the order of 0.3 g Γ^1. Nevertheless, we know that the problem of erosion and sedimentation has assumed great local significance in some areas and has caused much concern for the future of water and land resources development of such watersheds.

Had the Anambra basin soil conservation measures met with greater success, it might have become the nation's Tennessee Valley, considering the conservation efforts described in Section 3.9.4.2 above. However cultural and physical peculiarities as well as the excess population concentrations have discounted the results. Thus although so many measures were tried, it appears that they were neither properly coordinated nor seen as part of a comprehensive and integrated river basin management which transcends mere watershed protection.

There remains so much to be done therefore in the area of data collection and planned research on erosion and sediment transport in Nigeria, to enable us to understand better the processes and factors for proper scientific prediction and control of their effects. It is particularly important to obtain information on both phenomena for the post-construction periods of the water projects in the various basins to make possible the assessment of the influence of such controls on the sediment discharge regime of such river systems.

REFERENCES

Bagnold, R. A., 1954 Some flume experiments on large grains but little denser than the transporting fluids and their implications, *Inst. Civil Engnrs., Proc.,* pp. 174–205.

Bogardi, J. L., 1970, *Sediment Transportation in Alluvial Streams,* Research Institute for Water Resources Development, Budapest.

Chow, V. T., 1964, *Handbook of Applied Hydrology,* McGraw-Hill.

Colson, B. E., 1969, Surface-water resources of the Yobe River System, 1963–68, *Open File Report,* USGS.

Dorman, J. M., 1978, *Problems of Soil Erosion on the Jos Plateau Proceedings,* Nigerian Geographical Association Conference, Jos, pp. 16–75.

Floyd, B., 1965, Soil erosion and deterioration in Eastern Nigeria, *Niger. Geogr. J.,* 8(1), pp. 33–44.

Imevbore, A. M. A., 1975, The Chemistry of Lake Kainji Waters, August 1968 to September 1969, in, Imevbore, A. M. A., Adegoke, O. S. (eds)., *The Ecology of Lake Kainji,* pp. 82–102, Ile-Ife.

Jones, R. G. B., 1975, Central Nigeria Project Report on a Soil Conservation consultancy to Study Soil erosion problems on the Jos Plateau, *LRD Report No. 6.*

King, C. A. M., 1966, *Techniques in Geomorphology,* London, pp. 84–85.

Leopold, L. B., Wolman, M. G., and Miller, J. P., 1964, *Fluvial Geomorphology,* San Francisco, pp. 67–80.

Linsley, R, K., and Franzine, J. B., 1972, *Water Resources Engineering,* McGraw-Hill.

Motor Columbus Consulting Engineers and G. F. Appio and Associates, 1978, *Feasibility Report on Lokoja, Makurdi, and Ikom HEP Projects—Appendix 8: Suspended Sediment Load Measurements.*

MRT Consulting Engineers Ltd., 1978, *Water Supply Master Plan,* Kaduna State 6, Appendices.

NEDECCO, 1959, *River Studies—Niger and Benue,* North Holland Publ. Co., Amsterdam.

Ofomata, G. E. K., 1965, Factors of soil erosion in the Enugu area of Nigeria, *Niger Geogr. J.,* **8(1),** pp; 45–59.

Thorne, W., 1963, The land and water use, in Thorne, W., *Land and Water Use,* Washington D.C., p. 15.

Wischmeier, W. H., and Smith, D. D., 1958, Rainfall energy and its relationship to soil, *Am. Geophys. Union,* Trans. **39,** pp. 285–291.

Udo, R. K., 1970, *Geographical Regions of Nigeria,* Heinemann, London.

Tropical Agricultural Hydrology
Edited by R. Lal and E. W. Russell
© 1981, John Wiley & Sons Ltd.

3.10

Soil and Vegetation Development on Fresh Landslide Scars in the Mgeta Valley, Western Uluguru Mountains, Tanzania

L. LUNDGREN

3.10.1. INTRODUCTION

The present report on soil and vegetation development of fresh landslides in the Mgeta area, western Uluguru mountains, Tanzania, is a follow-up study of one of the DUSER-project (Dar-es-Salaam–Uppsala Universities Soil Erosion Research) studies carried out in Tanzania during the period 1968–1972. The results of the DUSER-project studies were published in a double issue of *Geografiska Annaler* (1972), **54 A (3-4)**, including fifteen papers, e.g. Temple (1972) and Temple and Rapp (1972).

3.10.2 THE STUDY AREA

The Uluguru mountains are situated approximately 200 km E of Dar-es-Salaam. Morogoro, administrative centre of Morogoro Region, lies on the northern foot-hills of the mountain massif. The mountains are of major importance as a water-catchment area, and important perennial rivers rise in the mountains. Dar-es-Salaam and Morogoro, as well as several industries and agricultural areas, are dependent on the water of these rivers. The mountain area has for a long time been affected by severe erosion, which influences the quality and the flow of the water in the rivers, as well as the crop yields in the mountains.

The rocks of the mountains are predominantly Precambrian. In the study area meta-anorthosites with gabbroic anorthosites dominate—the most important mineral being plagioclase feldspar with $>$ 50 per cent anorthite (Sampson and Wright, 1964). The normal soil to be expected on this rock type and under prevailing climatic conditions—rainfall 1000–2000 mm yr^{-1}—is deep yellow-red kaolinitic clay, which is also found in a very small area under undisturbed forest cover (Sampson, 1954). However, the deep soil has been washed off the steep slopes—mostly more than 20°—in the Mgeta Valley area, and the existing soil is shallow yellowish-whitish in colour and coarse textured—sandy-loams dominate.

The natural vegetation in the Mgeta Valley is submontane woodland at lower altitudes (below 1200 m alt.) and montane evergreen forest at higher altitudes (Pócs, 1976). Due to a very intensive small-scale agriculture, there is, however, almost no natural vegetation left, except in the Forest Reserve, which dates back to the time before the First World War (cf. Plate 11).

The cultivation of the mountains started about 200 years ago, when the Lugurus—originally plains people—cleared and cultivated the slopes until the soils became impoverished and then abandoned it. It is unlikely that any conservation measures whatsoever were followed in this initial phase of exploitation of a virgin environment (Temple, 1972). Reports on severe erosion problems in the mountains date from the beginning of this century.

During the German and British colonial administration period (1891-1961), several conservation schemes were carried out which tried to improve the situation (e.g. Bagshawe, 1930; Young and Fosbrooke, 1960 and Grant, 1965). On the whole these schemes failed, since 'they often were unsoundly based and unwisely implemented' (Temple, 1972). During the 1960s and early 1970s, erosion problems increased and serious landslides were recorded in 1968, 1969, 1970, and 1973 (Temple, 1972; Temple and Rapp, 1972; and Lundgren and Rapp, 1974). It has been suggested by others, e.g. Sternberg (1949) and Tricart and Cailleux (1965) that landsliding is the terminal phase of man-induced accelerated erosion.

3.10.3. BACKGROUND AND AIMS OF THE STUDY

On 23rd February 1970 a large landslide catastrophe took place in the Mgeta Valley after intensive rainfall—100 mm in less than 3 hours. The area was surveyed by Temple and Rapp (1972), who found that a total area of approximately 75 km^2 was affected by no less than 1000 landslides, most of which can be regarded as small debris slides which quickly turned into mudflows and joined the streams at the bottom of the slopes. Temple and Rapp (1972) also examined the economic and geomorphologic effects of the catastrophe. To follow up these studies, I have studied the soil and vegetation development of twelve of the slides (L1-L12) during the period 1970-1977. Two additional slides (L13-L14) were included in the study in April 1973, when roughly 100 new slides were triggered off in the same area.

The general aims of the present follow-up study were to investigate and, where possible, quantify

—continued erosion in the areas affected by the landslides
—topsoil development in the landslide scars
—recolonization of vegetation in the landslide scars
—land reclamation measures by the local farmers.

Plate 11 Photo of part of the Mgeta Valley taken in January 1977, i.e. seven years after the main landslide catastrophe. The slide in the lower right-hand corner is one of the investigated slides. In the lower left-hand corner old slide edges (i.e. pre-1970) are clearly visible. Several typical features of the land use are seen—fallow land, grazing land and annual crops (with ladder terraces) on steep slopes, the sharp boundary of the forest reserve (on top of the background), groups of eucalyptus planted around houses for fuel and as erosion control. (Photo Lill Lundgren, January 1977)

Facing page 228

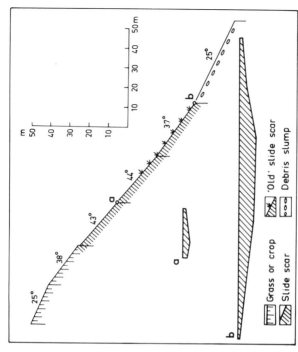

Plate 12 Photo and long and cross profiles of slide L12. The vegetation-covered central part of the slide coincides with the original slide from Feb. 1970, while the bare surfaces represent retrogressive slumping areas from April 1973. Note rill erosion in the exposed soil. (Photo Lill Lundgren, October 1973)

3.10.4. RESULTS

3.10.4.1. Types of Landslides and Mechanism of Slope Failure

According to Varne's terminology (1958, p. 29), the landslides from 1970 are debris slumps of which the majority continued as mudflows. 'The typical slide scar is 5–10 m wide, 10–50 m long, up to 2 m deep, wide near the upper end, narrowing in a wedge-like fashion downslope. Basically they are debris slides which due to their small thickness are called sheet-slides.' (Temple and Rapp, 1972). The same authors also establish: 'Most scar angles measured were within the interval 33.5° to 44°, which indicate critical angles for triggering sheet-slides . . . 60 per cent of the recorded slides had occurred on straight valley sides well below ridge crests. . .Many slides were close to stream sides on the lower steepening of slopes resulting from stream incision. None of the examples studied had resulted from flood erosion, however.' To judge from morphological evidence, e.g. dimensions and angles as well as soil characteristics, the slides of 1973 are of the same type as the ones from 1970, i.e. debris slides of the sheet-slide type.

The mechanism of slide initiation in 1970 was probably high pore water pressure built up in subsurface pore spaces and pipes due to heavy rainfall (Temple and Rapp, 1972). The mechanism is likely to have been the same in 1973. It has, however, not been possible to get information on the exact day of slope failures. According to the local people it occurred during heavy rainfall. Reports from a landslide catastrophe at Palu village, Morogoro River Catchment, in April, 1973, due to heavy rainfall (Lundgren and Rapp, 1974) support this theory. The Palu landslide was a bottle-slide, i.e. many metres deep and large—300 m long and 130 m wide in the upper part but 35 m wide in the lower part. It was triggered off after a week of heavy rainfall (about 50 mm day^{-1}). Although the type of landslide is different from the Mgeta type, it still indicates that instability prevailed in the mountains at the time in question.

3.10.4.2. Continued Erosion

Signs of *sheet erosion* were noted on all slides. Soil pedestals and shoestring washlines are evidence of this. Soil pedestals, that were used to measure minimum depth of sheet erosion, measured up to 12 cm four years after sliding.

Rill erosion occurred in all investigated slides seven years after sliding, except in those which were cultivated at this time. The rills converge downslope on the longer slides into deeper rills, or gullies.

Gully formation took place in four of the fourteen slides investigated. A gully up to 4 m deep was measured after seven years. In these four slides, gully erosion is so bad that reclamation activities will be very difficult.

Retrogressive slumping had occurred on six slides seven years after sliding. One slide triggered off in 1970, was enlarged to more than double its size in 1973 owing to slumping (Plate 12).

3.10.4.3. Topsoil Development

The surface soil layer of the landslides, which has been analysed for organic carbon content, bulk density, pH, texture, and plant available and reserve phosphorus and potassium contents, shows a weak development during that period. Changes can be seen for texture (clay content), organic carbon, and for available potassium (Figure 3.16). In all slides investigated 0.5, and 7 years after sliding the clay content has decreased except in L11. The mean value, calculated from all slides (except L11), has decreased from 13.4 to 7.7 per cent. This can be explained by the high erosion rate in the exposed scars. There is also a possible leaching of clay particles to deeper layers. An explanation of the increase of clay content in L11 and the small decrease in L4 (2 per cent) can possibly be the recultivation of these slides. The organic carbon content has increased in all slides (mean value changes from 0.16 to 0.44 per cent) and most obviously in L4 and L11, and are most likely due to cultivation. There is a tendency of increase of available potassium. All slides except L2, where no change has occurred, show an increase. Mean change is from 3.1 to 5.5 mg (100 g)$^{-1}$. Development of bulk density, pH, available phosphorus, reserve phosphorus and reserve potassium show no clear trends over the period. Compared to the corresponding topsoil values it is obvious that big differences still exist, e.g. in organic carbon (mean 0.4 per cent seven years after sliding compared to 1.8 per cent in the corresponding topsoil), bulk density (1.43 g cm^{-3} compared to 1.18 g cm^{-3}). and clay content (7.7 per cent compared to 14.5 per cent).

3.10.4.4. Recolonization by Vegetation

Recolonization by vegetation has been investigated in 2 X 2 m sample plots in the centre of the slide scars. The vegetation shows a clear succession from annual herbs to perennial grasses, though the total vegetation cover on average is still small, being approximately 25 per cent after seven years (Figure 3.17). Recolonization has mainly occurred through seed germination. Vegetation growth from 'floes' and slide edges seem to have played an unimportant role. Only species found in the neighbourhood, i.e. weeds and species from grazing land, have been found in the slides. Water availability has most likely been the major factor restricting vegetation establishment, since the coarse textured soils of the landslides readily dry out. Establishment of lichens on the fresh landslide scars was first noted four years after sliding. After seven years about 75 per cent of the slide scars were covered with lichens.

3.10.4.5. Reclamation by the Local Farmers

The most common practice was to leave the slides and the areas around the slides for grazing. On four of the fourteen slides investigated, tree seedlings, were planted.

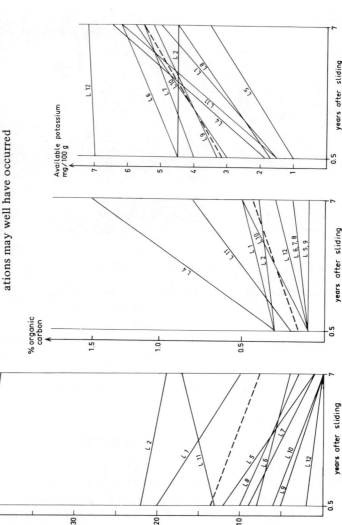

Figure 3.16 Changes in surface (0–10 cm) soil contents of clay (per cent), organic carbon (per cent), and plant available potassium (mg/100g) in the investigated land-slides. There is a general trend of increase in carbon and available potassium content and decrease in clay content from 0.5 to 7 years after sliding. Mean values are indi-cated with a broken line. (L 11 is excluded from mean clay content). Observe that the parameters were only measured 0.5 and 7 years after sliding. The straight lines therefore give an impression of a uniform rate of development although fluctu-ations may well have occurred

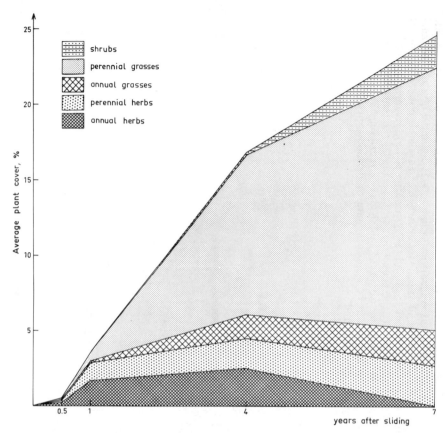

Figure 3.17 Average plant cover in per cent of different groups of plants over age in the 2 X 2 sample plots of the investigated slides. The straight lines of the curves between the investigation times 0.5, 1, 4, and 7 years after sliding give an impression of a uniform development, although fluctuations may well have occurred

On two of them the seedlings (*Cupressus lusitanica*) died after one to two years, most likely due to desiccation. The small trees (*Eucalyptus maidenii* and *Acacia mearnsii*) on the other two slides are threatened by gully erosion. On two of the investigated slides recultivation with ladder terraces was recorded—on one (L11) after 4 years and on one (L4) after 7 years.

3.10.5. DISCUSSION

Continued erosion, vegetation recolonization and topsoil formation in the landslides are not only closely interdependent processes, but their speed and direction in each individual case will depend on the specific interaction of land use and environmental factors. Bearing in mind this complexity of factors, the following discussion of the results under separate headings may give the impression of oversimplifying the ecological relations influencing the landslides. Still, for reasons of clarity, the same subdivision as used in the presentation of the results has been retained in the discussion, though the interaction of different factors is more firmly emphasized.

3.10.5.1. Continued Erosion

Sheet, rill, and gully erosion on the bare surface of the landslide scars, and retrogressive slumping of the slide edges, have been of such magnitude that the amounts of material eroded from the slide in some cases are as much as that removed in the initial landslide and in some even more. The relative importance of the different forms of erosion varies from slide to slide.

The protection against erosion, particularly the initial forms—splash erosion and raindrop impact—that may result from lichen development, which occurrred in many slides after 3–4 years, probably speeds up soil recovery and vegetation recolonization. The terraces prepared in L4 and L11 (when recultivation started) provide a protection against sheet erosion. In L11 the clay content has increased and in L4 the decrease is very small (cf. Figure 3.16). This can most likely be explained by limited sheet erosion. As Temple suggests (1972), the terraces are, however, perhaps the most serious erosional threat to cultivation on steep slopes in the Ulugurus because they increase the infiltration rate and thus speed up the building up of pore water pressure. Drainage, which increases the internal friction by lowering the water pressure is a conservation measure generally recommended, but as Haldeman (1956), from his studies in the Rungwe mountains, Tanzania, and Temple (1972) conclude, it is impracticable for several reasons, including financial considerations.

The pattern and degree of the continued erosion of the slides and of the affected area as a whole have, of course, been dependent on the rainfall pattern during the seven years this investigation has continued. It must be emphasized that the development of erosion on the slide scars would probably have been different if, for example, the intensities of the rains had been different, or if periods of drought had been different from normal conditions, etc. The differences in rainfall characteristics influence erosion not only directly, but also indirectly, since they influence vegetation establishment and topsoil development.

3.10.5.2. Topsoil Development

Only very weak topsoil development can be seen after seven years. The percentage of organic carbon has for example increased from 0.16 to 0.44 per cent (mean values), but with big differences between individual slides (cf. Figure 3.16). Bulk density development shows no clear trends. Mean clay content of the surface soils has decreased from 13.4 per cent to 7.7 per cent, mostly due to the high sheet erosion rate, but probably also as a result of leaching of clay particles. There are still

big differences between the surface soil in the slide scars and the corresponding top-
soils in the surrounding areas. Mean organic carbon content in the scars is for example 0.4 per cent, whereas it is 1.8 per cent in the corresponding topsoil. Bulk density is $1.43 \, g \, cm^{-3}$ compared to $1.18 \, g \, cm^{-3}$. and clay content 7.7 per cent compared
to 14.5 per cent. It thus seems to take a long time for the surface soil layer of the
landslides to develop features comparable to the surrounding topsoil, a development that is delayed by the high erosion rate of the bare soils.

In L4 and 11 where cultivation started about four years after sliding, the organic
carbon content increased from 0.3 per cent to 1.5 per cent in L4 and from 0.2 per
cent to 0.8 per cent in L11, as a result of mixing up of mulch with the mineral soil,
which is done when hoeing the terraces.

3.10.5.3. Recolonization of Vegetation

The plants of the slide scars have above all recolonized through seeds from the
surrounding area that have germinated in the scar. The microclimate on the bare
scar is comparable to that of cultivated land, i.e. a bare soil surface with high insolation, resulting above all in wide ranges of temperature and humidity. Species adapted to these environmental factors are the ones that germinated first on the slides.
The sample plots were initially dominated by annual herbs, but perennial herbs and
grasses—both annual and perennial—took over after one to two years. After about
four years the perennial grasses dominated and shrubs started to appear (cf. Figure
3.17). This succession of communities is similar, for example, to that one observed
by Olsson (1961) from recolonization of landslides in Sweden, although this is in a
temperate climate.

The average cover of plants (lichens excuded), about 25 per cent after seven
years, is less than that found in corresponding studies from Scandinavia, e.g. Olsson
(1961) and Resvoll (1903), who both found a denser vegetation cover after a comparable period. This is probably due to much better availability of water and to less
active continued erosion in the Scandinavian slides. On the whole, it is difficult to
compare different studies on recolonization speed, since the climatic conditions
play such an important role. In this study the years 1973 and 1974 were dry with
rainfalls of 841 mm and 358 mm compared with the mean rainfall of 1080 mm in
Mgeta Mizugu. These dry years which caused disastrous dought conditions over
great areas of Africa, most likely contributed to a slower recolonization than if
average rainfall conditions had prevailed. Drought is particularly detrimental to
plant survival on coarse-textured soils, since they easily dry out. Furthermore, the
dry season climate, with four dry months (on average) in the Mgeta area, makes
microclimatic conditions on the slide scars (dry and hot) quite unfavourable for
small seedlings to survive.

3.10.5.4. Land Reclamation

The slide catastrophe in 1970 occurred over large areas and a large number of
families were directly affected. In spite of the seriousness and economic implica-

tions of the event, only very general recommendations on suitable land reclamation measures were issued to those local farmers who asked for advice. The only recommendations offered were for setting aside for grazing, areas surrounding the slides, and for sowing elephant grass (*Pennisetum pupureum*) or for planting trees in the scar itself. Tree seedlings were given free of charge to the farmers. Setting aside the areas for grazing and planting trees seems to have been the only general activities, as observed in this investigation. But land left for grazing is burnt once a year, which leads to a decrease in the organic matter content of the soil and a consequent degradation of the topsoil structure. When the grassland surrounding the slide is burnt, even the sparse vegetation on the slides is sometimes burnt. Sowing elephant grass (*Pennisetum purpureum*) does not seem to have been practised. Tree planting on the other hand has been relatively common probably because trees provide a more useful crop than elephant grass. The tree species used—*Acacia mearnsii, Cupressus lusitanica,* and *Eucalyptus maidenii*—are the ones used by the Forest Department for timber and fuel production and are thus the only species available in the area. These species do not have the most desirable qualitites for erosion control. Neither are they adapted to survive on the very unfavourable slide scars. All *Cupressus lusitanica* seedlings planted died. *Acacia mearnsii* and *Eucalyptus maidenii* are better adapted and have a higher survival rate, but their continued survival is threatened by gully erosion. In my opinion, when trees are planted on the slide scars, species well adapted to the environment and with a root system and general appearance (e.g. more bush-like) suitable for erosion control must be used. Furthermore, the bare soil must be protected against continued erosion, e.g. by putting mulch on the slides. The mulch would also give the soil better moisture condition and provide organic matter to the mineral soil of the slides.

Recultivation and terracing of the slides had been carried out on L4 and 11 after about four years. Many other slides not investigated were also cultivated after this period. In the case of L11, only the slide scar was cultivated, a pattern that may seem very odd. It is not restricted to this slide only, but has been observed on several other slides. The most probable explanation is that the farmer found it easier to work the exposed soil compared with the grass-covered surroundings. In L4, both the slide scar and the surrounding area were treated in the same way.

3.10.6. CONCLUSIONS

Based on the results and discussion above, the following conclusions can be drawn after observing the slides over a seven-year period:

— The continued erosion has been very severe, including sheet, rill, gully, and slumping erosion.
— The development of topsoil characteristics has been rather slow during the seven years of study.
— Recolonization of vegetation shows a clear succession from annual herbs to perennial grasses and shrubs, though the total plant coverage is still small after seven years.

- Land reclamation measures have been haphazard and generally not effective.
- Continued erosion, topsoil development, recolonization of vegetation and land reclamation measures show different patterns in the individual slides.
- Successful land reclamation measures must aim at preventing continued erosion so that vegetation may become established and well-structured topsoil may develop.
- Finally, the decisive role of climate should be emphasized.

If, for example, periods of drought or rainfall intensities had differed from what they actually were during the seven years of study, erosion topsoil development and vegetation establishment patterns would probably have been different from those actually observed.

REFERENCES

Bagshawe, F. J., 1930, A report by the Land Development Commissioner on the Uluguru hills, *Land Development Survey Rept. 3.*, Tanzania National Archives; 61/378/4.

Grant, H. St. J., 1965, Uluguru land usage scheme; annual report for 1955, *Unpubl. rept.*, Tanzania National Archives:61/D/3/9.

Haldeman, E. G., 1956, Recent landslide phenomena in the Rungwe volcanic area, Tanganyika, *Tanganyika Notes Rec.*, **45**, 1-14.

Lundgren, L., and Rapp, A., 1974, A complex landslide with destructive effects on the water supply of Morogoro town, Tanzania, *Geogr. Ann.* **56A(3-4)**, 251-260.

Olsson, O., 1961, En vegetationsstudie för åldersbedömning av västsvenska jordskred, *Gothia*, **9**, 109-127.

Pócs, T., 1976, Vegetation mapping in the Uluguru mountains (Tanzania East Africa), *Boissiera*, **24**, 477-498.

Resvoll, T. R., 1903, Den nye Vegetation paa Lerfaldet i Vaerdalen, *Nyt. Mag. Naturv.*, **41(4)**, 369-396.

Sampson, D. N., 1954, printed 1956, The forest cap clays of the western Uluguru mountains, *Rec. geol. Surv., Tanganyika*, **4**, 64-69.

Sampson, D. N., and Wright, A. E. 1964, The geology of the Uluguru mountains, *Bull. geol. Surv. Tanzania*, **37**, 1-67.

Sternberg, H. O'Reilly, 1949, printed 1951, Floods and landslides in the Paraiba Valley, December 1948, Influence of destructive exploitation of the land, *Int. geogr. Congr.*, **3**, 335-364.

Temple, P. H., 1972, Soil and water conservation policies in the Uluguru mountains, Tanzania, *Geogr. Ann.*, **54A(3-4)**, 110-123.

Temple, P. H., and Rapp, A., 1972, Landslides in the Mgeta area, western Uluguru mountains; geomorphological effects of sudden heavy rainfall, *Geogr. Ann.*, **54A(3-4)**, 157-193.

Tricart, J., and Cailleux, A., 1965, *Introduction a la géomorphologie climatique*, SEDES, Paris, 306pp.

Varne, D. J., 1958, Landslide types and processes, In *Landslides and Engineering Practice* (Ed. E. B. Eckel), Highway Res. Board, USA, Spec. Rep., 29, N.A.S.-N.R.C. Pub., 544, 20-47.

Young, R., and Fosbrooke, H., 1960, *Land and Politics among the Luguru of Tanganyika*, Routledge and Kegan Paul, London, 212pp.

PART 4

Management and Catchment Hydrology

Tropical Agricultural Hydrology
Edited by R. Lal and E. W. Russell
© 1981, John Wiley & Sons Ltd.

4.1

Watershed Management as a Basis for Land Development and Management in India

K. G. Tejwani

4.1.1. INTRODUCTION

Watershed management implies rational utilization of land and water resources for optimum and sustained production with the minimum of hazard to natural resources. It essentially relates to soil and water conservation in the watershed which means proper land use and the protection of land against all forms of deterioration building and; it also implies maintaining soil fertility, conserving water for farm use proper management of local water for drainage, flood protection and sediment reduction, and the increase of productivity from all land uses.

The problem of soil and water conservation in India is rather alarming. Recent surveys have shown that 58 per cent of the agricultural land (80,000,000 ha), 33 per cent of the forest land (20,000,000 ha), 86 per cent of the culturable waste land (15,000,000 ha), 95 per cent of permanent pastures and other grazing lands (14,000,000 ha), 75 per cent of the fallow lands (15,000,000 ha), 24 per cent of land under trees and groves (1,000,000 ha) and 44 per cent of the geographical area of India (145,000,000 ha) are subject to severe erosion (Anonymous, 1968). The degradation of the forest and pasture lands in river catchments, including 2,300,000 to 4,000,000 ha of gullied lands contribute to the disastrous floods and sediment yield that pose the greatest threats to the well-being and economy of India.

For example a survey of 21 reservoirs has indicated that they receive sediment at the rate of 8.51 ha m $(100 \text{ km}^2 \text{ yr})^{-1}$ as against the designed inflow of 3.02 ha m $(100 \text{ km}^2 \text{ yr})^{-1}$. This represents 182 per cent more inflow of sediment than the designed inflow (Gupta, 1975).

4.1.2. CONSERVING LAND AND WATER AND INCREASING PRODUCTION ON AGRICULTURAL LANDS

4.1.2.1. Contour Cropping

It has been experimentally shown at Dehra Dun that contour cultivation reduces runoff and prevents soil erosion when compared with up and down cultivation

Table 4.1. Effect of contour cultivation on runoff and soil loss. Average of 4 years for Dhoolkot silty clay loam with 8% slope at Dehra Dun

	Up and down cultivation	Contour cultivation
Runoff (mm)	670	511
Runoff as percentage of rainfall	54.1	41.2
Soil loss (t ha^{-1})	28.5	19.3
Rainfall (mm)	1239 Crop—maize	

(Tejwani, Gupta, and Mathur, 1975). Contour farming is easy, simple, and economical.

Experiments at Ootacumund have shown that by adopting contour farming for potato, cultivated on 25 per cent slope, runoff was reduced from 52 to 29 mm and soil loss from 39.0 to 14.9 t ha^{-1} (Raghunath *et al.*, 1967). Apart from conserving the water and soil, contour cultivation conserves soil fertility and increases crop yields. Contour cultivation on alluvial soil at 2.2 per cent slope at Kanpur conserved 11.2 kg N, 10.2 kg P_2O_5, 44.8 kg K_2O, 557.4 kg CaO, 109.7 kg MgO, and 74.0 kg ha^{-1} humus in one season alone. Contour cultivation also produced 440 and 4176 kg of sorghum (*Sorghum vulgare*) grain and stover respectively more than up and down cultivation. Thus every centimetre of rainwater conserved produced 44.8 and 422 kg of grain and stover respectively more than up and down cultivation (Bhatia and Chaudhary, 1977). Dehra Dun and Kanpur have an average annual rainfall of 1674 and 805 mm respectively, 80 to 85 per cent of which is received in three months during south-west monsoon. It would thus appear that contour cultivation is effective both in low and high rainfall areas. However, in low rainfall areas contour cultivation not only conserves soil and moisture but the conserved soil moisture contributes to increased crop production. In high rainfall area the major gain is reduction of soil erosion while conserved moisture may not benefit the crop as soil moisture is not a limiting factor for crop yields.

4.1.2.2. Mechanical Measures of Erosion Control on Agricultural Lands

Mechanical measures, which include contour *bunding*, grading *bunding*, and bench terracing on steep slopes are adopted to supplement the agronomical practices when the latter alone are not adequately effective.

At Dehra Dun, it has been observed that as a result of field *bunding* of an agricultural watershed (54.63 ha area), there was a 62 per cent reduction in runoff amount and 40 per cent reduction in the peak runoff rate (Rambabu *et al.*, 1974).

At Dehra Dun a *bunded* agricultural watershed of 12 ha had less runoff and peak discharge than another *unbunded* watershed of 2.94 ha (Bansal and Husenappa, 1977).

Table 4.2. Monthly rainfall, runoff, and peak discharge of agricultural watersheds at Dehra Dun

Month, 1976	Rainfall (mm)	(mm)	Runoff (Percentage of rainfall)	Peak discharge (mm hr^{-1})
	Bunded sub-catchment—(12.08 ha)			
July	527.5	90.1	17.1	39.7
August	606.5	134.1	22.1	26.2
September	131.5	1.6	1.2	1.0
	Unbunded sub-catchment—(2.94 ha)			
July	527.5	144.4	27.4	47.5
August	606.5	271.4	44.8	53.1
September	131.5	20.5	15.6	23.1

Table 4.3. Effect of *bunding* on runoff, peak discharge, soil loss, and crop yield at Chandigarh. (Average of 6 years : 1972–1977)

	Unbunded	*Bunded*
Rainfall (mm)	764.3	764.3
Runoff (mm)	169.2	107.4
Runoff (%)	22.2	14.0
Annual peak discharge (m^3 ha^{-1})	0.206	0.120
Soil loss (t ha^{-1})	3.51	0.97
Maize yield (kg ha^{-1})^a	1541	1699

^{a} Average of 5 years after excluding the yield of 1974

Similar results have been reported from Chandigarh, where *bunding* of agricultural land reduced runoff, peak discharge and soil loss, and on an average conserved 62 mm of rainfall and 2.34 t ha^{-1} yr^{-1} of soil (Sud *et al.*, 1977) (Table 4.3).

Bunded area also gave 10.1 per cent more crop yield than *unbunded* area.

At Agra, it was observed that field *bunding* of an agricultural watershed (22.3 ha) reduced the runoff by 45 per cent. In addition to the soil erosion control, the field *bunding* due to moisture conservation is reported to have increased grain and stover yields of pearl millet by 1.0 and 5.0 q ha^{-1} respectively (Anonymous, 1971; 1978).

On steeply sloping and undulating lands, intensive farming can be practised only with bench terracing. In rainfed areas bench terracing is recommended on slopes from 6 to 33 per cent. It is recommended for slopes of less than 6 per cent for irrigated agriculture.

Investigations conducted in South India by Das *et al.* (1970) indicated that different lengths of bench terraces had no significant effect on runoff, soil loss, and potato yield. However terraces longer than 100 m had less soil moisture reserve than short length terraces. The runoff and soil loss on a 25 per cent slope increased with an increase in the longitudinal terrace gradient from 0.15 per cent to 0.84 per cent. However, maximum runoff with 0.84 per cent gradient was only 0.9 per cent of the rainfall and soil loss was only 0.3 t ha^{-1}, both of which are negligible. The differences in potato yield from bench terraces of various grades were also not significant.

4.1.2.2.1. Peak Rate of Runoff

If appropriate measures are adopted for safe disposal of water runoff, its storage and recycling provide an opportunity for supplemental irrigation. The prediction of peak runoff rate and amount for design purposes for Ambala Siwaliks have been facilitated by the use of the modified rational formula: $q = CIA^{0.73}$ (Erasmus and Bansal, 1965).

where q = runoff, ft^3 sec^{-1}

C = runoff coefficient

I = rainfall intensity for duration equal to time of concentration (T_c), in hr^{-1}

A = Area of catchment, acres

The value of 'C' for different conditions in India are shown in Table 4.4.

Peak runoff rates are also estimated by the use of nomograms prepared on the basis of: (i) rational formula (ii) Cook's method, and (iii) Hydrologic soil cover complex method, for small watersheds up to 300 ha in area (Gupta *et al.*, 1970). A comparison of experimental and predicted peak runoff rates has indicated that the prediction based on the hydrological soil cover complex method was closely associated with the experimentally measured data (Gupta *et al.*, 1969).

Runoff volume: Information on runoff volume is necessary for estimation of storage capacity and for designing structures for disposal of surplus water. Wasi Ullah and Rambabu (1970), established an empirical relation between peak runoff rate and the expected volume of runoff:

$$Q = 0.14A^{0.33} q$$

where Q = Volume of runoff, ha m

A = Area of watershed, ha

q = Peak discharge, m^3 sec^{-1}

Based on the hydrologic soil cover complex method, weekly yield of runoff for 15 stations located in different agro-ecological regions of India were computed for different levels of probabilities (Gupta *et al.*, 1971). An equation and nomograph have been devloped for estimation of annual runoff (cm) for Nilgiris hills in South India for watersheds of 1 to 400 km^2, different shapes, drainage densities, relief and temperature (Das *et al.*, 1971).

$$Q = \frac{1.911\,(P)^{1.44}}{T_m^{1.34}\,A^{0.0613}}$$

Table 4.4. Value of C for Rational Formula

	Land use		
	Cultivated	Pasture	Forest
General (Gupta *et al.*, 1970)			
With above average infiltration rates usually sand or gravel	0.29	0.15	0.10
With average infiltration rates. No clay pans loam and similar soils	0.40	0.35	0.30
With below average infiltration rates; heavy clay soils or soils with clay pan near the surface; shallow soils above impervious rock	0.50	0.45	0.40
Specific for Doon Valley (Gupta *et al.*, 1969) Agricultural small watersheds with benches	0.18 to 0.25 Average 0.22	–	–
Forest watersheds	–	–	0.22 to 0.28 Average 0.25
Chandigarh (Erasmus and Bansal, 1965)			
Unculturable hilly lands	0.70		

where Q = Annual runoff, cm
T_m = Mean annual temperature, $°C$
A = Catchment area, km^2
P = Annual rainfall, cm

Runoff storage: The losses due to seepage and evaporation from the storage tanks should be minimized. Seepage losses are generally more in alluvial soils. Annual losses due to seepage and evaporation have been estimated to be 1.8 to 4.0 m of water for Dehra Dun and Chandigarh regions of northern India (Agnihotri *et al.*, 1971; Erasmus *et al.*, 1971; Bansal and Husenappa, 1977). Seepage losses may be slight from dugout ponds in vertisols. Simple and economical techniques for storage of water in farm ponds are not yet available. With high demand for supplemental irrigation, high investments for lining the ponds with bricks or concrete may be justified. For example, Sastri *et al.* (1975) observed in the semi-arid regions of Bellary that supplementary application of 5 cm of harvested water increased the sorghum grain yield from 22 to 29 t ha^{-1} and stover yield from 52 to 57 t ha^{-1}. Similar investigations for wheat in Dehra Dun showed 57 per cent increase in grain yield (Singh and Bhushan, 1977).

Table 4.5. Yield (kg ha^{-1}) of wheat as influenced by supplemental irrigation at Dehra Dun (Singh and Bhushan, 1977)

Treatment	1974–1975		1975–1976		1976–1977		Average	
	Grain	Straw	Grain	Straw	Grain	Straw	Grain	Straw
Supplemental irrigation								
Control	3776	7140	1249	2925	1110	5366	2045	5144
Presowing irrigation (5 cm)	4540	7061	2088	4889	2981	5741	3203	5897
Irrigation at crown root initiation (5 cm)	4671	7140	1999	4659	3039	3498	3236	5099
Presowing irrigation+ irrigation at crown root initiation (10 cm)	4660	7240	2955	6940	4078	8535	3898	7572
SE_m (±)	146	566	163	379	192	199		
CD at 5%	439	–	564	1309	863	688		

4.1.3. CONSERVING LAND AND WATER AND INCREASING PRODUCTION FROM LANDS NOT SUITED FOR AGRICULTURE

Land capability classes V, VI, VII, and VIII have limitations of slope, erosion, stonniness, rockiness, shallow soils, wetness, flooding, climate, etc., which make them generally unsuited for agricultural crops and limit their use largely to pastures, forests, wild life, and cover. The importance of vegetation cover on soil and water loss is evident from the data presented in Table 4.6.

Trees play a multifarious role in effective utilization of precipitation. By intercepting rainfall trees reduce the raindrop impact by dissipating the energy. Tejwani *et al.* (1975) reported that as much as 14 to 20 per cent rainfall may be intercepted by *Acacia spp.* and *Shorea robusta*. Depending on the canopy structure some species may intercept even more (Table 4.7). For example, *Acacia nilotica* can intercept about 60 per cent more rainfall than *Dalbergia sissoo*. The interception increases with the increase in canopy cover and with age (Singh and Prajapti, 1974).

4.1.3.1. Closure of Areas to Biotic Influences

Regrowth of natural flora on eroded lands constitute hardy annual and unpalatable grasses rather than the desirable climax associations that normally exist in the prevailing soil-climatic environments. For example, in Gujarat ravines *Aristida spp.* and *Themeda triandra* colonize the eroded lands in place of the desirable *Dichanthium cenchrus* association (Tejwani *et al.*, 1961). Investigations on ravine lands of Vasad and Kota indicated that protection against grazing resulted in a gradual replace-

Table 4.6. Effect of vegetative cover on runoff and soil loss (Tejwani *et al.*, 1975)

| | Centre | | |
	Dehra Dun	Agra	Kota
Soil	Inceptisol	Inceptisol	Vertisol
Slope (per cent)	8	2	1
Runoff (% of rainfall)			
Cultivated fallow	35.9	NO[a]	17.3
Grass cover	2.5	17.3	10.7
Arable rowcrop (maize/ sorghum)	41.2[b]	18.7[c]	15.7[c]
Soil loss (t ha^{-1})			
Cultivated fallow	44.0	NO[a]	3.7
Grass cover	1.3	2.1	0.4
Arable rowcrop (maize/ sorghum)	19.3[b]	3.1[c]	2.4[c]

[a]NO = Not observed [b] maize [c] sorghum

Table 4.7. Effect of tree species and their age on interception of rainfall. These observations were made at Agra, in northern India (Singh and Prajapati, 1974)

| | Rainfall Intercepted (%) | |
Year	*Acacia nilotica*	*Dalbergia sissoo*
1969	26.4	11.3
1970	12.8	5.1
1972	17.4	12.9
1973	20.4	17.1
1974	28.1	19.6
Average	21.0	13.2

ment of *Aristida funiculata* and *Themeda triandra* by *Apluda mutica, Erempogon faveolatus, Heteropogon contortus, Dicanthium annulatum*, and *Cenchrus spp*. As a result, the runoff and soil loss progressively decreased with the improvement of the natural vegetation and the yield of desirable grass species increased (Tejwani *et al.*, 1961; Singh, 1971, 1972; Singh and Verma, 1971). The closure to biotic interference may also lead to regeneration of trees adapted for those environments (Table 4.8).

With closure to grazing the number of trees and shrubs increased from 1960-1961 (year of closure) to 1973-1974. Among the tree species *Azadirachta indica, Prosopis spicigera*, and *Heloptelia integrifolia* increased consistently in number. Pradhan

Table 4.8. Effect of closure on regeneration of tree species at Vasad, Gujarat (Pradhan and Vasava, 1974)

Tree species	Tree population in the year			
	1960–1961	1962–1963	1965–1966	1973–1974
Acacia nilotica	49	55	50	30
Acacia eburnea	Nil	4	20	11
Acacia senegal	23	42	81	259
Azadirachta indica	20	20	80	133
Holoptelia integrifolia	Nil	6	18	69
Prosopis spicigera	60	74	98	88
Cassia auriculata	57	44	63	44
Cymnosporea montana	3	6	6	64
Zizyphus species	101	98	321	331
Total	313	349	737	1028
Percentage increase over 1960–1961	–	11	135	228

(1977) reported that in 1973-1974 there were 6563 trees of more than 50 cm girth at breast height, stocking of 80 trees ha^{-1}.

Similar results have been obtained at Agra where in a period of ten years of closure edible grasses increased from 44.6 per cent to 48.5 per cent and non-edible decreased from 42.9 to 30.1 per cent (Prajapati, 1974).

4.1.3.2. Fodder–fuel Plantations

In India, as in other developing countries, the fodder fuel needs are increasing rapidly. In this connection an experiment was initiated at Vasad in 1960-1961 to investigate if fodder and fuel-cum-timber trees species (*Acacia nilotica*) can be grown together without reduction in the yield of fodder.

Results presented in Table 4.9 indicate that there was no difference in the forage yields of *Cenchrus ciliaris* or *Dichanthium annulatum*, when grown as monoculture or in association with a tree species. Moreover the mean forage yield increased by 101 per cent in 1964-1965 over the yield obtained in 1960-1961. This increase was partly due to the effect of closure and protection of the area and partly due to the rejuvination of pasture species. Fuel-cum-fodder plots that were planted to *A. nilotica* did not differ in forage yield from the plots of pure grass during the initial 6 years since the shading by tree species was less during this initial period. Prajapati and Joshie (1977) reported a yield potential of 2 t ha^{-1} yr^{-1} of fuel in a twenty-year-old plantation of *Acacia nilotica* which was grown as a protective forest on black soil in ravines in semi-arid region of Kota. It has been observed

Table 4.9. Forage (green) yield (Kg ha^{-1}) from fodder-fuel plantation at Vasad (Dayal, 1974)

Treatment	Year							Average
	1960–1961	1961–1962	1962–1963	1963–1964	1964–1965	1965–1966	1966–1967	
Pure grass block of *Cenchxus ciliaris*	2790	4180	6135	4919	6318	5192	5224	4965
Pure grass block of *Dicanthium-annulatum*	3948	3636	5873	4697	7277	5585	6193	5316
Fuel-cum-grass block with *C. Ciliaris*	3375	4378	3314	6423	6754	4887	4938	4867
Fuel-cum-grass block with *D. annulatum*	3227	4277	5656	5493	6505	5827	5945	5276
Average	3315	4244	5589	5728	6663	5363	5575	5106
Percentage increase in yield over 1960–1961	—	28.0	68.6	72.8	101.0	61.8	68.2	

at Dehra Dun (Mathur and Joshie, 1972) that a yield potential of about 11 t ha^{-1} yr^{-1} of grass *Chrysopogon fulvus* (at 0.75 m $\times$ 0.75 m) and 97 trees ha^{-1} of fuel from *Dalbergia sissoo* tree (at 9.14 m $\times$ 9.14 m) is attainable when grown on Class V and VI waste land in Doon valley. In the Siwalik hills, the estimated yield of a mixed plantation of *Eucalyptus hybrid* for fuel and *Eulaliopsis binata* (a fodder and industrial grass used for rope and paper making) was standing fuel stock at 55.35 t ha^{-1} and an average air dry grass yield of 5.57 t ha^{-1} yr^{-1} was obtained in 4 yr. When *Eulaliopsis binata* was grown alone it gave an average yield of 14.49 t ha^{-1} yr^{-1} of air dry grass in 4 years. Such high yields were obtained by planting tall plants of *Euclayptus hybrid* (average initial height of seedlings 2.74 m) making tie ridges 15 cm high and at interval of 2 m $\times$ 2 m (Mishra and Sud, 1978).

4.1.4. EVALUATION OF SOIL AND WATER CONSERVATION MEASURES

4.1.4.1. Peak Rate of Runoff and Sediment Control

Patil and Sohoni (1969) reported that contour *bunding* increased yield of crops in semi-arid tropical rainfed areas of Maharashtra state. Since the runoff losses decreased, there was a trend of increasing ground water level in wells.

Investigations conducted on a 2-ha watershed in the Siwalik ranges of Chandigarh showed that the total runoff and peak rate of runoff were reduced by 60.4 per cent and 61.1 per cent respectively as a result of the following soil and water conservation measures: (i) construction of earthen debris basins, (ii) earthen pondage banks, (iii) staggered contour trenches and (iv) planting with *Eucalyptus hybrid* and *Acacia catechu* (Kaushal *et al.*, 1975). Similar results have been reported from Dehra Dun where a reduction of 28 per cent in runoff and 73 per cent in peak rate of flow were achieved be reforestation of a small watershed (Mathur *et al.*, 1976). Another series of experiments conducted at Chandigarh indicated that whereas soil and water conservation practices of contour trenching and afforestation reduced the runoff by 41 per cent, annual burning, over-grazing, and cutting of trees plus over-grazing increased runoff by 69 per cent, 88 per cent, and 71 per cent respectively (Gupta *et al.*, 1974).

Studies conducted at Chandigarh in the Siwalik region (Mishra *et al.*, 1977) on a watershed of 20 ha showed that with appropriate soil and water conservation measures the rate of sediment was reduced from 80 t ha^{-1} yr^{-1} to 6 to 7 t ha^{-1} yr^{-1} within four years. Thereafter the rate of sediment has gradually decreased to only 2.9 t ha^{-1} yr^{-1}. Runoff was reduced from 23.5 in 1964 to about 10 per cent. Peak rate of discharge was reduced from 0.065 in 1964 to 0.034 m^3 sec^{-1} in 1977. At Vasad, it has been shown that when 1422 tonnes of sediment from an untreated catchment of 67 ha were delivered to a treated catchment of 72 ha, all the sediment was either contained or retained in the latter as a result of soil and water conservation measures (Nema and Kamanwar, 1976). Jalote and Malik (1974) reported complete stabilization of 270 ha of severely eroded land by afforestation in the catchment of Gomti river.

The results of the treatment of some of the large reservoir catchments have been very encouraging. For example, in 16 sub-watersheds of Damodar Valley, the sediment load has been reduced from 0.30 to 0.019 ha m km^{-2}. In five major reservoirs catchments (Bhakra, Machkund, Panchet, Maithon, and Hirakud), their treatment resulted in a decrease in the sediment load by 16.3 to 42 per cent (Patnaik, 1977). Pathak (1974) reported that sediment load of 0.18 ha m km^{-2} was reduced to 0.15 ha m km^{-2} when 770 ha (0.25 per cent of the total area) of agricultural land were treated with soil and water conservation measures and 16000 ha (5.2 per cent of the total area) were afforested in the Ram Ganga catchment.

4.1.4.2. Increase in Production

While sediment control and water conservation are necessary for management of reservoirs, it is essential that the production of food, fibre, fuel, timber, etc., is also sustained. The Indian experience reinforces the belief that soil and water conservation measures are economical in the long run. The beneficial effects of soil moisture conservation by construction of *bunds* and levelling of land have been extensively demonstrated in the alluvial plains of UP where 35 per cent, 63 per cent and 98 per cent increase in crop yield was obtained by *bunding*, levelling, and *bunding*-cum-levelling respectively (Khan, 1962). Bunding increased yields of *Setaria*, cotton and sorghum by 18, 11, and 17 per cent in large field trials in Madras State (Kanitkar *et al.*, 1960). As a result of contour *bunding* Tamhane *et al.* (1967) reported 25 per cent increase in crop yield in the Maharashtra state; 35.6 per cent and 25.4 per cent increase in yield of sorghum and pearl millet respectively in Tamilnadu; 20 per cent increase in yield of groundnut in Tamilnadu; 21.4 per cent, 15.0 per cent, 19.7 per cent, and 13.9 per cent increase in yield of wheat, gram, maize, and pearl millet respectively in Punjab state.

Afforestation trials carried out in the ravines at Vasad have shown that *Dendrocalamus strictus* (bamboo), *Tectonia grandis* (teak), *Dalbergia sissoo* (*shisham*) and *Euclayptus camaldulensis* are very promising (Table 4.10).

Aggarwal *et al.*, 1977, carried out an economic evalution of 17 years old fodderfuel plantations of *Chrysopogan fulvus* and *Eulaliopsis binata* grasses and *Dalbergia sissoo* and *Acacia catechu* trees. The results presented in Table 4.11 indicate that *Eulaliopsis binata* grass with *Acacia catechu* produced the highest gross income R 35498/- per hectare). The association of *Chrsopogon fulvus* grass with and without trees produced net income of R 9110/- (mean of 7 *Chrysopogon* treatments) and R 11447/- per hectare respectively. Similarly growing *Chrysopogon fulvus* with *Dalbergia sissoo* and *Acacia catechu* fetched an income of R 9141/- (mean of 4 *D. sissoo* treatments) and 8670/- (mean of 3 *A. catechu* treatments) per hectare respectively. Spacing of the trees did not have any significant effect on the growth or yield of *Chrysopogon fulvus* grass. On an average *Dalbergia sissoo* and *Acacia catechu*

Table 4.10. Net annual income from various tree species at Vasad

Species	Rotation (years)	Cost of production (R rotation^{-1} ha^{-1})	Gross income (R rotation^{-1} ha^{-1})	Net annual income (R ha^{-1})	Remarks
Dalbergia sissoo[a]	30	3380	23356	666	
Dendrocalamus strictus[a]	30	3380	44475	1370	
Eucalyptus camaldulensis[a] (3 rotations)	24	12824	13478	444	
Tectonia gradis[b]	15	1300	17500	1080	For poles

[a]Singh, 1972; [b] Dayal *et al.*, 1974

produced an income of R 4077/- (mean of 4 *D. sissoo* treatments) and R 1202/- (mean of 4 *A. catechu* treatments) per hectare respectively. *Dalbergia sissoo* at 9.15 m × 9.15 m spacing and *Acacia catechu* at a spacing of 4.55 m × 4.55 m resulted in the maximum income of R 8684/- and R 2652/- per hectare respectively. The data indicated that for fuel purposes, trees at a closer spacing of 4.55 m × 4.55 m or 9.15 m × 9.15 m would be more profitable than when grown at wider spacings. The benefit: cost (*b:c*) ratio was more than unity for all treatments investigated. From these *b:c* ratios, it was inferred that *Dalbergia sissoo* at 9.15 m × 9.15 m + *Chrysopogon fulvus* for fuel and fodder; *Chrysopogon fulvus* alone for fodder, and *Acacia catechu* and *Eulaliopsis binata* for fuel and fibre requirement are economically justified to be grown in the wasted lands of Doon Valley.

4.1.5. WATERSHED MANAGEMENT AND YEAR 2000

While it is very heartening that there is an on going and active soil and water conservation programme on watershed basis in India and a strong and stable infrastructure for executing it, if one looks ahead to the year 2000 one feels concerned and staggered by the terrific pressures which will be exerted on the land–water–plant systems by the human and cattle population in India. For example, human population, which increased from 361,000,000 in 1951 to 548,000,000 in 1971 will be 931,000,000 by the year 2000; the net area *per capita* which decreased from 0.9 ha in 1951 to 0.6 ha in 1971, will decrease to 0.35 ha. Though the cultivated area increased from 119,000,000 ha in 1951 to 140,000,000 ha in 1971, yet the *per capita* availability of land for production of food, fibre, and other needs shrank from 0.33 to 0.292 ha and this will further decrease to 0.175 ha by the year 2000 (Anonymous, 1976a; 1976b). To the growing pressure of human population on the

Table 4.11. Economic evaluation of fodder fuel plantation at Dehra Dun (Aggarwal et al., 1977)

Grass species	Tree species	Tree spacing	Total of establishment and maintenance cost for 17 years (R ha^{-1})	Summed up value of income from grass (Rs ha^{-1}) for 17 years (R ha^{-1})	Total income from trees for 17 years (R ha^{-1})	Benefit:cost ratio
C. fulvus	–		3796	11447	–	3.015
C. fulvus	D. sissoo	4.55 m × 4.55 m	11136	8941	2711	1.044
C. fulvus	D. sissoo	9.15 m × 9.15 m	6835	10089	8684	2.746
C. fulvus	D. sissoo	13.75 m × 13.75 m	5518	9378	2638	2.177
C. fulvus	D. sissoo	18.30 m × 18.30 m	4657	9355	2275	2.497
E. binata	A. catechu	4.55 m × 4.55 m	11136	35498	2652	3.425
C. fulvus	A. catechu	9.15 m × 9.15 m	6835	8657	1140	1.433
C. fulvus	A. catechu	13.75 m × 13.75 m	5518	8328	649	1.629
C. fulvus	A. catechu	18.30 m × 18.30 m	4657	9026	367	2.016

Table 4.12. Projection of soil and water conservation development programme

Period	Achievement of development programme (million ha)	Expenditure (million R)	Remarks
Total up to the end of Fifth Plan (31.3.78)	21.9	5081	
Projection of Development Programme			
Sixth plan (1978–1983)	9.9	4660	
Seventh plan (1983–1988)	13.5	6345	At the price level of the sixth plan
Eighth plan (1988–1993)	17.0	7990	As above
Ninth plan (1993–1998)	20.5	9635	As above
Tenth plan (1998–2003)	24.0	11280	As above
Total up to 2003	106.8	44991	

Table 4.13. Projection of need of trained man power

Period	Officers — Dehra Dun and Ootacamund Centres	Assistants — Bellary, Kota, Ootacamund, and Hazaribagh Centres
Total manpower trained up to the end of fifth plan (31.3.78)	1076	3494
Projection of training needs		
Sixth plan (1978–1983)	2050	10250
Seventh plan (1983–1988)	2800	14000
Eighth plan (1988–1993)	3500	17500
Ninth plan (1993–1998)	4200	21000
Tenth plan (1998–2003)	4900	24500

available land resources has to be added the growing pressure of livestock population. The latter increased from 264,400,000 cattle units in 1951 to 312,200,000 cattle units in 1972. These dense population pressures on land result in cultivation of marginal lands and destruction of forests and thus aggravate the present rates of the degradation of the production base. Already 145,000,000 ha of land representing 44 per cent of the geographical area of India are subject to severe erosion. If development costs were fixed from the sixth plan onwards and provision were made for an increase of 3,500,000 ha in every five-year plan, it would be possible to implement soil and water conservation measures on some 107,000,000 ha of land by the end of the tenth five year plan in the year 2003 (Table 4.12).

By the year 2003 about 37,000,000 ha of land which are now in need of conservation measures will still remain to be treated. Furthermore the lands that have been and are being treated now will be in need of retreatment. To achieve this level of development it will be necessary to train sufficient manpower for this purpose (Table 4.13).

It is essential to plan for training the man power in advance if the programmes of soil and water conservation are to succeed.

4.1.6. ACKNOWLEDGEMENT

The author would like to thank the Central Soil and Water Conservation Research and Training Institute, Dehra Dun, UP (ICAR), India for its support.

REFERENCES

References marked with an asterisk(*) are cited from the *Annual Report of the Central Soil and Water Conservation Research and Training Institute*, Dehra Dun for the corresponding year.

Aggarwal, M. C., Rambabu, and Joshie, P., 1977.*

Agnihotri, Y., Raghubir, and Mishra, P. R., 1971.*

Anonymous, 1968, *Report of the Working Group for Formulation of Fourth Five Year Plan Proposal on Land and Water Development*, Ministry of Food, Agriculture Community Development and Cooperation, Department of Agriculture, Government of India, New Delhi.

Anonymous, 1971*, Effect of bunding on the hydrological behaviour of watershed −Agra.

Anonymous, 1976a, *Report of the National Commission on Agriculture,* Ministry of Agriculture and Irrigation, Government of India, New Delhi.

Anonymous, 1976b, *India: Habitat-76.* The UN Conference on Human Settlements, Vancouver, Department of Science and Technology, Government of India, New Delhi.

Anonymous, 1978, *Report of the Working Group on Soil Conservation and Land Reclamation for 1978-79 to 1982-83,* Ministry of Agriculture and Irrigation, Government of India, New Delhi.

Bansal, R. C., and Hussenappa, V., 1977.*

Bhatia, K. S., and Chaudhary, H. P., 1978, Runoff and erosion losses and crop yields from sloping and eroded alluvial soils of Uttar Pradesh in relation to contour farming and fertilization, *Soil Conserv. Dig.,* **5(2)**, 16−22.

Das, D. C., Kurian, K., Lakshmannan, V., *et al.,* 1971.*

Das, D. C., Raghunath, B., Murthy, S. S., and Pooranchandran, G., 1970, Optimum length of bench terraces at Ootacamund (Presented at the *Eighth Annual Convention of Indian Society of Agricultural Engineers,* Ludhiana).

Dayal, R., 1974, *Report on Achievement of Soil Conservation Research Demonstration and Training Centre, Vasad* (unpublished data of various workers).

Dayal, R., Pradhan, I. P., and Vasava, S. S., 1974.*

Erasmus, I. I., and Bansal, R. C., 1965, A runoff prediction in watershed planning in Ambala Siwaliks, *Indian Forester,* **91(8)**, 548–552.

Erasmus, I. I., Gupta, R. K., and Kaushal, R. C., 1971.*

Gupta, G. P., 1975, Sediment production status report on data collection and utilization, *Soil Conserv. Dig.,* **3(2)**, 10–21.

Gupta, R. K., Mishra, P. R., *et al.*, 1974.*

Gupta, S. K., Rambabu, and Nayal, M. S., 1969, Comparison of peak discharge rates by various methods, *Harvester,* **2(3)**, 150–154.

Gupta, S. K., Rambabu, and Rawat, N. S., 1971, *Weekly Expected Runoff at Various Stations in India at Different Percent Chance* (Unpublished data).

Gupta, S. K., Rambabu, and Tejwani, K. G., 1970, Nomographs and important parameters for estimation of peak rate of runoff from small watersheds, in *India J. Agr. Eng.,* **7(3)**, 25–32.

Jalote, S. R., and Malik, O. P., 1974, Farm forests of Rehmankhera, *Soil Conserv. Dig.,* **2(2)**, 59–69.

Kanitkar, N. V., Sirur, S. S., and Gokhale, D. H., 1960, Dry farming in India, *Indian Counc. Agr. Res.,* New Delhi.

Kaushal, R. C., Dayal, S. K. N., and Mishra, P. R., 1975.*

Khan, A. D., 1962, Measurement of increase in productivity by adopting soil and water conservation practices. III. *J. Soil Water Conserv. India,* **10(3 and 4)**, 25–33.

Mathur, H. N., and Joshie, P., 1972*.

Mathur, H. N., Rambabu, Joshie, P., and Singh, B., 1976, Effect of clear felling and reforestation on runoff and peak rates in small watershed, *Indian Forester,* **102(4)**, 219–226.

Mishra, P. R., Kaushal, R. C., Dayal, S. K. N., and Prabhushankar, C., 1977.*

Mishra, P. R., and Sud, A. D., 1978.*

Nema, J. P., and Kamanwar, H. K., 1976.*

Pathak, S., 1974, Role of forests in soil conservation with special reference to Ramganga watershed, *Soil Conserv. Dig.,* **2(1)**, 44–47.

Patil, R. G., and Sohoni, D. K., 1969, Long term economic benefits of soil conservation programme, *J. Soil Water Conserv.,* India, **17(1 and 2)**, 22–26.

Patnaik, N., 1977, Soil Conservation works in river valley projects. *Soil Conserv. Dig.,* **5(2)**, 53–61.

Pradhan, I. P., 1977.*

Pradhan, I. P., and Vasava, S. S., 1974.*

Prajapati, M. C., 1974, *Report on Achievements (1955–74) of Soil Conservation Research and Demonstration Training Centre, Agra* (Unpublished data of Prajapati, M. C., and Sajwan, S. S.)

Prajapati, M. C., and Joshie, P., 1977.*

Raghunath, B., Sreenathan, K., Das, D. C., and Thomas, P. K., 1967, Conservation evaluation of various land use practices on moderately steep sloping lands in Niligris Part I & II (Presented in the *Sixth Annual meeting of Indian Society of Agricultural Engineering, Bangalore*).

Rambabu, Bansal, R. C., and Srivastava, M. M., 1974.*

Sastri, G., Rao, D. H., and Ram Mohan Rao, 1975.*

Singh, B., 1971, A comparative study on economics of various soil conservation-cum-grass land improvement practices in ravine lands II, *Indian Forester,* **97(7)**, 387–391.

Singh, B., 1972, Economics of afforestation of ravine lands, *Proc. Symp. Man-made Forests in India, Dehra Dun,* **II**, 26–32.

Singh, B., and Verma, B., 1971, A comparative study on economics of various soil conservation-cum-grassland improvement practices for rejuvenating forage production in ravine lands I (Forage Production), *Indian Forester,* **97(6)**, 315–321.

Singh, G., and Bhushan, L. S., 1977.*

Singh, J. P., and Prajapati, M. C., 1974.*

Sud, A. D., Sadhu, Singh, and Mishra, P. R., 1977.*

Tamhane, R. V., Khemchandani, H. T., and Kulkarni, G. A., 1967, *Critical Review of the Assessment of Soil Conservation Measures in India,* Ministry of Food, Agriculture Community Development and Cooperation, Government of India, New Delhi.

Tejwani, K. G., Gupta, S. K., and Mathur, H. N., 1975, *Soil and Water Conservation Research,* Indian Council of Agricultural Research, New Delhi.

Tejwani, K. G., Srinivasan, V., and Mistry, M. S., 1961, Gujarat can still save its ravine lands, *Indian Farming,* **11(8)**, 20–21.

Wasi Ullah, and Rambabu, 1970, Runoff estimates based on rainfall retention relationship, *Indian Forester*, **96(2)**, 89–101.

APPENDIX I: Vernacular names and their English equivalents

Vernacular name	English equivalent name
Jowar	*Sorghum vulgare*
Bajra	*Pennisetum typhoides*
Rabi	Crop season after south west monsoon in India; crops grown during October–April season
Kharif	Crop season during south west monsoon in India; crops grown during June–October season
Bund	Terrace
Bunding	Terracing

Tropical Agricultural Hydrology
Edited by R. Lal and E. W. Russell
© 1981, John Wiley & Sons Ltd.

4.2

Rainy Season Cropping on Deep Vertisols in the Semi-arid Tropics – Effects on Hydrology and Soil Erosion

J. Kampen, J. Hari Krishna, and P. Pathak

4.2.1. INTRODUCTION

Vertisols are one of the major soil orders found in the Semi-Arid Tropics (SAT). The SAT, are defined by ICRISAT (International Crops Research Institute for the Semi-Arid Tropics) as those areas where the average monthly precipitation exceeds the potential evapotranspiration in at least two and at most seven months. However, the rainfall patterns are erratic with frequent rainless periods even within the rainy season. Annual rainfall at ICRISAT for example was 346 mm in 1972-1973 and 1137 mm in 1975-1976; the coefficient of variation is 26 per cent (Virmani *et al.*, 1979). Much of the rains occur in high intensity storms resulting in runoff and severe erosion particularly under conditions of limited vegetative cover.

Large areas of deep Vertisols in India (about 18,000,000 ha) and also in West Africa are currently fallowed during the rainy season and crops such as sorghum or wheat are grown on residual soil moisture in the post-rainy season. In India, yields are low due to lack of moisture at the time when crops mature. The practice of rainy season fallowing with repeated cultivation to control weeds has been traditionally adopted by farmers because of the risks involved in rainy season cropping and a lack of appropriate land and water management technology which would facilitate working these sticky clay soils during the rainy season.

The Vertisols (Usterts) referred to in these investigations are fine calcareous montmorillinitic isohyperthemic members of the family of Typic Chromusterts. They will be referred to as Vertisols in this paper. These Vertisols are high in montmorillinitic clay (50 to 64 per cent) and undergo pronounced shrinkage during drying, resulting in large cracks that close only after prolonged rewetting. These soils become hard when dry and sticky when wet. Drainage during wet periods in the rainy season can be a serious problem. The Vertisols investigated in this study have slopes from 0.5 to 3 per cent; runoff and erosion are serious problems, particularly under rainy season cultivated fallow. The soils are high in bases, including calcium, magnesium, and potassium, and the pH ranges from 7.5 to 8.6. Under semiarid conditions, the soils are low in organic matter and are usually deficient in nitrogen, phosphorus, and sometimes zinc.

An important assumption, which has been strengthened by experience gained at ICRISAT in the past few years, is that the farming systems presently found in the areas of concern cannot be significantly improved through the introduction of better technology with regard to any one single production factor alone (Thierstein and Kampen, 1978). The climatic and soils environment, the traditional cropping systems, the present yield levels and the behaviour of farmers exposed to risk, create a situation in which improvement of one component of the production system very rarely results in adequate returns for the long run. Thus, several components of technology must be identified and integrated in such a manner that synergistic effects make the concurrent introduction feasible and rewarding while long term stability is assured. A clearly visible impact on agricultural production is required; deferred, or small gains will not capture the imagination of farmers and will impede acceptance.

In many of the Vertisol areas of the SAT the average annual rainfall would seem to be sufficient for one or sometimes two good crops per year; for example at Hyderabad the average rainy season precipitation between 1 June and 31 October is 691 mm. However, the average cereal yields that farmers presently obtain in good rainfall years may range from only 500 to 800 kg ha^{-1}; when the late rains fail, yields will be even lower. Thus, methods to more effectively utilize the total available rainfall and the soil in these regions are envisaged to have a large potential impact on agricultural production and also on the conservation of the natural resource base.

4.2.2. RESOURCE UTILIZATION ON VERTISOLS

The present farming systems found on most Vertisols in the SAT are characterised by inefficent use of the total available precipitation and low productivity. In ICRISAT's research program on improved resource utilization, the central objective is to make the best use of the rain that falls on a given area (Kampen *et al.*, 1974). In order to study water as an input, natural watersheds (small agricultural catchments) were chosen as a unit for research (Krantz *et al.*, 1978). Also, since water is the major natural constraint to agricultural development in the SAT, it is envisaged that the watershed or drainage basin will, in time, become the focus for resource development and utilization. Several land and water management techniques are simulated on Vertisol watersheds, e.g. flat planting, broadbed-and-furrow cultivation, field, contour and graded bunds, runoff collection and supplemental irrigation, etc. Alternative cropping systems are superimposed on these treatments, these consist of intercrops of maize or sorghum with pigeonpea, sequential crops of maize and chickpea, sole crops of sorghum in the rainy or in the post rainy season, etc. A distinction is made between 'existing' or 'traditional', and 'improved' levels of technology in case of varieties, fertilization, farm equipment, management, etc. All cultural operations are executed using animals for draft power because these are a major power source of farmers in the SAT today. Thus, these watersheds are operational-scale 'pilot plants' where the integrated effect of alternative systems of farming on productivity, resource use, and conservation can be monitored (Krantz, 1979). Results from operations research on the integration of new technology into economically feasible farming systems are used to arrive at viable technology, for research priority setting, and as feedback into specific research projects (Kampen, 1979).

4.2.2.1. New Farming Systems on Vertisols

Important components of the improved natural resource management system for deep Vertisols, developed at ICRISAT in collaboration with scientists of the National Research Programs (Kampen, 1979; Krishnamoorthy *et al.*, 1977) include:

Dry sowing: An important factor which has facilitated cropping the deep Vertisols during the rainy season has been the realization that crops can be sown dry (shortly ahead of the rainy season) in areas where the rains commence fairly reliably and where there is a good probability of follow-up rains to ensure the establishment of the germinating crop (Virmani *et al.*, 1979). One serious problem encountered was the difficulty of preparing a good seedbed during the dry season, when tillage on these hard, clayey soils is difficult. However, the development of new tillage techniques, more suitable equipment and the occurrence of 1 or 2 prerainy season rains which moisten the surface soils in most years made adequate land preparation ahead of the rainy season feasible (Plate 13).

Animal-drawn precision equipment: Operations research in 1973 and 1974 clearly indicated that in order to provide the necessary speed and precision and to improve the efficiency of draft animal use, improved animal drawn equipment was required (Thierstein and Kampen, 1978). It was therefore attempted to adapt a tool carrier that could provide the same vertical and horizontal precision as tractor mounted equipment, at lower speed but at only a fraction of the cost. These tools consist of a frame on two rubber-tyred or steel wheels with a toolbar to which various implements (moldboard and disc ploughs, harrows, planters, cultivators, fertilizer applicators, scraper, cart body, etc.) can be attached (Plates 13-14). The use of animal-drawn, wheeled carriers and modified equipment resulted in considerable improvement of this segment of farming systems on Vertisols (ICRISAT, 1976; ICRISAT, 1978).

Broadbed-and-furrow system: The system found to be most successful both to facilitate cultural operations and to control excess water and erosion is a broadbed-and-

Narrow beds and furrows are adapted to 75 cm rows only (e.g. maize)

Broadbeds and furrows are adapted to many row spacings

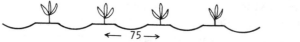

A maize crop

A sorghum or millet crop

A groundnut or chickpea crop

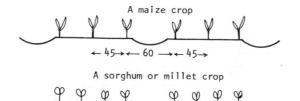

A pigeonpea/sorghum intercrop or
a pigeonpea/maize intercrop.

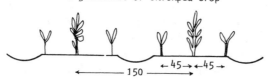

Figure 4.1 Alternative cropping systems and row arrangements on broadbeds (150 cm). All dimensions in centimetres

furrow system 150 cm wide (Figure 4.1); the broadbed is about 100 cm wide and the sunken furrow about 50 cm. Initial experience with narrow, graded ridges of 75 cm width at ICRISAT indicated instability of this system on lighter soils such as Alfisols; the ridges were insufficiently wide to prevent 'breaching' which can be catastrophic for any soil and water conservation system (ICRISAT, 1976). The 75-cm ridges were also found to have very limited flexibility to accommodate the wide range of crops grown in the SAT. With the 150-cm broadbeds, it is possible to plant two, three or four rows at 75-, 45-, and 30-cm row spacings respectively (Figure 4.1). Having semi-permanent furrows for draft animals or tractors to follow results in reduced soil compaction where plants grow, which in turn facilitates and speeds up dry season land preparation. Comparisons of flat planting with cultivation on broadbeds indicate substantial yield advantages of the graded broadbed system on deep Vertisols (Table 4.14). In 1976, values of rainfed crop yields in field scale experiments were about 800 rupees per hectare (R ha^{-1}) higher on broadbeds at a 0.4 to 0.6 per cent slope than under flat cultivation at that slope; with supplemental irrigation the difference in yields were similar. Narrow graded ridges were superior to flat planting, however yields were generally less than on broadbeds (Table 4.15).

Erosion and runoff are lower in the broadbed system than with flat planting on deep Vertisols because the excess water is led off the land at a controlled velocity in many furrows rather than in concentrated streams down the steepest slope (Table 4.15). Surface drainage during wet periods is also facilitated. The data obtained at ICRISAT indicate that the optimum slopes along the furrows on deep Vertisols are in the range of 0.4 to 0.8 per cent (Kampen and Krishna, 1978; Krishna and Hill, 1979). Steeper slopes resulted in accelerated erosion, while lower slopes did not provide sufficient surface drainage during wet periods (ICRISAT, 1974; ICRISAT, 1978).

The broadbeds and furrows are so arranged that excess water discharges into grassed waterways, which if required, may lead to a runoff-collection facility consisting of a dug 'tank' or small earth dam (Figure 4.2). The collected water may be used in the dry (and cool) season for supplemental irrigation at crop establishment or later. In brief, the advantages of the graded broadbed-and-furrow system observed in operational-scale research on watersheds including the following:

— Reduces soil erosion
— Provides surface drainage
— Transfers to the broadbed the top soil, with its associated organic matter which is removed from the furrow bottom between adjacent broadbeds
— Reduces soil compaction in the plant zone
— Is adaptable to supplemental water application
— Can be laid out on a permanent basis
— Is easily maintained with minimum tillage
— Facilitates land preparation
— Reduces the power and time requirements of agricultural operations
— Provides furrows for animals to follow
— Is adaptable to many row spacings.

4.2.3. RESOURCE CONSERVATION AND PRODUCTIVITY UNDER ALTERNATIVE MANAGEMENT SYSTEMS

4.2.3.1. Runoff and Erosion

Investigations on runoff and soil loss during high intensity, long duration storms on operational-scale research watersheds have shown that total runoff, peak runoff rates and soil erosion are much lower on Vertisol watersheds using cropped broadbeds than for other management treatments. This is particularly

Table 4.14. Mean monetary values of flat cultivation and the semi-permanent broadbed-and-furrow system on Vertisol watersheds using improved technology[a] in 1976 and 1977. All values in R ha^{-1} (1\$ = 8 R)

Ws. No.	Land Manag.	Year	Intercrop			Sequential crop			Means	
			Maize	Pigeonpea	Total	Maize	Chickpea	Total	Crop systems	Both years
1, 2, 3A	Beds	1976	2840	2080	4920	2730	950	3680	4300	
1, 2, 3A	Beds	1977	2270	2770	5040	2880	2400	5280	5160	
Means										4730
3B, 4B	Flat	1976	2530	1680	4210	2300	570	2870	3540	
3B, 4B	Flat	1977	2450	1810	4260	2790	2200	4980	4620	
Means										4080
LSD (0.05)										280
CV(%)										9.2

[a]Improved technology: Improved varieties, (Maize Deccan Hybrid 101 in sole crops, SB23 in intercrops; pigeonpea ICRISAT and chickpea local; fertilization with 75 kg ha^{-1} of 18-46-0 and 67 kg ha^{-1} of N top-dressed. All cultural operations executed with improved animal-drawn equipment, fertilizers band-placed, minimum insect control (if required).

Plate 13 Primary tillage on a deep Vertisol executed during the dry season using an animal-drawn, multipurpose toolbar

Plate 14 **Planting** on a broadbed-and-furrow system on Vertisols before the rainy season, using the animal-drawn multipurpose toolbar

Facing page 262

Plate 15 Parshall flume and water level recorder, during a runoff event on a rainy season fallowed deep Vertisol watershed (BW4C)

Plate 16 A rainy season fallowed deep Vertisol watershed immediately after high intensity rainfall and runoff (BW4C). Note substantial erosion and deposition within the cultivated fields

Table 4.15. The effect of land management[a] on runoff, erosion, and yields of sequentially cropped maize and chickpea on deep Vertisols (ICRISAT, 1979).

Land treatment	Runoff (mm)	Erosion (kg ha^{-1})	Maize	Chickpea Unirrigated	Chickpea Irrigated[b]	Total value (R ha^{-1}) Unirrigated	Total value (R ha^{-1}) Irrigated
Flat planting	141	240	2740	490	610	3170	3360
Narrow ridges	77	110	3240	450	650	3540	3860
Broadbeds	110	170	3170	740	940	3940	4260

[a] The furrow in all ridged treatments were maintained at a 0.6% grade; flat planting was executed at the same slope.
[b] One supplmental irrigation (75 mm) at crop establishment.

true when compared to the traditional cultivated fallow (Plate 15, Table 4.16, Table 4.17, and Table 4.18). Fallowing also results in the profile becoming saturated early in the growing season and then substantial runoff begins to occur. When a crop is grown during the rainy season, water is continuously being withdrawn from the profile and surface runoff is reduced. Data collected on profile moisture (to 187 cm depth) indicate that in most years the total quantity of available moisture at sowing time of the post-rainy season crop (late September) is similar on soils that have been cropped compared to those fallowed in the rainy season. Thus land cropped during the rainy and post-rainy season (June to March) to maize and pigeonpea only contained 30–40 mm less water in the profile at the end of the dry season than land fallowed during the rains (June to September) and cropped to sorghum from October to March.

Thus, the practice of rainy season fallowing under conditions of a relatively dependable late season rainfall, does not generally result in an improved moisture environment for post-rainy season crop growth compared to the technique of growing a crop during the rainy season. As expected, erosion is more severe under cultivated fallow conditions than when a rainy season crop is grown; this difference becomes very evident at high rainfall intensities (Figure 4.3). Under fallow conditions significant erosion was observed to occur within the cultivated fields (Plate 16). In some years (without high intensity, long duration rainfall) the quantities of eroded soil measured at watershed outlets were relatively small (Table 4.18).

4.2.3.2. Rainfall Utilization

A parametric simulation model has been developed to predict storm runoff from these small Vertisol watersheds (Krishna, 1979; Krishna and Hill, 1979). Computed

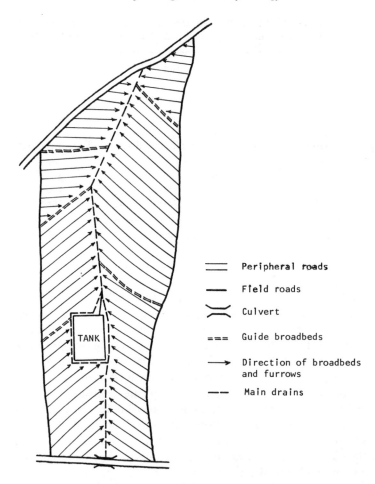

Figure 4.2 Schematic drawing of a watershed in broadbeds and furrows with a runoff storage facility

runoff values compared very well with the measured values; the estimated runoff values for the two treatments (rainy season cropping on ridges with a post-rainy season crop and rainy season fallow) for 1974, 1975, and 1976 are summarized in Table 4.19.

For the cultivated rainy season fallow treatment, on average (over these three years) about 42 per cent of the precipitation was lost as evaporation from the soil between June and October (Krishna, 1979). In an average year this would amount to 290 mm (the average long term rainy season precipitation is 691 mm). The average profile soil moisture accretion was 23 per cent, this is equivalent to about 160 mm in an average year. Rainfall from November through February is on average 46 mm, therefore a total of approximately 200 mm would be available for crop use.

Table 4.16. Rainfall and runoff on a Vertisol watershed with a broadbed-and-furrow system at 0.6 per cent slope (BW1) and a rainy season fallowed watershed (BW4C). (ICRISAT, 1979)

Date	Rainfall[a]	Runoff	
		Cropped	Fallow
	(mm)	(mm)	(mm)
June 15	34	0	0.2
July 4	37	0.7	1.8
17	54	0.5	3.2
Aug 10	75	0.3	25.4
22	30	0	3.2
23	15	0	4.4
24	5	0	2.9
28	21	0	3.6
Sept 1	14	0	5.7
Small storms (5)	52	0	2.6
Total	349	1.5	53.0

[a]The total rainfall from the 42 rainy days of the season (June–October) was only 519 mm. The runoff was unusually low in 1977, due to below-normal total precipitation and lack of high-intensity storms.

The total rainfall on average received during the period June through February is 737 mm. Thus, the effectively utilized rainfall would amount to only about 27 per cent.

Much greater effective use of the rainfall can be achieved when graded ridges are introduced to facilitate rainy season cropping, because in many years sufficient residual soil moisture will be available to grow a second crop in the post-rainy season. Runoff in this treatment on average amounted to only about 14 per cent approximately 100 mm in an average rainy season. The actual evapotranspiration in the rainy season would amount to approximately 45 per cent or 310 mm. The profile moisture accretion amounts to about 22 per cent of the rainfall, or 155 mm in an average year. With additional rainfall (46 mm) during the post-rainy season growing period, the total amount available for crop use is about 510 mm or almost 70 per cent of the seasonal rainfall in an average year.

4.2.3.3. Runoff Collection and Use

The runoff can be collected for utilization during the post-rainy season. If all runoff were collected and no losses occurred, the total water available for crop use during these three years would have been about 610 mm and the effectively used rainfall would have increased to 82 per cent (Krishna, 1979).

Table 4.17. Rainfall and runoff on a cropped deep Vertisol with broadbed-and-furrow system (BW1) and on a rainy season fallowed watershed (BW4C) (ICRISAT, 1978)

		Runoff	
Date	Rainfall[a]	Deep Vertisol cropped (BW1)	Deep Vertisol fallow (BW4C)
	(mm)	(mm)	(mm)
June 17	36.8	5.0	2.0
18	24.7	2.1	1.9
21	24.7	3.3	4.4
July 12	44.2	0.7	16.5
16	33.6	1.9	10.5
18	32.2	10.9	21.5
Aug 7	27.2	0.1	10.1
8	14.6	0.3	5.9
13	52.7	6.8	28.6
14 and 15	219.5	170.0	190.6
19	16.7	1.5	10.7
21	17.3	–	5.8
22	65.5	49.5	58.9
23	13.5	9.3	10.8
28	25.9	8.2	16.5
Small storms	57.0	2.9	15.4
	706.1	272.5	410.1

[a]Runoff producing storms only; total seasonal rainfall (June–October) was 1120 mm.

The collection of surface runoff during periods of excess rainfall, and its subsequent use during dry periods in the rainy season or early in the dry season, would markedly decrease the risks involved in rainfed agriculture. The deep Vertisols rarely require supplemental irrigation for the rainy season crop.

However, supplemental water can always be used on a post rainy-season crop and growing a second crop can sometimes be facilitated by a small initial quantity of water (Table 4.15). Additional evidence on the yield increasing effect of relatively small quantities of supplemental water (50 to 100 mm) has also been collected for sorghum in the post rainy season (ICRISAT, 1978; ICRISAT, 1979).

Thus, the potentials for use of collected surface or groundwater as a back-up resource need to be further explored (Kampen et al., 1974). The direct effect of improved water utilization technology appears substantial in years of ill-distributed rainfall. A similar effect may be expected in producing a second crop in the dry season. A decreased risk in production will provide the basis for greater assured profitability, and result in more certain responses to other inputs such as improved

Table 4.18. Rainfall, runoff, and soil loss (t ha^{-1}) measured at watershed outlets of cropped deep Vertisol watershed (BW1) and rainy season fallow deep Vertisol watershed (BW4C) from 1973 to 1978

Year	BW1				BW4C			
	Rainfall (mm)	Runoff (mm)	Peak rate (m^3 sec^{-1} ha^{-1})	Soil loss (t ha^{-1})	Rainfall (mm)	Runoff (mm)	Peak rate (m^3 sec^{-1} ha^{-1})	Soil loss (t ha^{-1})
1973	697.0	51.2	0.03	3.0	734.6	58.7	0.06	3.9
1974	810.4	116.1	0.20	1.3	806.9	223.4	0.22	6.8
1975	1041.6	162.2	0.06	0.7	1055.0	253.2	0.15	2.1
1976	687.3	73.1	0.09	0.8	710.1	238.1	0.16	9.2
1977	585.6	1.5	0.01	0.1	585.9	53.0	0.06	1.7
1978	1125.2	272.5	0.11	3.4	1116.7	410.1	0.15	9.7

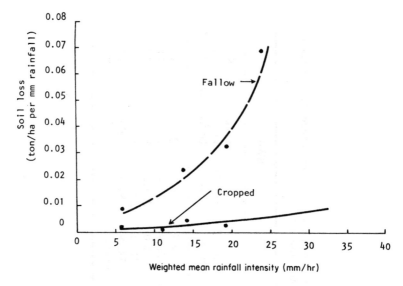

Figure 4.3 Effect of rainfall intensity on soil loss in a fallowed and a cropped watershed on a deep Vertisol (ICRISAT, 1976)

Table 4.19. Estimated and measured runoff on a double cropped narrow ridged Vertisol watershed (BW1) and a rainy season fallowed Vertisol watershed (BW4C) during 1974, 1975, and 1976 (All values in mm)

Watershed	BW1			BW4C		
Year	1974	1975	1976	1974	1975	1976
Total rainfall	776	965	647	774	966	666
Computed runoff	112	155	70	204	281	192
Measured runoff	114	157	72	210	250	209

seeds and fertilizers. This means greatly increased yields during years of adequate and well-distributed rainfall and thus, better opportunities to 'harvest the good years'.

4.2.3.4. Rainfall Productivity

The productive use of available water is critically important in the SAT. Rainfall productivity (RP) has been defined as the agriculrual production (in kg ha^{-1} or the monetary equivalent) in relation to the seasonal precipitation (in cm); it is the product of effectively used rainfall and the water-use efficiency (Kampen, 1979).

The crop yields obtained illustrate the potentials to arrive at vastly increased rainfall productivities. In the rainy season fallow treatment in 1975, with traditional technology (local implements and animal draft power, local varieties, minimal insecticide use and farmyard manure at 50 cartloads ha^{-1}) a safflower yield of 260 kg ha^{-1} was obtained, giving an RP of 4 R ha^{-1} cm^{-1}, and sorghum gave an RP of 16 R ha^{-1} cm^{-1}. In contrast to these figures, an RP of 58 R ha^{-1} cm^{-1} was obtained using improved technology (improved animal-drawn implements, improved varieties, adequate plant protection, graded ridges and furrows, chemical fertilizers 100 kg ha^{-1} of 22-57-0 at planting and 58 kg ha^{-1} N side dressed) with maize as the rainy season and chickpea as the post-rainy season crop. Similar gains in RP were obtained in other years (Kampen and Krishna, 1978).

4.2.4. COOPERATIVE ON-FARM RESEARCH

These improved farming systems on Vertisols are currently being tested and adapted in cooperative on-farm studies with the Indian Council of Agricultural Research. In its cooperative research programme on real farms, ICRISAT does not envisage the demonstration effect or the implementation of new technology as a major goal. Those activities are the exclusive responsibility of the national organizations for research and extension. However, it is expected that the results of such cooperative research projects will greatly facilitate later implementation of new technology across large regions of the SAT by adapting the new technology to what farmers require, by identifying bottle-necks, and through the development of guidelines on appropriate village and farmer level organizational structures (Kampen, 1979).

In view of the qualitative differences in the factor endowments of farmers compared to research stations, the performance of technology in farmers' fields may differ from what has been observed at research stations. Testing under real world conditions can provide information on the actual economic viability of watershed-based technology. It is therefore important that the technology be studied on farmers' fields so that constraints to adoption can be identified and specifically addressed by further research to improve the technology (Kampen and Doherty, 1979).

On-farm research provides for participation of farmers in development of technology. Ideally, agricultural research does not start on research stations it starts on farms. Only in this setting will feedback to the research programs be generated. Such feedback is considered essential to assure the continuing relevance of technological solutions to the priority problems of farmers.

Research on watershed-based systems in on-farm situations also has an advantage in that critical questions related to the initiation of group action among farmers who operate plots on a watershed area can be addressed. Indeed, these issues only express themselves in an on-farm situation. With average farm sizes in the Indian SAT around one to two hectares and with farmers having several plots comprising this area, often in locations which are spatially separated and with watershed sizes generally many times larger than this, it is imperative that operational guidelines for eliciting group action be defined.

It is envisaged that the information generated from these cooperative research projects will have impact far beyond the actual locations of project execution. Improved understanding of the technology generating process for appropriate natural resource development and use and of effective organizational structures identified as prerequisites to the effective implementation of new farming systems, will facilitate agricultural development not only in India; the results of such research may, in the long run, well affect research development and technology implementation in the entire SAT.

REFERENCES

ICRISAT (International Crops Research Institute for the Semi-Arid Tropics), 1974, *ICRISAT Annual Report 1973-74,* Hyderabad, India.

ICRISAT, 1976, *ICRISAT Annual Report 1975-76,* Hyderabad, India.

ICRISAT, 1978, *ICRISAT Annual Report 1976-77,* Hyderabad, India.

ICRISAT, 1979, *ICRISAT Annual Report 1977-78,* Hyderabad, India.

Kampen, J., 1979, Watershed management and technology transfer in the semi-arid tropics, *International Symposium on Development and Transfer of Technology for Rainfed Agriculture and the SAT Farmer,* ICRISAT, Hyderabad, India.

Kampen, J., and Doherty, V. S., 1979, Methodologies to attain improved resource use and productivity, *(Report on a Cooperative ICAR–ICRISAT On-farm Research Project),* ICRISAT, Hyderabad, India.

Kampen, J., *et al.,* 1974, Soil and water conservation and management in farming systems research in the SAT, *International Workshop on Farming Systems,* ICRISAT, Hyderabad, India.

Kampen, J., and Krishna, J. H., 1978, Resource conservation, management and use in the semi-arid tropics, *American Society of Agricultural Engineers, Summer Meeting,* Logan, Utah, USA.

Krantz, B. A., 1979, Small watershed development for increased food production, Leaflet, ICRISAT, Hyderabad, India.

Krantz, B. A., Kampen, J., and Virmani, S. M., 1978, Soil and water conservation and utilization for increased food production in the semi-arid tropics, ICRISAT, Hyderabad, India.

Krishna, J. H., 1979, Runoff prediction and rainfall utilization in the semi-arid tropics, *Ph.D Dissertation,* Utah State University, Logan, Utah.

Krishna, J. H., and Hill, R. W., 1979, Hydrologic investigations on small watersheds at ICRISAT, presented at the *Combined Meeting of the American and Canadian Societies of Agricultural Engineering,* Winnipeg, Canada.

Krishnamoorthy, C., Chowdhury, S. L., Anderson, D. T., and Dryden, R. D., 1977, Cropping systems for maximizing production under semi-arid conditions, *Indian Council for Agricultural Research,* All India Coordinated Research Project for Dryland Agriculture, Hyderabad, India.

Thierstein, G. E., and Kampen, J., 1978, New Farming Systems for Agriculture in the Semi-Arid Tropics, American Society of Agricultural Engineers Summer Meeting, Logan, Utah, USA.

Virmani, S. M., Shivakumar, M. V. K., and Reddy, S. J. 1979, Climatological features of the semi-arid tropics in relation to the Farming Systems Research Program, International Crops Research Institute for the Semi-arid Tropics, *Proceedings of International Workshop on the Agro-climatological Research Needs of the Semi-arid Tropics,* Hyderabad, India (In press).

Tropical Agricultural Hydrology
Edited by R. Lal and E. W. Russell
© 1981, John Wiley & Sons Ltd.

4.3

Sand Dune Fixation in Tunisia by Means of Polyurea Polyalkylene Oxide (Uresol)

M. De Kesel and D. De Vleeschauwer

4.3.1. INTRODUCTION

Desertification, mostly described as a more or less irreversible reduction of the plant cover (Le Houérou, 1968) is the single and perhaps the most important hazard that is threatening the low rainfall areas of the world today. In the not so distant past, Tunisia was covered by a woody plant protective cover ranging from shrubs to tall trees. In the centuries that followed and especially during the Roman rule, the abuse of the land and its vegetative cover, together with the severe climatic conditions, led to the destruction of a considerable part of the free cover. Today, the resulting picture of the havoc and destruction which has taken its toll throughout the ages is that of huge, practically completely denuded areas of shifting sand dunes, which form continuous belts throughput the agricultural areas and in the hinterland (Plate 17 and 18).

In the Kairouan region, an area of 20,000 ha has been invaded by shifting dunes, following the historical floods of the autumn of 1969. To protect the holy town Kairouan and its neighbourhood against the moving sand, the forest service of Tunisia started sand dune works which had two main objectives:

(1) fighting wind and water erosion and soil degradation by maintaining and creating a protective vegetative cover;
(2) production of new sources of wood through afforestation.

In the past few years, many methods have been tried out for the stabilization and afforestation of the drifting dunes. One of these methods is the conventional method consisting of the creation of artificial wind breaks made from palm branches or asbesto-cement corrugated plates (Ben Salem, 1974) (Plate 19). Those barriers are 200 m to 300 m apart and they are raised when needed. This treated area is then divided in small plots 30 m X 20 m each, by means of hedges of tamarisk. In these plots, a fast-growing tree species is planted.

Table 4.20. Particles of size distribution of the
soil at Oulled Dhifallah

Fractions μm	%
0–100	1.2
100–200	20.5
200–500	72.4
> 500	5.9

Table 4.21. Physico-chemical and chemical composition of the sand

pH H_2O	E.C.[a] (1/5) μS	CaCO$_3$ %	Organic Matter %	Total N mg N($100g^{-1}$)	Ca	Mg	K (p.p.m.)	Na	P	Fe
8.5	45	0	0.08	10.9	200	27	20	15	35	23

[a]The electrical conductivity of a suspension of one part of soil in five parts of water.

Besides the conventional method, other methods, based on a chemical fixation of the sand surface, have been tried out. Trials with bitumen emulsions proved to be unsatisfactory (Ramadan D'jaj, 1974). Other materials such as polyacrylamide and Curasol, which were very effective against water erosion under laboratory conditions, were not successful when used in the field (Gabriels, 1972; Mhiri, 1978). Another method of sand dune fixation consists of using sand stop, a material mainly composed of lignin. This method also proved to be unsuccessful because, as soon as the sand dried out, the coating broke away and eroded (Ramadan D'jaj, 1974).

The purpose of the present investigation was to evaluate the use of polyurea for the fixation of sand dunes near Kairouan, in comparison with the traditional method.

4.3.2. METHODS AND MATERIALS

The experiment was conducted in collaboration with the Ministry of Agriculture of Tunisia (Directorate for Forestry); the experimental site was situated near El Ala and was called the Oulled Dhifallah, about 100 km south-west of Kairouan. This area is subject to severe wind erosion, so that both the houses and the indispensable water wells are threatened by burial under the sand mass. The yearly precipitation is between 200 and 300 mm; the soil is sandy with a coarse texture (Table 4.20), a low organic matter content and a high pH (Table 4.21).

Plate 17 Shifting sand dune

Plate 18 Olive-tree roots exposed by wind and water erosion

Facing page 274

Plate 19 Artificial wind barrier made from asbesto-cement corrugated plates

Plate 20 Hydrophilic version of Uresol (left): hydrophobic version of Uresol (right)

Plate 21 Spray boom

Plate 22 Spray gun

Plate 23 Uresol skin of approximately 5 mm thickness

Plate 24 Left and upper middle treated with Uresol: right underneath is untreated

4.3.2.1. The Uresol Polymer

The polyurea prepolymer is composed of a polyfunctional polyol with terminal isocyanate groups. The polyol (R_1) used is a polycondensate of a mixtute of a hydrophilic ethylene oxide and a hydrophobic propylene oxide, their ratio depending on the degree of hydrophilicity or hydrophobicity required. The constituent polymers are

$$\text{Hydrophilic ethylene oxide } RO - (CH_2 - CH_2 - O)_n - CH_2 - CH_2$$
$$- OH \leftrightarrow 20 < n < 65$$

$$\text{Hydrophobic propylene oxide } RO - (CH_2 - \overset{\overset{\displaystyle CH_3}{|}}{CH} - O) - CH_2 - \overset{\overset{\displaystyle CH_3}{|}}{CH}$$
$$- OH \leftrightarrow 15 < n < 50$$

The actual pre-polymer used is soluble in water and is hydrolysed to give polyurea following the reaction

$$\text{prepolymer} + \text{water} \longrightarrow \text{carbamic} \longrightarrow \text{amine} + CO_2$$

$$O = C = N - R_1 - N = C = O + 2H_2O \rightarrow HO - \underset{\underset{\displaystyle O}{\|}}{C} - NH - R_1 - NH - \underset{\underset{\displaystyle O}{\|}}{C} - OH$$

$$\rightarrow H_2N - R_1 - NH_2 + 2CO_2,$$

amine + prepolymer   polyurea polymer

$$H_2N - R_1 - NH_2 + O = C = N - R_1 - N = C = O \qquad (R_1 - NH - \underset{\underset{\displaystyle O}{\|}}{C} - NH -)_n$$

The commercial product Uresol is a liquid and remains soluble until it is dried, when it becomes insoluble and can then aggregate the soil particles.

4.3.2.2. Laboratory Trials

Using the rainfall simulator as described by Gabriels and De Boodt (1975) the effectiveness of polyurea (Uresol) as soil conditioner was evaluated in comparison with bitumen and polyacrylamide. Table 4.22 summarizes the characteristics of these three chemicals. The mode of action of the polyacrylamide and bitumen emulsions have already been described in detail by Huylebroeck (1973) and De Boodt (1970) (Plate 20).

The solutions or emulsions of the chemicals were sprayed on air-dry sand from Oulled Dhifallah in small soil trays (20 × 30 cm). The trays were set at a slope of 20 per cent, and the soil was subjected to one or two hours of simulated rainfall at an intensity of about 40 mm hr^{-1}. Splash erosion was considered as criterion for evaluating the stability against surface erosion. The tests were replicated.

Table 4.22. Characteristics of the soil conditioners

Basic compound	Form	% Active material	Hydro-philicity	Application gl^{-1} m^{-2} (active material)	Trade name	Company
Bitumen emulsion	Emulsion	50	Hydrophobic	75	Humofina Bitumen	Labofina Belgium
Polyacrylamide	Solution	16	Hydrophilic	20	Humofina PAM	Labofina Belgium
Polyurea	Solution	73	Hydrophilic	50	Uresol	PRB-Recticel Belgium

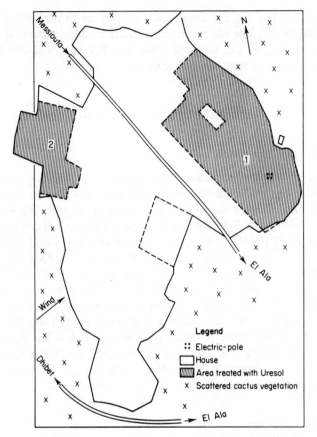

Figure 4.4 Layout of the experimental site (Scale 1:4,500)

4.3.2.3. Field Trial

For the afforestation of the sand area Oulled Dhifallah (60 ha), *Acacia cyanophylla* was chosen, because this plant has many advantages:
- rapid, early growth
- longevity
- dense crown
- excellent wind firmness
- thornless
- very resistant against drought
- valuable wood product.

One-year-old acacias were planted at a density of one plant per six square metres. Immediately after planting (beginning January 1979), eighteen out of the sixty hectares (Figure 4.4) were treated with a solution of 70 g Uresol in one litre of water per square metre. The soil conditioner was applied on to the soil by means of a spray boom (Plate 21), but when natural obstacles hindered smooth spraying with this spray boom, a spray gun was used (Plate 22). The remaining area (48 ha) was treated in the traditional way, however without tamarisk hedges.

4.3.3. RESULTS AND DISCUSSION

4.3.3.1. Laboratory Experiment

Since the soil conditioner must stabilize the sand dune until the vegetation is established, it must preserve the stability of the soil during this period of establishment. Thus the stabilized surface must be resistant to raindrop impact. This was evaluated using a rainfall simulator. The results of this test using three soil conditioners, bitumen, polyacrylamide (PAM), and polyurea (Uresol) are summarized in Table 4.23.

Splash erosion was reduced by as much as 90 per cent for PAM, 88 per cent for bitumen and 91 per cent for Uresol, which are excellent results. After a drying period of 2 days, the same soils were again subjected to a storm of 1 hour at the same intensity of 40 mm hr^{-1}; the results are summarized in Table 4.24.

During this second storm, the reduction of splash erosion was 70, 76, and 85 per cent for respectively PAM, bitumen, and Uresol, which is less than in the first run. Only for the Uresol treatment, was the soil surface still intact at the end of this second raintest, whereas for the PAM and bitumen treatments, small parts of the soil surface started to erode.

4.3.3.2. Field Experiment

4.3.3.2.1. Evaluation after 41 days

(a) Weather. Exceptionally, there was no rainfall in January; the first rain was only registered twenty days after planting, this initial dry spell was of course harmful to the growth of the young acacia. The total rainfall during the observed period amounted to 54 mm. In February, 19 sandstorms were registered with wind speeds varying from 70 up to 130 km hr^{-1} at a height of 2 m.

(b) Soil conditions. In the non-treated area, 20 cm sand had been blown away. This lowering of the soil surface could be clearly seen at the boundary of the treated plots, because these plots looked like an elevated table land compared with the adjacent eroded ones. Despite the many sandstorms, the Uresol film was still intact, and the thickness of the stabilized skin was about 5 mm (Plate 23). However the treated area 1 (Figure 4.4) was covered with a thin film of loose sand blown in from the adjacent untreated part (Plate 24).

Table 4.23. Splash erosion after 1 hour simulated rainfall

	Splash g	Amelioration %	Evaluation
Control	36		
PAM	3.6	90	Excellent
Bitumen	4.2	88	Excellent
Uresol	3.1	91	Excellent

Table 4.24. Splash erosion after 2 hours simulated rainfall

	Splash g	Amelioration %	Evaluation
Control	30		
PAM	9.1	70	Very good
Bitumen	7.3	76	Very good
Uresol	4.4	85	Excellent

Table 4.25. Evaluation of the vegetation after 41 days

Area	Number of plants per 100 m^2	Plants with shoots (%)	Uprooted plants (%)
Treated	15	70	0
Untreated	3	30	65

(c) Vegetation. The observations concerning number and appearance of the *Acacia cyanophylla* plants are summarized in Table 4.25.

From the original 15 plants per 100 m^2 only 3 were left on the untreated area, whereas on the treated part, all plants were still there. Another great difference was noticed in the number of uprooted plants: none on the treated part compared with 65 per cent on the untreated part.

As a result of the bombardment of the young plants by the blowing sand grains the leaves and stems facing the wind direction turned completely brown. But green shoots budded from the base of 70 per cent of the plants on the treated plots compared with 30 per cent of the untreated due to the higher moisture content of the soil under the stabilized surface.

Table 4.26. Evaluation of the vegetation after three months

Area	Number of plants per 100 m^2	Plants with shoots (%)	Uprooted plants (%)
Treated	13	75	3
Untreated	4	20	70

4.3.3.3. Evaluation after Three Months

(a) Weather. The extremely low temperature during the first two months after planting had strongly hindered the growth of the plants. The onset of the rains was much delayed; and the sand storms, reaching 130 km hr^{-1} lasted until mid-April.

(b) Soil conditions. Part of the treated site was covered with loose sand. In the most elevated parts and where no mechanical protection was present, the Uresol skin had been partly torn away.

(c) Vegetation. The observations concerning the number and appearance of the *Acacia cyanophylla* are summarized in Table 4.26.

As a consequence of the very serious loss of plants on the untreated area 7000 new acacia were planted in early March, at an average density of 15 per 100 m^2. In mid-April only 4 per 100 m^2 were alive compared with 13 original plants per 100 m^2 on the treated areas. At this time only 3 per cent of the plants on the treated areas were uprooted and 75 per cent had green shoots compared with 70 per cent uprooted and only 20 per cent with shoots on the untreated. The mean height of the acacia on the treated soil was 40 cm compared to 30 cm for the untreated area.

Another effect of the treatment was in the density of spontaneous vegetation: in the treated area, 20 per cent of the soil was covered compared to 10 per cent on the untreated. The following species were observed:

— *Retama retam*
— *Dactylis cynodum*
— *Orysopsis* (did not exist before the treatment).

4.3.4. CONCLUSIONS

A sand surface treated with the Uresol polymer solution showed a very good improvement in its resistance onto water erosion after two hours of simulated rainfall in the laboratory. In the field three months after treatment, it was found that the young trees make faster growth on the soil-stabilized plots. It is absolutely necessary to combine mechanical barriers (living or dead hedges and verges) with a soil stabilizer in order to get optimum results. The planting or sowing must only be carried out when the soil moisture and the temperature are favourable, for it is important that the young plants make sufficient growth to stabilize the soil surface by the time the soil stabilizer has degraded.

Finally we believe that this polymer has a great potential for improving the living conditions in many parts of the world.

REFERENCES

Ben Salem, 1974, *Proc. Inter-regional Training Centre on Heathland and Sand Dune Afforestation FAO/DEN/TF/ 123,* Country statement Tunisia, 206–208.

De Boodt, M., 1970, New possibilities for soil conditioning by means of diluted bitumenous emulsions, *FAO, ECA, W.R./70/4(b),* 11, 15 May 1970.

Gabriels, D., 1972, Response of different soil conditioners to soils, *Proc. Int. Symp. on Soil Conditioning,* Ghent, Belgium, 1014–1034.

Gabriels, D., and De Boodt, M., 1975, A rainfall simulator for soil erosion studies in the laboratory, *Pedologie,* **XXV,** 2, 80–86.

Le Houérou, 1968, La desertisation du Sahara septentrional et des steppes linitrophes, PBI, *Coll. de Hammamet, 26 p, London et Ann. Alger, de Geogr., no. 6,* 2–27, Alger.

Huylebroeck, J., 1973, De intcrakties tussen polymeren en bodemmineraleu in verband met bodemkonditionering. *Ph.D. Thesis,* University of Ghent.

Mhiri, 1978, Service des Forets—Ministere de l'Agriculture, *Personal Communication.*

Ramadan D'jaj M., 1974, Proc. Inter-regional training centre on Heathland and Sand Dune Afforestation FAO/DEN/TF 123, *Country Statement Libya,* 187–190.

Tropical Agricultural Hydrology
Edited by R. Lal and E. W. Russell
© 1981, John Wiley & Sons Ltd.

4.4

Effect of Reclamation of Alkali Soils on Water Balance

V. V. Dhruva Narayana and I. P. Abrol

4.4.1. INTRODUCTION

The soils in which exchangeable sodium is so high as to adversely affect plant growth are termed alkali or sodic soils. Alkali soils occur extensively in the Indo–Gangetic plains in north India (Figure 4.5) and are generally confined to areas with a mean annual rainfall between 550 and 1000 mm. Preliminary estimates (Abrol and Bhumbla, 1971) indicate that these soils occupy approximately 2,500,000 ha in the Indo–Gangetic plains alone. Large chunks of alkali soils are usually interspersed with productive normal soils. Increasing pressure on land resources and need for increased food production require that alkali soils, hitherto lying barren, are reclaimed for crop production. Investigations at the Central Soil Salinity Research Institute (CSSRI) at Karnal, since its establishment in 1969 have shown that these soils can be economically reclaimed through a series of steps involving land shaping, application of amendments like gypsum, growing rice during rainy season, adoption of proper agronomic and cultural practices and storage and utilization of rain water in the catchment (Abrol *et al.*, 1973; Narayana, 1979). This paper is a case study of the impact of reclamation measures on the water balance components of a representative alkali soil area.

4.4.2. SOIL AND CLIMATIC CHARACTERISTICS

The research farm of the Central Soil Salinity Research Institute at Karnal is representative of alkali soil catchments of the Indo–Gangetic region. Much of these areas were lying uncultivated for the last 80 years or so (Plate 25) due to the accumulation of excess exchangeable sodium in the soil profile.

These salt affected regions have a flat topography with gentle to very gentle slopes (slopes less than 0.5 per cent). The low lying areas often get flooded during

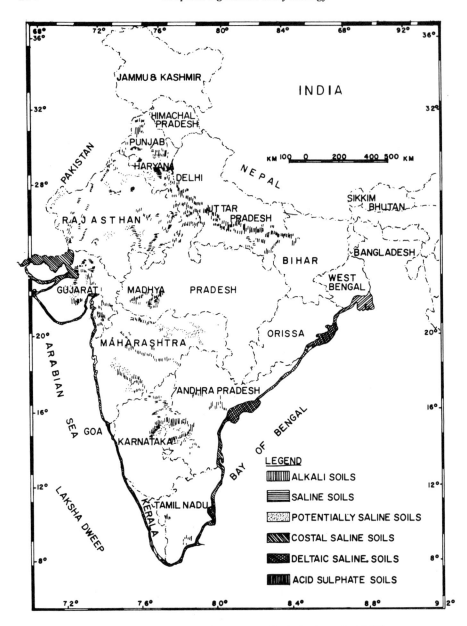

Figure 4.5 Salt affected soils in India (Abrol *et al.*, 1973)

rainy season and this water stagnates for prolonged periods even after the cessation of rains.

The climate of these areas is characterized by hot summer and cool winter. The

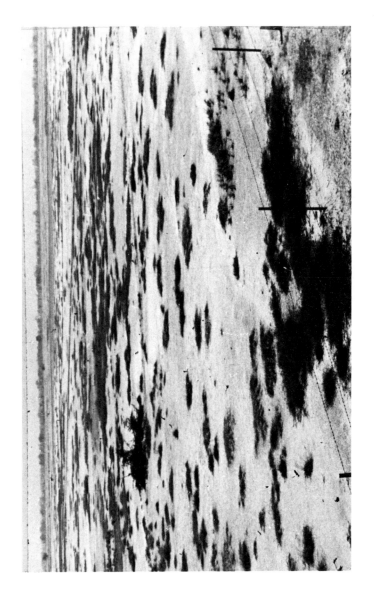

Plate 25 A view of alkali land

monsoon rains are received during the period from July to September. There is also a large variation in the temperature during the year. These climatic features give rise to two distinct cropping seasons in a year, viz. Kharif (summer) i.e. June to October and Rabi (winter) i.e. November to April. Both the annual rainfall and its monthly distribution are highly variable and the temperature is far less variable from year to year (Table 4.27).

The average annual rainfall is about 700 mm. Nearly 90 per cent of this is received from the south-west monsoon during the months of June to September. The average annual open pan evaporation value is approximately 1900 mm (Table 4.27) thus giving a net annual water deficit of about 1200 mm. Generally, this water deficit is experienced in all the months except during July and August.

The maximum two day rainfall and the length of dry spell of five year return period are estimated to be 20 cm and 34 days respectively (Table 4.28). However at 75 per cent confidence level (Figure 4.6), the occurrence of dependable rain (greater than 10 mm) in any five day interval during June–September period is limited only to certain spells during August. Such a situation calls for greater emphasis on the conservation of rainfall in this region not only from a drainage point of view, but also to meet the water requirements of crops during the dry spells of the growing season.

Physico-chemical characteristics of alkali soils of the Indo–Gangetic plains have been summarized by Abrol and Bhumbla (1977) as follows:

1. Excess soluble salts are present, chiefly, in the surface 30 to 60 cm soil depth. Sodium carbonate and bicarbonate form an appreciable portion of the soluble salts. The soils have a high pH (up to 10.5 in 1:2 soil water suspension).
2. Excess sodium carbonate in the soil causes the precipitation of calcium in the soil solution and thus increases the exchangeable sodium percentage (ESP) to a very high level. ESP is usually high throughout the profile depth, and the values of 80 to 90 are quite common.
3. Soils are generally medium textured (sandy loam to clay loam) in the surface layers and medium to heavy (clay loam to clay) in the subsurface layers.
4. Soils are calcareous and often have a zone of accumulation of calcium carbonate in amounts large enough to have a calcic horizon. The zone of accumulation of calcium carbonate may indicate the zone of fluctuating water table.
5. Illite is the dominant clay mineral.
6. Soils are highly dispersed and are extremely impermeable to water and air.

Some of the physico-chemical characteristics of an alkali soil profile are presented in Table 4.29.

Excess exchangeable sodium exerts a primary influence on soil physical properties including soil water behaviour (Abrol *et al.*, 1978; and Acharya and Abrol, 1978) and this in turn influences the various components of the hydrologic cycle. From the catchments with such soil characteristics, the runoff produced by storms of even a five year return period can be very high. The measured peak runoff values

Table 4.27. Climatic data for Karmal

Year	Annual				June–October (Kharif)				November–March[a] (Rabi)			
	Rain fall (mm)	Evaporation (mm)	Mean Temperature Maximum (°C)	Minimum (°C)	Rain fall (mm)	Evaporation (mm)	Mean Temperature Maximum (°C)	Minimum (°C)	Rain fall (mm)	Evaporation (mm)	Mean Temperature Maximum (°C)	Minimum (°C)
1971	520	2044	28.3	16.9	461	851	31.4	23.5	44	605	23.7	9.0
1972	811	2314	29.5	16.9	724	1032	33.4	23.2	62	474	22.3	9.0
1973	689	1909	29.3	17.1	629	724	33.4	24.0	17	494	23.0	8.1
1974	473	2091	30.2	16.1	426	876	34.0	22.7	72	446	22.8	7.9
1975	609	1871	29.6	16.4	638	774	32.8	23.3	65	401	23.4	8.4
1976	954	1753	29.8	16.6	871	799	33.5	23.1	30	434	24.6	7.9
1977	710	1612	30.2	16.9	597	688	33.3	23.4	176	378	23.2	9.0
1978	897	1582	29.8	16.8	741	687	33.2	23.3	148	348	23.4	9.1
Average	718	1900	29.6	16.7	636	804	33.1	23.3	84	448	23.4	8.6

[a]Up to March month of the next year.

Table 4.28. Maximum storm rainfall and dry spells of different return periods in Karnal region (Narayana *et al.*, 1978)

Duration of event	Return period in years					
	1.01	2.33	5	10	25	100
Maximum 1-day rainfall (cm)	4.1	12.0	15.2	18.3	22.1	28.2
Maximum 2-days rainfall (cm)	5.1	15.5	20.1	23.8	28.5	35.5
Maximum 3-days rainfall (cm)	6.1	17.1	21.9	25.8	30.7	38.1
Maximum 4-days rainfall (cm)	6.7	17.9	22.8	26.8	31.8	39.4
Maximum dry spell in monsoon season (days)	15	28	34	39	45	54

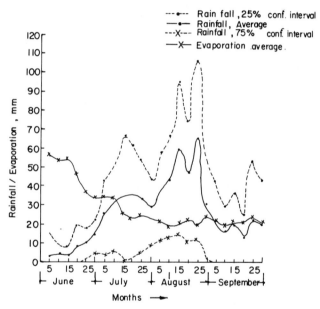

Figure 4.6 Five days' mean evaporation and mean rainfall with 25 per cent and 75 per cent confidence levels of rainfall (Narayana *et al.*, 1978)

reported by Sharma and Sehgal (1972) from these catchments are in the range of 13 to 26 m³ sec⁻¹ km⁻².

Fluctuations in groundwater table are associated with the rainfall patterns. A sharp rise in water table is recorded during the rainy season and this is followed by a gradual recession (Figure 4.7). However, due to large scale utilization of ground water in the region, there is a general declining trend in the groundwater levels. The

Table 4.29. Soil analysis of the experimental site

Depth cm	Mechanical composition (%)			CEC me $(100g)^{-1}$	$CaCO_3$ (per cent)	pH^a	Composition of saturation extract (mg l^{-1})								
	Clay	Silt	Sand				CO_3^{-2}	HCO_3^{-1}	Cl^-	SO_4^{-2}	Ca^{++}	Mg^{++}	Na^+	K^+	ESP
0–9	12.0	25.0	59.9	8.8	2.6	10.3	280.0	89.2	56.0	18.4	0.8	0.8	440.6	0.3	93
9–30	20.7	24.5	51.1	12.5	3.3	10.5	15.6	26.4	7.2	6.9	1.2	2.1	50.6	0.2	95
30–74	24.5	28.2	45.9	13.6	3.9	10.2	2.4	9.0	3.6	0.2	1.8	0.2	12.8	0.4	88
74–109	20.2	22.5	43.7	16.6	3.4	9.9	2.4	6.6	2.4	0.2	0.6	0.3	10.3	0.3	88

[a] pH was measured on 1:2 soil–water suspension

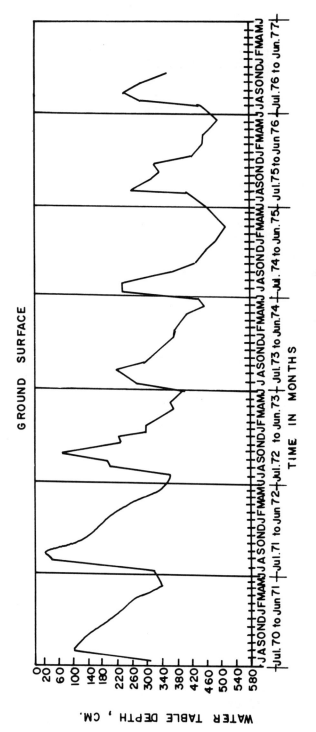

Figure 4.7 Ground water fluctuations on the CSSRI Farm, Karnal

quality of groundwaters in the area is uniformly good thoughout the year with electrical conductivity of about 500 micromho cm^{-1}.

4.4.3. RECLAMATION PROCEDURES

4.4.3.1. Package of Practices

The following steps are adopted for the reclamation of alkali soils.

(i)   Land levelling and bunding (small earthen ridges on the field peripheries)
(ii) Application of gypsum at the rate of up to 15 t ha^{-1} depending on initial soil sodicity. Gypsum is broadcast on the soil surface and shallow cultivation is done so that gypsum mixes with the soil up to 10 cm depth only. Gypsum is to be applied only once and the application is not to be repeated.
(iii) Growing of rice crop during the rainy season (June–October) and wheat during the winter season (November–April) with appropriate cultural and agronomic practices.

In this procedure, it is clearly emphasized that application of gypsum alone is not sufficient for good crop yields. It is important to follow proper cultural and management practices. These include choice of suitable crops and varieties in a rotation, correct time of transplanting and sowing of crops, maintaining adequate plant population, use of adequate amounts of fertilizers including micronutrients, adoption of proper water management practices, and timely plant protection measures, etc. (Abrol *et al.*, 1973).

It has been shown (Mehta and Abrol, 1974; Mehta *et al.*, 1975) that, by adopting these practices, moderate to good yields of rice (3–4 t ha^{-1}) can be obtained in the first year of reclamation and good yields in the second and subsequent years. However it usually takes two to three years before good yields of wheat can be obtained. It has been further shown (Chhabra and Abrol, 1977; Abrol and Bhumbla, 1979) that rice acts as a reclamation crop and including this crop in a rotation further improves the soil physical properties including infiltration rates.

4.4.3.2. Rainwater Management

The rainwater management procedure recommended for these soils consists of a three-tier system, with the following features (Narayana, 1979).

1. Collection of part of the rainfall in the cropland until such time and extent as will not be harmful to the crop.
2. Directing the runoff (after storage in step 1) from various parts of the catchment into the dugout storage ponds of sufficient capacity located in the

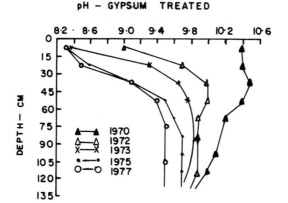

Figure 4.8 Changes in soil pH at different depths
in a gypsum treated plot as a result of continuous
cropping (Singh and Abrol, 1977)

lowest portions of the catchments. The stored water is utilized for irrigation
in the adjacent lands by pumping during the dry spells of the monsoon
season.
3. The remaining excess water is then discharged into the shallow surface drains
   provided on a regional basis.

Field studies conducted on the storage of rainwater in the cropland and in the
dugout ponds indicated that rainwater up to 15 cm of storm rainfall could be
stored within the bunded fields without affecting the rice yields adversely. Further,
storage of rainwater in the dugout ponds during heavy storms and utilizing the same
for irrigation during the intervening dry spells of the rainy season serve the dual
purpose of drainage and irrigation.

4.4.4. HYDROLOGIC EVALUATION

Reclamation measures outlined above bring about significant changes in the soil and
hydrologic characteristics of a watershed and these in turn affect the hydrologic
balance of the area.

4.4.4.1. Soil Physical Characteristics

Reclamation measures bring about gradual improvement in the physical properties
of the soil profile. Initially prevailing high levels of exchangeable sodium are replaced
with more favourable calcium resulting in the improvement of water retention and
transmission properties of soils (Figure 4.8). An idea of the extent of improvement
in this regard could be obtained from Table 4.30, where the variation of intake
rates of sodic soils at different stages of reclamation are shown. It could be seen

Table 4.30. Effect of reclamation on the infiltration characteristics (Abrol, 1977)

| Soil | ESP | | Basic intake rate (After-72 hours) (mm day^{-1}) | Cumulative intake in 20 hours (mm) |
	Depth range (0–15 cm)	Depth range (15–30 cm)		
Unreclaimed sodic soil	95	90	0.5	10
Reclaimed sodic soil (After gypsum application and taking 2 rice crops)	16	68	18	49
Normal soil unaffected by sodicity	–	–	50	104

that the basic intake rate increased from almost a negligible value to about 18 mm day^{-1} just in a period of one year after reclamation had begun. If rice is included in the cropping sequences, the rates will further improve and would approach those of normal soils after 6 to 7 years. Similarly effect of the degree of soil improvement as judged from soil ESP on the cumulative intake by soils is presented in Table 4.30. These data indicate that with reclamation increased amount of water will pass through the profile and enhance the soil moisture storage in the root zone besides contributing to the ground water recharge.

Reclamation of the alkali soils, through application of gypsum, also affects the soil–water evaporation (Figure 4.8).

During June–September periods when high intensity storms are intervened by prolonged dry spells, most of the rainfall is lost as surface runoff due to poor water absorption and storage and is usually collected in the depressional areas. During the dry spells, whatever little water is absorbed, in the surface soil layers, is lost rapidly as evaporation. As these soils get reclaimed, much of rainfall could be absorbed within the soil profile due to favourable changes in soil properties.

4.4.4.2. Catchment Retention Storage

Catchment retention storage is generally computed by measuring the rainfall and the net runoff (Narayana, 1977). Under improved land management systems (as at the CSSRI research farm) the catchment retention storage also includes the water stored in the crop land and in the dugout ponds. The water stored in the dugout ponds may be recycled for irrigation during the dry spells. An example of the reten-

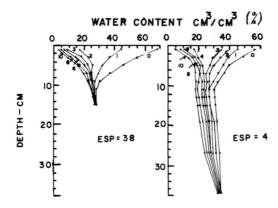

Figure 4.9 Soil water contents of the profiles initially and on 1–10 days under high evaporative demand for soils of ESP 38 and 4 (Acharya and Abrol, 1978)

Table 4.31. Rainwater storage during June–September for the CSSRI research farm (Narayana *et al.*, 1979)

| Year | Rainfall (mm) | Runoff (mm) | Water retained | | Water retained as percentage of rainfall |
			Dugout ponds (mm)	Crop land (mm)	
1972	695	121	80	494	82.5
1973	589	94	10	395	84.5
1974	379	5	16	358	99.0
1975	605	12	52	540	98.5
1976	867	100	122	645	88.0
1977	589	56	87	446	94.0

tion storage in a catchment at CSSRI from 1972–1977 is shown in Table 4.31. It is obvious that retention storage was maintained fairly high and the average crop yields and the soil properties (as indicated by ESP) of the area under reclamation also improved.

Singh *et al.* (1977) also studied the relationship between the individual storm rainfall and the retention storage (*RI*) for the years 1976 and 77 in a relatively large catchment of 1200 ha which is under reclamation. The *RI* values for 1977 (Figure 4.10) are much higher than those of 1976 particularly for higher storm values thus confirming that more rainfall is retained in the catchment due to reclamation.

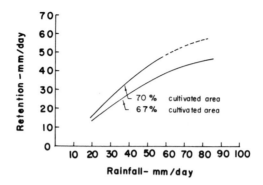

Figure 4.10 Changes in the catchment retention storage due to progressive reclamation
(Singh *et al.*, 1979)

4.4.4.3. Runoff Characteristics

The effect of reclamation on the runoff characteristics of the watershed can be
evaluated by different methods. The runoff from a watershed may be gauged before
and after adopting the reclamation measures on two comparable catchments—one
may be kept as control and the other subjected to reclamation measures. Alternatively, runoff measurements could be recorded at the same gauging station every
year after reclamation measures were implemented in the watershed and then
compared.

Typical runoff hydrographs recorded on CSSRI farm are shown in Figure 4.11.
These hydrographs show that the reclamation measures would decrease the values
of peak and volumes of runoff for a given storm intensity. An example of runoff
hydrographs produced by the same storm in comparable normal and unreclaimed
watersheds are shown in Figure 4.11(d). The values of peak and volume of runoff
from the normal catchment are 0.43 and 0.25 times the corresponding values from
the unreclaimed watershed.

4.4.5. WATER BALANCE ANALYSIS

Water balance analysis of any area is based on the equation of continuity of mass
expressed as:

Input = Output ± changes in watershed storage.

In this analysis the measured or estimated inputs (I) include precipitation and
runoff or seepage from other sources. The outputs (O) are runoff, evapotranspiration, and changes in soil water storage. Such an analysis will reflect the effects of
reclamation measures on different components of the hydrologic cycle. A water
balance analysis of CSSRI research farm for the June–September period (1972-77)
is presented in Table 4.32. (Narayana *et al.*, 1979; Narayana and Gupta, 1974).

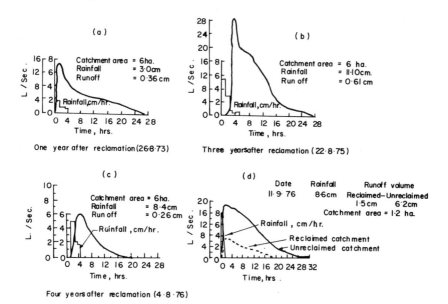

Figure 4.11 Gauged runoff hydrographs from alkali and soil watersheds (Narayana and Singh, 1976)

4.4.5.1. Changes in Hydrologic Components

Changes in the ratios of runoff to rainfall, water table levels and the groundwater recharge are presented in Figure 4.12. It is seen that the ratio of runoff to rainfall decreased progressively. The rise in ground water levels, during June–September declined from 1972 to 1974 probably due to excessive ground water pumping. Differences between the pre- and post-monsoon ground water levels after 1974 have stabilized approximately at a value of 2.40 m. This corresponds to a replenishment of about 24 cm of water to ground water storage (Narayana *et al.*, 1974). This trend of ground water fluctuation may also be confirmed from the changes in the values of water surplus, during the period of the study. To start with the water surplus available for local recharge or the difference between the total input and the output (Table 4.32) declined from about 26 per cent of the seasonal rainfall in 1972 to zero level in 1974. This may be due to relatively higher runoff losses from the area in the first season of reclamation. It has later on increased up to 40 per cent by 1977 probably due to consistently good rainfall and increased catchment infiltration rates in the area. These are the direct consequences of the various reclamation measures in the study area.

The significant aspects of these observations are that favourable water balance has been attained even while the probable recharge from outside the study area has declined to nearly zero level and with a generally decreasing trend of ground water levels in the region (Narayana *et al.*, 1978; Narayana and Singh, 1976). Moreover,

Table 4.32. Water balance analysis of the research farm (June–September)

Particulars		1972	1973	1974	1975	1976	1977
Precipitation	(Cm)	69.5	58.9	38.4	60.4	86.7	58.9
Canal seepage[a]	(Cm)	4.0	4.0	4.0	4.0	4.0	4.0
Total output (I)	(Cm)	73.5	62.9	42.4	64.4	90.7	62.9
Runoff	(Cm)	12.1	9.4	0.5	1.2	10.0	5.5
Evapotranspiration	(Cm)	34.2	37.2	37.9	35.8	42.6	25.2
Addition to soil moisture storage	(Cm)	9.0	8.0	6.8	8.8	10.9	7.5
Total output (O)	(Cm)	55.3	54.6	45.2	45.8	63.5	38.2
Water surplus for local recharge $S = (I\text{-}O)$	(Cm)	18.2	8.3	−2.8	18.6	27.2	23.7
	%	26	14	0	31	31	40
Computed input to ground water (G)	(Cm)	29.1	18.7	17.1	23.9	25.5	22.8
	%	42	32	45	40	29	39
Probable recharge from outside the area (G–S)	(Cm)	10.9	10.4	19.9	5.3	−1.7	−0.9
	%	16	18	52[b]	8	0	0
Runoff (R)	%	17	16	1	2	12	9

% These are values expressed as percentage of rainfall.

[a] Value adopted from Narayana and Gupta (1974).

[b] This unusually high value is due to the non-operation of the Haryana state tubewells in the vicinity of the study areas during this season.

the reclaimed area crop yields also continuously improved with corresponding improvement in the soil properties. The reclamation of alkali soils, will, thus, not only improve the agricultural production but also change the water balance favourably to minimize the drainage needs of such areas and to meet some of the water needs for reclamation itself. Most of the water needs could be met from within the area by increased rain water utilization and induced groundwater recharge.

REFERENCES

Abrol, I. P., 1977, Exchangeable sodium and soil water behaviour, *Proceedings Indo-Hungarian Seminar on Management of Salt Affected soils,* Central Soil Salinity Research Institute, Karnal, pp. 103–119.

Abrol, I. P., and Bhumbla, D. R. 1971, Saline and alkali soils in India—their occurrence and management, *World Soil Resour. Rep. FAO Rome,* **41**, 42–51.

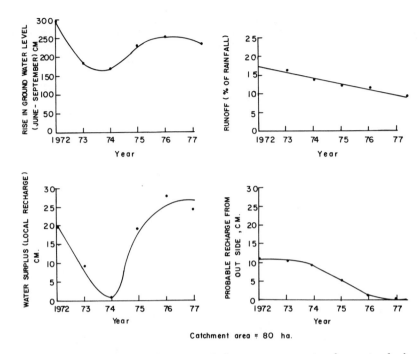

Figure 4.12 Changes in the water balance components of a watershed under reclamation (Singh *et al.*, 1979)

Abrol, I. P., and Bhumbla, D. R., 1977, Sodic soils of the Indo-gangetic plains in India–Characteristics, Formation, and Management, *Paper presented at FAO Expert consultation on Identification and reclamation of Salt Affected Soils and Secondary Salinization,* December 5-7, 1977, Rome.

Abrol., I. P., and Bhumbla, D. R., 1979, Crop responses to differential gypsum applications in a highly sodic soil and the tolerance of several crops to exchangeable sodium under field conditions, *Soil Sci.,* **127**, 79–85.

Abrol., I. P., Dargan, K. S., and Bhumbla, D. R., 1973, Reclaiming alkali soils, Central Soil Salinity Research Institute, *Karnal. Bull.,* **2**, 58p.

Abrol., I. P., Saha, A. K., and Acharya, C. L., 1978, Effect of exchangeable sodium on some soil physical properties, *J. Indian Soc. Soil Sci.,* **26**, 98–105.

Acharya, C. L., and Abrol., I. P., 1978, Exchangeable sodium and soil water behaviour under field conditions, *Soil Sci.,* **125**, 310–319.

Chhabra, R., and Abrol., I. P., 1977, Reclaiming effect of rice grown in sodic soils, *Soil Sci.,* **124**, 49–55.

Mehta, K. K., Yadav, J. S. P., and Abrol, I. P., 1974, Reclamation of alkali soils makes impact on Sangrur farmers, *Indian Farming,* **25(1)**, 7–8.

Mehta, K. K., and Abrol., I. P., 1975, In village Kachhwa alkali fields turn productive, *Indian Farming,* **25(9)**, 17–20.

Narayana, V. V. Dhruva, 1977, Reclamation of alkali soil reduces flood problems in the catchment, *Indian Farming,* **27(2)**, 7–9.

Narayana, V. V. Dhruva, 1979, Rain water management for low land rice cultivation in India, *Journal of the Irrigation and Drainage Division, ASCE,* **105(IR 1)**, 87–98.

Narayana, V. V. Dhruva, and Gupta, S. K., 1974, Analysis of canal seepage Paper presented at XII annual convention of I.S.A.E. Bhubaneswar, 5–7 March, 1974.

Narayana, V. V. Dhruva, Gupta, S. K., and Joginder, Paul, 1973, Analysis of water table fluctuations for the study of aquifer properties. *International Syposium on Ground Water Resources, Madras,* **I(II-1)**, 29–40.

Narayana, V. V. Dhruva, Gupta, S. K., and Tiwari, A. K., 1978, Rainfall and runoff analysis for rain water management in agriculture, *Proceedings of the Symposium on Hydrology of Rivers with Small Catchments,* C.B.I.P. New Delhi, 11, 23–28.

Narayana, V. V. Dhruva, Kalra, V. D., and Tiwari, A. K., 1979, Evaluation of alkali soil catchment responses to reclamation through water balance analysis, *ICID Bulletin,* **28(1)**, 50–55.

Narayana, V. V. Dhruva, and Singh, O. P., 1976, Hydrologic responses to various reclamation measures in alkali soils, *Proc. National Symposium on Hydrological Problems Related to the Development of Power and Industries,* I.I.T. Kanpur III, 105–114.

Sharma, D. R., and Abrol, I. P., 1973, Quality of some drain waters in Haryana, *J. Indian Soc. Soil Sci.,* 21, 349–353.

Sharma, R. P., and Sehgal, S. R., 1972, Design criteria for drainage projects in Punjab, Symposium on water logging causes and measures for its prevention, *C.B.I.P. Publication,* **118(1)**, 111–122.

Singh, O. P., Narayana, V. V. Dhruva, and Tiwari, A. K., 1979, Analysis of hydrologic changes taking place in alkali soil watersheds during reclamation, *Canadian Agricultural Engineering,* **21(2)**, 147–150.

Singh, S. B., and Abrol, I. P., 1977, Effect of gypsum on long term changes in soil properties and crop growth, *Annual Report, CSSRI,* Karnal 132001, India, 174pp.

Tropical Agricultural Hydrology
Edited by R. Lal and E. W. Russell
© 1981, John Wiley & Sons Ltd.

4.5

Impact of Intensive Silviculture on Soil and Water Quality in a Coastal Lowland

RICHARD F. FISHER

4.5.1. INTRODUCTION

The hope of many forestry organizations in the tropics is to convert natural forest to highly productive domesticated forests of fast growing species. The experience of declining productivity in second and subsequent rotations of domesticated forests in Australia (Florence, 1967) has caused us to become concerned that such a fate might befall similar forests in the tropics. To study this we have isolated some experimental watersheds in the coastal lowlands of Florida.

4.5.2. STUDY SITE

The study area is located in the coastal lowlands of western Florida. This area is derived from Pleistocene marine deposits and extends in terraces from sea-level to about ten metres elevation. The terrain is flat and is occupied by forests of recent origin. These consist of slash pine (*Pinus elliottii* var. *elliottii* Engelm.), mixed slash pine-longleaf pine (*Pinus palustris* Mill.), or mixed pine-hardwoods, interspersed with a variety of wetland types. The soils are predominantly poorly drained, sandy, inceptisols. These siliceous soils are highly weathered with low cation exchange capacities. The organic fraction of the surface soil and the forest floor layers contain most of the nutrient reserves.

The N content of coastal plain soils is low in comparison to most forests (Table 4.33). Phosphorus values for coastal plain soils range from 2 to 10 kg ha^{-1}, based on double acid extraction of the top 30 cm of mineral soil. Although such values are generally considered to be lower than those found elsewhere regardless of forest type, direct comparisons of extractable P with published values for other forest soils are difficult due to the differing methods of extraction used (Ballard, 1973; Pritchett, 1976).

Table 4.33. Estimated total nitrogen content in the rhizosphere mineral soil of diverse forest types

	kg N ha^{-1}
Subtropical pine forest	
Florida, USA	1200
Temperate pine forest	
Mississippi, USA	3000
Boreal pine forest	
Canada	3380
Tropical rain forest	
Costa Rica	1750
Temperate broadleaf	
Illinois, USA	6800

Table 4.34. Nitrogen and phosphorus reserves in a natural secondary growth slash pine forest on subtropical lowland

Component	kg N ha^{-1}	kg P ha^{-1}
Tree		
foliage	22	2
branches	16	1
stem	58	5
Total tree	96	8
Understory	107	5
Forest floor	51	2
Total above ground	254	15
Rhizosphere mineral soil	2560	2
Total site	2814	17

Because these soils are low in nutrient reserves, largely associated with the organic fraction, productivity of coastal plain slash pine forests is dependent on the maintenance of a conservative nutrient cycling system. The high frequency with which slash pine responds to fertilizers applied to these wet sandy soils (Pritchett and Smith, 1974) attests to their lack of soil nutrient reserves and the consequences of disrupting the conservative cycling through harvest removal and intensive site preparation.

The primary sources for cycling of nutrients within these ecosystens include: (a) the living biomass, (b) decomposing organic matter, (c) soil mineral reserves, and (d) precipitation and fixation inputs (Switzer and Nelson, 1972). Analysis of the data in Table 4.34 indicates that, of these primary sources of nutrients available, 7 per cent of the N is immobilized in the living aerial biomass with an additional 2 per cent contained in forest floor material. These N values range from one-half to three-fourths of those reported for loblolly plantations in the upper coastal plain and Piedmont (Switzer and Nelson, 1972; Wells and Jorgensen, 1977). Although 90 per cent of the site N reserves are contained in the rhizosphere mineral soil, the majority of this is associated with the organic matter fraction of these highly weathered sands. Consequently, any major perturbation of the vegetation component can result in the depletion of the source of this soil organic fraction, and, ultimately, cause a reduction in three of the four primary nutrient sources listed above.

A similar analysis with respect to P indicates that approximately 80 per cent of the available P is immobilized in the above ground organic fraction. This strongly amplifies the contention of Pritchett and Wells (1978) that lower coastal plain slash pine forests are highly inelastic and sensitive to disturbance.

4.5.3. METHODS

Two rectangular watersheds of approximately 400 hectares were isolated by a system of roads and drainage ditches. The roads were sloped away from the ditch adjacent to the watershed so that those ditches carried no runoff from the road surface. All of the runoff from each watershed was collected into a single ditch and monitored by a culvert weir. Water samples from these weirs were collected systematically for analysis of nutrient content.

Throughout each watershed 18 sets of porous cup soil solution sampling tubes were installed at 10, 25, and 100 cm depths. These tubes were sampled seasonally to monitor the movement of nutrients downward. These and the weir water samples were analysed for pH, NO_3-N, NH_4-N, total N, PO_4-P, K, Ca, Mg, Al, and suspended solids.

After a short calibration period one of the watersheds was harvested, site prepared and planted. The site was chopped by a rolling drum chopper, felled by crawler tractor mounted shears, and skidded tree length with rubber tired tractors. After harvest the site was again chopped and then bedded with crawler tractors using bedding plows. After the beds had settled for several months the treatment watershed was planted with slash pine seedlings at a 2 X 3 m spacing.

4.5.4. RESULTS AND DISCUSSION

4.5.4.1. Harvest Losses

Clearcutting and slash disposal not only remove a substantial proportion of the total biomass from the site, but significantly alter the distribution and subsequent availability of organically bound nutrient reserves.

Estimated percentages of nutrients removed from the site, as a function of harvesting intensity, are presented in Table 4.35. A low intensity harvest, which

Table 4.35. Percentage of above ground nitrogen and phosphorus site reserves removed with the various components during harvest and site preparation in coastal lowland

	N (%)	P (%)
Stems only	2	29
Whole tree	4	44
Understory	4	27
Forest floor	2	12
Total biomass	10	83

removes only merchantable stems from the site, results in an estimated loss of 2 per cent of the total site of N reserves. Pritchett and Wells (1978) estimated that probably no more than one per cent of the elements contained in this mineral soil-organic complex becomes available annually through biological mineralization unless the site is disturbed. Consequently, one intensive harvest removal alone in these lower coastal plain slash pine forests, in the absence of any other form of site disturbance, represents a loss of N equivalent to an amount that would normally become available through mineralization of the organically bound soil reserves over a 10-year period. Such intensive harvest losses of N are also equivalent to the cumulative quantity of N recycled through southern pine systems over a 4-year period (Switzer and Nelson, 1972).

4.5.4.2. Site Preparation Losses

One of the important components in the nutrient cycling process in managed slash pine forests is the residual organic material left after harvesting. This can include foliage, branches, understory vegetation, and forest floor material. The quantity of residue remaining depends not only on the degree of harvest removal, but also on the intensity of site preparation. This residual slash and forest material can contain, 200 kg ha^{-1}, or 7 per cent, of the total site N reserves, and 10 kg ha^{-1} P, in coastal plain pine ecosystems (Table 4.34). These elements subsequently become available through decomposition and mineralization during early stages of stand development.

The variety of site preparation methods available entail a series of treatments designed to reduce competing vegetation, promote decomposition of organic matter and release of nutrients, improve soil aeration and water relations, facilitate planting and ensure seedling survival and early growth (Shoulders and Terry, 1978). The particular methodology selected depends upon landowner objectives and varies from a minimum reduction of residual organic matter by chopping to virtually complete removal of residual organic matter by burning, rootraking, and blading into windrows (Balmer and Little, 1978).

Table 4.36. Estimated average annual losses on nitrogen and phosphorus, due to runoff, from a natural forest and a young plantation of slash pine in the coastal lowlands

	NO_3-N	NH_4-N	Total N	PO_4-P	Total P	Suspended solids
			kg ha^{-1} yr^{-1}			
Natural stand	0.1	0.2	4.8	0.04	0.2	44
Plantation						
1 year old	0.2	1.8	10.9	0.30	0.8	550
2 years old	0.1	0.4	9.6	0.01	0.2	178
3 years old	0.1	0.2	6.5	0.01	0.2	40

Nutrient losses resulting from minimal site preparation primarily are restricted to leaching and runoff. Although the latter vary considerably with site conditions, for the sites examined here probably less than one per cent of the original N reserves and considerably less P would be lost following minimal intensity site preparation. In contrast, maximum intensity site preparation, which involves some combination of burning with blading, root-raking, discing, and bedding, virtually eliminates the above ground organic matter and nutrient reserves, particularly when coupled with maximum harvest removal. The consequences of such maximum intensity organic matter depletion, as reported by Haines *et al.* (1975) and Pritchett and Wells (1978), are a substantial reduction in initial survival and early height growth. Of greater importance are the long-term consequences resulting from removal of substantial proportions of the nutrient reserves. Since the majority of soil nutrients are organically bound in these coastal plain sands, harvesting and site preparation methods (which deplete the supply of soil organic matter) potentially can reduce the exchange capacity of these soils and substantially impoverish one primary source of nutrient reserves. For example, Haines *et al.* (1975) and Maki (1976) reported that the depletion of surface and soil organic matter, as a consequence of burning, root-raking, and scalping of the soil, damaged site quality sufficiently to reduce wood yield in planted loblolly pine by 1000 ft^3 acre (70 m^3 ha^{-1}) 19 years after establishment.

From a management perspective, the complete removal of logging slash and litter from the site described in Table 4.34 would require the equivalent application of 120 to 200 kg ha^{-1} yr^{-1} of ammonium nitrate and 10 to 20 kg ha^{-1} yr^{-1} of concentrated superphosphate fertilizers over a 5 year period to restore the site to its initial nutrient condition.

4.5.4.3. Runoff and Erosion Losses

The loss of nutrients from coastal slash pine forests *via* runoff and erosion has largely been ignored. For the first three years after harvesting and reforestation N and P losses averaged a two-fold increase (Table 4.36). The largest proportional

Table 4.37. Nutrients in the upper 25 cm of soil before and one year after site preparation on coastal lowland sites

| | N | P | K | Ca | Mg | Al |
	%			p.p.m.		
Natural	0.21	1.8	27	13	3	420
Planted	0.12	2.5	24	40	3	400

loss was associated with the soluble ammonium fraction. The majority of the N and P lost following regeneration was associated with the suspended particulate matter (52 per cent of which was organic) removed from the surface by the frequent flooding this site experiences. The highest values for runoff losses of both elements were observed during first year following harvest and site preparation.

Although increases in both N and P in the runoff were observed following reforestation, the area was fertilized at planting with 40 kg N ha^{-1} and 50 kg P ha^{-1}, thus some of the losses undoubtedly included the fertilizer addition. Even with this additional source of nutrients contributing to runoff losses, the quantity of soluble N lost, particularly in the nitrate fraction, is generally lower than that reported for other disturbed forested ecosystems (Swank and Douglass, 1977; Aubertin and Patric, 1974).

The low soluble N and P values in the runoff are not necessarily serious losses in relation to the total site nutrient reserves; however, they do illustrate the relatively low availability of nutrients in these forested ecosystems and the need to maintain a conservative nutrient cycling system.

4.5.4.4. Soil Nutrient Changes

A comparison of the extractable nutrients in forest soils in relation to soil solution content before and after harvesting and site preparation provides an indication of the rate of nutrient mobilization, availability and potential for loss through leaching and runoff.

The amounts of extractable nutrients in the soil before and one year after harvest and intensive site preparation are shown in Table 4.37. Although fertilizer was applied (40 kg N and 50 kg P ha^{-1}) at time of planting, the quantity of total N decreased. Part of any change in this system is due to rapid mineralization and solublization of N in the residual organic fraction on the soil surface. This mobilized fraction of N has the potential to be lost from the system by leaching, runoff, or denitrification if rapid revegetation of the site does not occur. In contrast to N, the extractable P in the mineral soil increased following harvest and intensive site preparation. Undoubtedly, this was the result of fertilization and subsequent binding of the applied P by the organic and Al fractions of the soil. The concentration of

Table 4.38. Soil solution nitrogen and phosphorus concentrations before and after preparation of coastal lowland sites

	p.p.m. NO_3-N			p.p.m. NH_4-N			p.p.m. P		
Depth—cm	10	25	100	10	25	100	10	25	100
Natural	0.01	0.03	0.00	0.2	0.4	0.04	0.03	0.02	0.03
0 year	0.21	0.37	0.01	2.4	3.0	0.04	0.13	0.05	0.04
1 year	0.13	0.12	0.00	3.2	4.3	0.01	0.02	0.01	0.02
2 year	0.02	0.11	0.01	0.1	2.3	0.02	0.04	0.04	0.04
3 year	0.01	0.03	0.01	0.2	0.9	0.02	0.03	0.03	0.02

other elements (Table 4.37) remained relatively unchanged with the exception of Ca. The increased Ca content of the soil again suggests that rapid mineralization of residual organic matter occurred after site disturbance.

An examination of the soil solution concentrations of N and P on this site (Table 4.38) shows, in fact, a large increase in soluble N in the A horizon, particularly in the ammonium fraction, shortly after site disturbance. Nitrogen appeared to be readily available during the first two years following site disturbance. However, in the third year the N availability returned to that observed for the undisturbed condition. To what extent the decline in soil solution N is the result of leaching losses, uptake by reinvading vegetation, or denitrification is unknown. Leaching does not appear to be important since no change in concentration of either N or P occurred at the 100 cm depth.

The lack of any appreciable change in soluble P is further evidence of the soil's P fixation capacity. Also, this underscores the problems encountered during attempts to alleviate P deficiencies in these sites in terms of elemental sources and their rates of application (Pritchett and Smith, 1974).

4.5.4.5. Soil Physio-chemical Modifications

Mechanical disturbances of forest soils during harvesting and site preparation can have both beneficial and detrimental effects on soil physical and chemical properties (Shoulders and Terry, 1978). The direct effects are upon the soil–water–air systems that affect the ability of roots to effectively penetrate and utilize the soil matrix. Soil compaction from the use of heavy machinery can reduce its porosity and the rate of water infiltration and increase erosional losses. Soils may require a decade or more to recover naturally from such impacts (Patric and Reinhart, 1971; Dickerson, 1976). For coastal plain soils in North Carolina, Shoulders and Terry (1978) reported that soil porosity changes resulting from site preparation persisted up to 6 years after bedding and resulted in an increase in soil aeration and percolation, and a decrease in resistance to root penetration and soil moisture storage. The resultant changes were considered beneficial to pine growth. However, Schultz (1976)

Table 4.39. Physical characteristics in the upper
25 cm of natural and disturbed coastal lowlands

	pH	Organic matter %	Bulk density (g cm^{-3})
Natural	3.9	7.0	0.82
1 year	4.5	4.8	0.99
2 year	4.3	3.1	0.96
3 year	4.1	4.3	0.92

reported that beneficial site preparation effects on a sandy soil in Florida persisted for less than 3 years. Severe damage to the structural and textural properties of forest soils will also retard root penetration, microbial activity, nutrient, and moisture availability, and the subsequent growth of the new tree crop. Whether site preparation improves soil physical properties or causes structural damage depends upon the soil type and moisture conditions at the time of perturbation (Dickerson, 1976).

Bulk density and organic matter content, factors that are closely linked to the textural structure of forest soils, were examined following harvesting and site preparation (Table 4.39). Harvest, chopping, and bedding activities reduced the organic matter content in the top 30 cm of soil. The substantial reduction (55 per cent) in organic matter in the soil following site preparation represents a depletion of the soil nutrient reserve storage potential.

The effects of harvesting and site preparation on bulk density and soil pH appeared to be relatively minor with a tendency to return to base line conditions during the observational period. These changes probably had little effect on the ability of roots to penetrate the soil or absorb nutrients and the data correspond favourably with the findings of Schultz (1976).

4.5.4.6. Hydrological Changes and Water Quality

Increased stream flow, as a result of forest canopy removal, has been reported for other areas of the south (Hibbert, 1967; Douglass and Swank, 1972). This increase has been attributed to the reduction of transpirational surface area, a reduction of interception and reevaporation, and decreased soil water storage capacity. Few studies of stream flow response to harvesting systems have been conducted in the lower coastal plains, partly due to the difficulty of establishing definable watersheds in an area of low relief and deep permeable soils.

During the first year after site preparation stream flow increased only 5 cm above control. During the second year runoff from the young plantation area was only 14

Table 4.40. Quality of stream discharge from natural and intensively managed coastal sites

Quality parameter	Natural	Annual means (mg/l)		
		1 year	2 year	3 year
NO_3-N	0.01	0.05	0.02	0.01
NH_4-N	0.02	0.43	0.07	0.03
Total N	0.85	2.69	1.50	0.09
PO_4-P	0.00	0.07	0.00	0.00
Total P	0.03	0.28	0.03	0.03
pH	3.9	4.2	4.3	4.1
Solids	6	137	28	8

cm more than from the undisturbed area, and by the third year the runoff from the two areas was similar. While this followed reported patterns, the magnitude of increase in stream flow following canopy removal was considerably lower than the 25 to 30 cm increase predicted by Douglass and Swank (1972) for areas having greater topographic relief. These small differences in water yield between undisturbed and reforested areas in the coastal flatlands can be explained on the basis of the storage capacity of cypress ponds, which occupy about 30 per cent of the land area, the ponding that occurs between the beds as a result of occulsion of the furrows with logging debris, low topographic relief and possibly deep infiltration.

The major impact of silvicultural operations on surface runoff occurred during periods of wet weather when saturated soil conditions prevailed. This was particularly evident during the dormant season when evapotranspiration was low. However, the dormant season, which ranges from 50 to 100 days, has only about 30 per cent of the annual rainfall. Also, the cold, saturated soil conditions inhibited mineralization processes and resulted in lower nutrient loading of the discharged waters during this period.

The quality of water issuing from managed forest watersheds in the lower coastal plain has been the subject of much debate based on a minimal amount of information. Comparisons of water quality parameters for the undisturbed and an adjacent reforested watershed area in the coastal lowlands are shown in Table 4.40. These data show that by the second year following harvesting most water quality parameters have returned to near background levels. The notable exception here is nitrate which is not affected by the harvesting operation. For some parameters, particularly sedimentation and colouration, water quality changes were of short duration and returned to near background conditions within 4 to 6 weeks after cessation of site disturbance activities. The overall impact of silvicultural operations on the hydrological properties of the coastal pine forest, at least for sandy soils, appears to be much less than in areas having greater relief and shallow soils.

Table 4.41. Nitrogen depletion on coastal lowlands sites with different silviculture practices

	Bole harvest minimum preparation	Total harvest maximum preparation
	$(kg\ ha^{-1})$	
Reserves	2050	2050
Inputs	237	237
Atmospheric	(125)	(125)
N_2 fixation	(112)	(112)
Removals	169	550
Losses	97	97
Leaching	(22)	(22)
Runoff	(75)	(75)
% nitrogen depletion per 20-year rotation	1.2%	20.0%

4.5.5. CONCLUSIONS

Intensive harvesting and site preparation of coastal plain slash pine forests can lead to significant losses of organic matter and, consequently, of nutrient reserves. Data presented here indicate that actual nutrient losses are low if conventional harvesting and minimal site preparation systems are used. These losses are usually recovered from atmospheric inputs and biological N-fixation. However, maximum harvest removal of aerial biomass coupled with intensive site preparation practices can result in a serious depletion of soil organic matter and associated nutrient reserves from which recovery on a short-rotational basis is incomplete.

Using the balance sheet approach to place nutrient flow into a management perspective, Table 4.41 shows a generalized input and loss of nitrogen for a 20 year rotation of slash pine. If only merchantable stems are harvested and the site is reforested with minimum disturbance, about 10 per cent of the toal N reserves would be removed. Such a reduction in N would easily be replaced over the next rotation by atmospheric inputs and biological fixation. However, an estimated 26 per cent of the total N reserves for the rotation are removed by complete tree harvest and an intensive site preparation that includes burning and blading or scalping of the forest floor. Atmospheric and biological N-fixation inputs during the next 20 year cycle will restore less than 50 per cent of these lost reserves. This example appears to be reasonable in comparison to similar studies on other southern pine sites (Switzer and Nelson, 1972 and 1973; Wells and Jorgensen, 1977). Obviously, these sites can not tolerate many repetitions of such activities without an appreciable decline in productivity.

REFERENCES

Aubertin, F. M., and Patric, J. H., 1974, Water quality after clearcutting a small watershed in West Virginia, *J. Environ. Quality.*, **3**, 243–249.

Ballard, R., 1973, Extractability of reference phosphates by soil test reagents in absence and presence of soils, *Soil Crop Sci. Soc. Fla. Proc.*, **33**, 169–174.

Balmer, W. E., and Little, N. G., 1978, Site preparation methods, in *Proceedings: A Symposium on Principles of Maintaining Productivity on Prepared Sites*, Missouri State University, pp. 60–64.

Dickerson, B. P., 1976, Soil compaction after tree length skidding in northern Mississippi, *Soil Sci. Soc. Amer. J.*, **40**, 965–966.

Douglass, J. E., and Swank, W. T., 1972, Streamflow modification through management of eastern forests, *USDA For. Serv. Res.*, Pap. SE-94, 15 pp.

Florence, R. G., 1967, Factors that may have a bearing upon the decline of productivity under forest monoculture, *Aust. For.*, **31**, 51–71.

Haines, L. W., Maki, T. E., and Sanderford, S. G., 1975, The effect of mechanical site preparation treatments on soil productivity and tree (*Pinus taeda* L. and *P. elliottii* Engelm. var. *elliottii*) growth, in *Forest Soils and Forest Land Management*, B. Bernier and C. H. Winget, eds. Laval Univ. Press, Quebec, pp. 379–395.

Hibbert, A. R., 1967, Forest treatment effects on water yield, in *International Symposium on Forest Hydrology*, W. E. Sopper and H. W. Lull, eds. Pergamon Press, N.Y. pp. 527–543.

Maki, T. E., 1976, Impact of site manipulation on the Atlantic Coastal Plain, *Proc. Sixth So. For. Soils Workshop*, pp. 108–114.

Patric, J. H., and Reinhart, K. G., 1971, Hydrologic effects of deforesting two mountain watersheds in West Virginia, *Water Resour. Res.*, 7, 1182–1188.

Pritchett, W. L., 1976, Phosphorus fertilization of pine in the Atlantic Coastal Plain, *Proc. Sixth So. For. Soils Workshop*, pp. 52–56.

Pritchett, W. L., and Smith, W. H., 1974, Management of wet savanna forest soils for pine production, *Fla. Agric. Expt. Sta. Tech. Bull.*, 762, 22 pp.

Pritchett, W. L., and Wells, C. G., 1978, Harvesting and site preparation increases nutrient mobilization, in *Proceedings: A Symposium on Principles of Maintaining Productivity on Prepared Sites*, Missouri State University, pp. 98–110.

Schultz, R. P., 1976, Environmental change after site preparation and slash pine planting on a flatwoods site, *USDA For. Serv. Res. Pap. SE-156*, 20 pp.

Shoulders, E., and Terry, T. A., 1978, Dealing with site disturbances from harvesting and site preparation in the lower coastal plain, in *Proceedings: A Symposium on Principles of Maintaining Productivity on Prepared Sites*, Missouri State University, pp. 85–97.

Swank, W. T., and Douglass, J. E., 1977, Nutrient budgets for undisturbed and manipulated hardwood forest ecosystems in the mountains of North Carolina, in *Watershed Research in Eastern North America*, Vol I, Chesapeake Bay Center for Environmental Studies, pp. 343–364.

Switzer, G. L., and Nelson, L. E., 1972, Nutrient accumulation and cycling in loblolly pine (*Pinus taeda* L.) plantation ecosystems: the first twenty years, *Soil Sci. Soc. Amer. Proc.*, **36**, 143–147.

Switzer, G. L., and Nelson, L. E., 1973, Maintenance of productivity under short rotations, in *Proceedings International Symposium on Forest Fertilization*, FAO-IUFRO Paris, pp. 365–389.

Wells C. G., and Jorgensen, J. R., 1977, Nutrient cycling in loblolly pine: silvicultural implications, in *TAPPI Forest Biology/Wood Chemistry Conference*, Madison, pp. 89–93.

Tropical Agricultural Hydrology
Edited by R. Lal and E. W. Russell
© 1981, John Wiley & Sons Ltd.

4.6

A Research Project on Hydrology and Soil Erosion in Mountain Watersheds in Sri Lanka

N. W. HUDSON

4.6.1. THE NATIONAL BACKGROUND

Agricultural research in Sri Lanka has traditionally concentrated on plantation crops, tea, rubber, and coconut. The increasing population and increasing need to reduce imports of food supplies now requires a new approach with more emphasis on food production. The present need has been defined as 'to formulate improved land use patterns, and to increase productivity through better management, requiring a thorough knowledge of the inter-relationships of soil, landscape, and hydrology in the wet zone'.

In the low, dry region, present and possible farming systems are fairly well understood. There is considerable room for the expansion of irrigation and further settlement, but increased agricultural production in that zone will be inhibited by the reluctance of people to move to it from the more densely settled areas.

In the mid-country, i.e. between 300 and 1000 m, there is both the greatest need and the greatest opportunity for more intensive farming systems. The population is already there, and there are substantial areas now under marginal plantation crops (particularly tea) which are suitable for diversification into more intensive use.

The topography, the climate, and the land use patterns in the mid-country are all complex. A single catchment of only 5000 hectares can include high land at 2,500 m with 2,500 mm of rainfall on tea plantations, then mixed farming in the middle altitudes with lower rainfall, then rubber plantations down to perhaps 500 m, and finally rice in the valley bottoms.

In order to understand these complex systems (and particularly the inter-relationships between the components) we need to study the water balance of these mixed-up catchments in order to see how much potential exists for intensification.

Later it will be required to monitor the effect of changes in the land use patterns, but the first requirement is to gain a basic understanding of the present systems.

As part of of a larger FAO project which seeks to strengthen the work of the Department, this research project will seek to obtain this basic understanding of the soil and water regime of mixed-use watersheds in Sri Lanka. The programme is operated by the Division of Land and Water Use, Department of Agriculture. The FAO input is the provision of imported measuring equipment, and specialist advisers.

4.6.2. METHODOLOGY

Within separate catchments of up to 5,000 ha it was required to measure rainfall, runoff, consumptive use by crops, groundwater movement, changes in soil moisture, and hence to compute the hydrological balance. At the same time we need to measure soil erosion from the land, and sediment movement in the streams and rivers in order to obtain a picture of soil movement. The runoff and soil loss data will be obtained for (a) the whole catchment, (b) smaller subcatchments within the main catchment, and (c) small runoff plots. This will allow the various kinds of land use to be evaluated separately, and also to see how they interact.

The two catchments of Phase I are:

1. Nanu Oya, about 5000 ha in agro-ecological Zone WM2, as defined in 'Agro-Ecological Regions of Sri Lanka' by the Land and Water Use Division, Department of Agriculture, Peradeniya. The land use is mainly mixed gardens, largely spices, with some tea on the high ground, and paddy in the valley bottoms. (Figure 4.1.3).
2. Hanguranketa, about 2000 hectares in agro-ecological Zone 1M3. The land use is tea at elevations from 2500 m to 1500 m—intensive arable farming at middle altitudes, mainly tobacco, and paddy in the valleys. There are several irrigation schemes. (Figure 4.14).

4.6.3. MEASUREMENTS

4.6.3.1. Runoff

Runoff will be gauged at the main outlet from each watershed, and each main sub-catchment, at rated stations, using existing structures where possible, e.g. railway and road bridges and culverts. In Nanu Oya, there are eight subcatchments and the estimated 10-year flood is shown in cusecs in Figure 4.15. One subcatchment of approximately 700 hectares is selected for detailed measurements using Parshall flumes, H-flumes, and rated stations, with continuous automatic stage recorders as shown in Figure 4.16.

At a still smaller scale, runoff is measured from 1 hectare single-crop plots. The

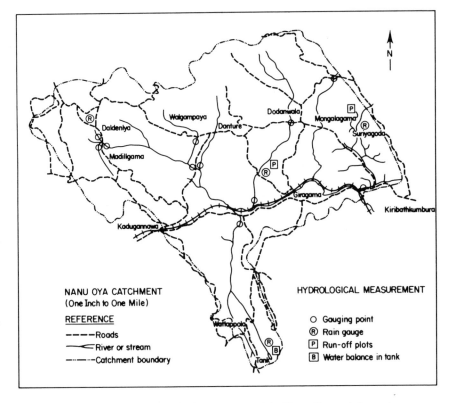

Figure 4.13 Hydrological measurement in Nanu Oye catchment

experimental design of these is 3 replications X 3 treatments. The treatments are—

(a) mixed garden with spice trees
(b) tea with good management
(c) tea with poor management

Even smaller runoff plots (0.002 ha) will measure the effect of slopes and land use. The experimental design is 3 treatments X 3 slopes X 3 replications = 27 plots. The treatments are:

(a) mixed garden with heavy canopy
(b) mixed garden with light canopy
(c) tea under good management

4.6.3.2. Soil Erosion

Soil movement will be measured at all the runoff measuring points. At the rated stations on main streams, and at the flumes and weirs on tributaries, the method

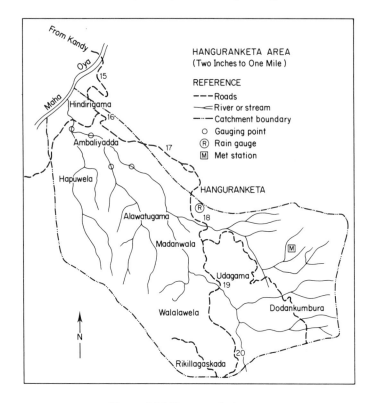

Figure 4.14 Hanguranketa area

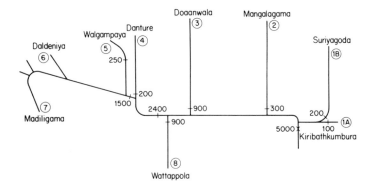

Figure 4.15 Nanu Oya experimental catchment. Schematic representation of estimated 10-year maximum flood flows

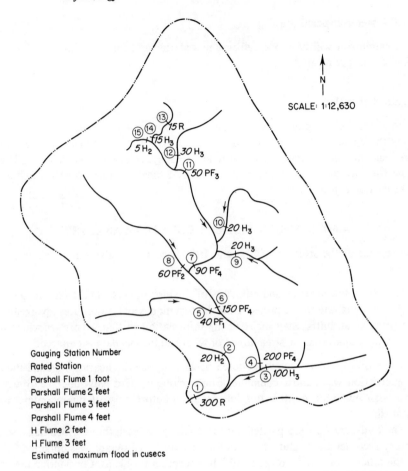

Figure 4.16 Sketch map of Suriyagoda sub-catchment with gauging points and estimated 10-year flood flows

will be *ad hoc* sampling during floods. On the 1-ha and 0.002-ha plots collecting tanks with divisors will provide daily measurement of soil loss.

If more detailed information is required on, for example, soil loss under different levels of management, or on different soil types, these will be studied using portable plots with artificial rainfall simulators.

4.6.3.3. Soil Moisture

Changes in the water storage in the soil will be monitored by a network of water-table tubes, and regular sampling of the soil profile above the water-table. Initially, this will be based on gravimetric sampling, but the possibility of using a neutron probe is being considered.

4.6.3.4. Meterological Data

Each catchment will have one full agro-met station and a network of recording and non-recording rain gauges.

4.6.3.5. Other Data

Good aerial photography is available, and good topographic maps down to 1:12630. Using these and field surveys, detailed base maps of each catchment have been prepared showing on separate maps the relief, geology, soil types, and land use. These allow the measurement of the areas of each of these variables in each catchment and subcatchment.

4.6.4. RESULTS—EXAMPLES OF STORM ANALYSIS

The records of the storm on 14th June 1979 have been analysed in detail, in order to:

(a) check the accuracy and reliability of the instruments and the recording staff.
(b) demonstrate to the project staff the techniques and procedures involved.
(c) make an initial assessment of the information so that we can see where the programme can be streamlined or where additional data is required.

The storm of 14th June was chosen because it was the heaviest rain and the instrumentation was almost complete. The cyclone of 23rd November 1978 would have been much more interesting but occurred before most of the recorders were installed.

The hydrographs were plotted for the discharge of each flume which had a satisfactory recorder trace. Out of 13 stations a usable hydrograph was obtained for 9. Of the other 4, one was drowned and 3 had recorder problems. Considering the miscellaneous collection of recorders used in order to get the experiment going, and considering the lack of experience of the staff, this is a highly satisfactory result. A selection of the hydrographs is attached as Figures 4.17–4.22.

Table 4.42 shows the volume of runoff during the storm, the percentage of rainfall lost as runoff, the maximum rate of runoff, and the flashiness expressed as time to peak flood and length of recession curve.

Some interpretive comments may be made on these results, but much more data is required before we can make authoritative statements about the effect of topography, land use, etc.

(a) the percentage runoff figures are all believable but several points need to be looked at when analysing later results. The percentage runoff from catchment 9 is high at 31.5 per cent, but it corresponds to a high maximum runoff (4.66 mm hr^{-1}) and long duration considering the catchment size. The land use distribution is not unusual, so some other explanation must be sought.

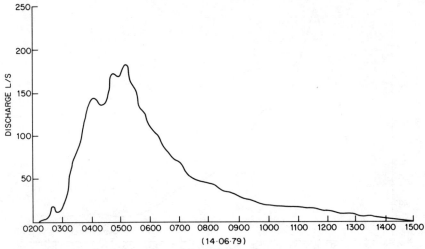

Figure 4.17 Flume 14-H₃

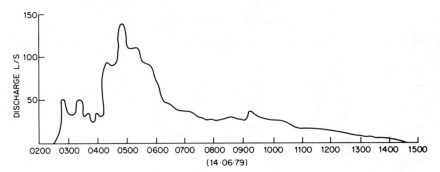

Figure 4.18 Flume 12-H₃

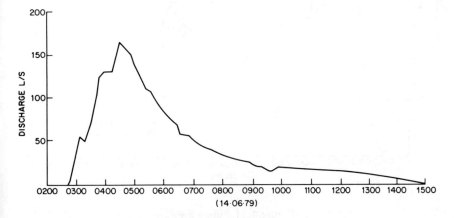

Figure 4.19 Flume 10-H₃

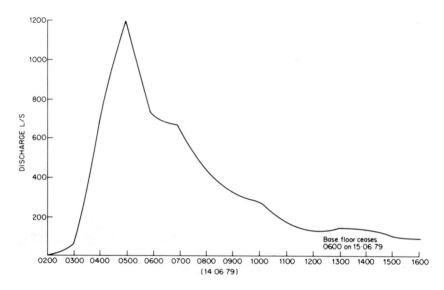

Figure 4.20 Flume 7-PF$_4$

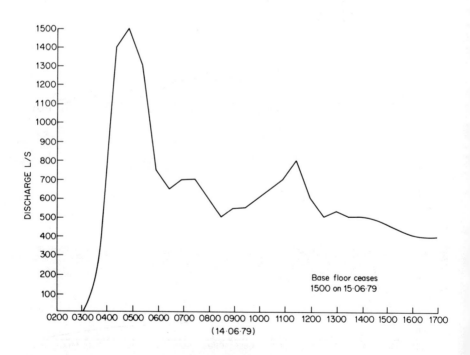

Figure 4.21 Flume 5-PF$_4$

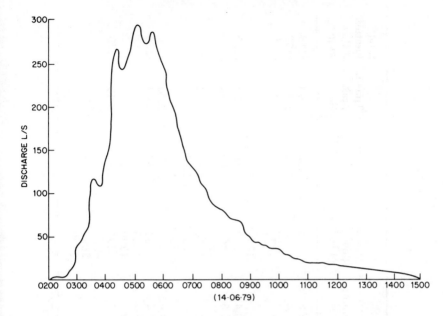

Figure 4.22 Flume 3-H$_3$

(b) catchment 4 also shows a high percentage runoff (27.4) but low maximum rate (1.68 mm hr^{-1}) which appears inconsistent. However, it has a considerably higher percentage of paddy which may be the explanation.

(c) The land use pattern as shown in Table 4.43 shows that the pattern is remarkably similar throughout Suriyagoda, and this is confirming what one would expect from observation. When a similar analysis has been made of soil type, this may explain some of the variations.

(d) However, the overall soil and water movement is already clear. The generally dense cover of this intensive mixed land use leads to low rates of runoff averaging around 10 per cent and negligible soil loss from the catchment as a whole. A very different pattern can be expected to emerge from the Hanguranketa catchment.

Table 4.42. Suriyagoda subcatchments, storm of 14th June 1979

Station number	Flume	Area (ha)	Total flow (m³ × 10³)	Runoff (mm)	Rainfall (mm)	Runoff (%)	Maximum flow (l sec⁻¹)	Maximum runoff (mm hr⁻¹)	Time to peak (hr)	Total duration of flow (hr min)
14	H3	41	2.34	5.7	97.0	5.8	185	1.62	3.00	1215
15	H2	5	0.36	7.2	97.0	7.4	26	1.91	1.30	0800
12	H3	26	1.46	5.6	97.0	5.7	140	1.94	2.30	1130
Total							351			
11	PF3	72			97.0		(350)	(1.76)	2.50	1220
10	H3	25	2.00	8.0	95.0	8.4	165	2.36	3.30	1700
9	H3	42	12.46	29.9	95.0	31.5	540	4.66		
Total							(1055)			
7	PF4	165	19.87	12.1	95.0	12.7	1200	2.62	3.00	2800
8	PF2	40					(250)	(2.24)		
Total							(1450)			
6	PF4	210	39.74	18.9	95.0	19.9	(1440)	2.47	3.00	4100
5	PF1	60	15.63	26.1	95.0	27.4	280	1.68	4.00	3200
Total							1720			
4	PF4	284	25.00	8.7	95.0	9.1	1800	2.28	3.00	1250
3	H3	62	3.82	6.2	95.0	6.5	295	1.71		
2	H2	7		nil			nil			
Total		353					2095	2.14		
1	R	384					(2100)	(1.96)		

Figures in brackets are estimates

Table 4.43. Summary of land use in Suriyagoda
subcatchments

Subcatchment	Land use Tea (%)	Paddy (%)	Cover (%)
14	–	19	81
15	33	8	58
12	14	9	77
Total			
11	7	15	78
10	48	8	44
9	21	17	62
Total			
7	17	17	66
8	25	20	55
Total			
6	18	17	65
5	11	36	53
Total			
4	17	20	63
3	19	21	60
2	–	11	89
Total			
1	16	21	63

Note: This is a summary of a more detailed study
which has 4 classes of tea and 4 subclasses of cover
crops

PART 5

Engineering Structures

Tropical Agricultural Hydrology
Edited by R. Lal and E. W. Russell
© 1981, John Wiley & Sons Ltd.

5.1

Engineering Structures for Erosion Control

F. W. BLAISDELL

5.1.1. INTRODUCTION

Erosion-control structures are needed on agricultural watersheds because vegetation and supporting soil and water conservation practices used alone often are not sufficient to prevent erosion due to concentrations of runoff or flows of long duration. Frequently permanent structures of stone, steel, concrete, or masonry are required to supplement conservation practices.

Erosion control structures are also required for grade stabilization in both intermittent and perennial streams. Many of the types of structures used on agricultural watersheds are also suitable for use in larger watercourses; many structures developed for erosion control in agricultural watersheds can be scaled up and used in small streams and in rivers.

Because water causes erosion, control of water also helps to control erosion. It is, therefore, appropriate to include water control structures in this discussion of erosion control structures.

In this chapter, are discussed in general terms, many, but by no means all, the types of engineering structures available for the control of erosion by water. It will suffice here to explain the characteristics and use of the various structures, so structures appropriate to the reader's problem can be selected for further consideration. The reader can then refer to the cited references to select the most appropriate structure and to obtain detailed information on the hydraulic design of the selected structure.

5.1.2. LINED CHANNEL STRUCTURES

Several channel structures that have a continuous lining will be considered in this section. The linings may be vegetation, riprap, or such non-erodible linings as metal, concrete, or masonry. The velocity of the water in the channel may be subcritical or supercritical. The channel slope may be flat or steep. Discharges may vary over a wide range. Each factor must be considered and evaluated by the designer.

Three types of lined channels will be discussed: (1) vegetation-lined channels; (2) riprap-lined channels; and (3) chutes.

5.1.2.1. Vegetation-Lined Channels

Because engineering is used to design vegetation-lined channels, vegetation-lined channels are here considered to be engineering structures for erosion control. Vegetation is an excellent and economical channel lining where it can be used.

Vegetation increases the stability of a channel in earth by creating a zone of relatively low velocity at and near the channel surface. This low velocity zone moves the plane of maximum fluid shear stress from the channel bed to the top of the grass lining. Also, it provides a barrier to the impact of turbulent vortices on the channel bed. Uniform, dense stands of flexible, long vegetation are best for creating an effective low-velocity zone. The vegetation must be able to survive and thrive in the channel it is to protect. The degree of protection by vegetation varies with the kind of vegetation, the uniformity of coverage, and the maintenance provided. Soil type, flow frequency and duration, and drainage also affect the performance of vegetation-lined channels.

Design information is available for vegetation-lined channels having slopes no steeper than 10 per cent and velocities up to 2.4 m sec^{-1}. Most of this information is based on research reported by Ree and Palmer (1949). Plate 26 shows a channel being tested at the Water Conservation Structures Laboratory, Stillwater, Oklahoma, USA. The plant species tested are those adapted to Southeastern and South Central United States (36 degrees north latitude, humid, and semihumid). The characteristics of the vegetation have been described by Ree and Palmer (1949) and by Palmer, Law, and Ree (1954; 1966). These descriptions should permit the adaptation of the Unites States vegetation design criteria to tropical vegetation having similar characteristics. Important plant properties are the length of the plant, the root system, plant count per unit area or completeness of coverage, the ability of the vegetation to reduce erosive velocities near the bed to non-erosive velocities when the stems are erect at low flows, and flexibility of the stem and ability of the vegetation to bend over and reduce the roughness coefficient at high flows.

Updated handbooks for channel design were prepared by Palmer, Law, and Ree (1954) for US customary units and Palmer, Law, and Ree (1966) for metric (International System of Units, SI) units. Chow (1959) summarizes the Palmer, Law, and Ree results and presents their design curves. Jacobson (1961) has written a chapter on waterway design. The most detailed information on grassed waterway design, construction, and maintenance is the chapter by Coyle (1969).

5.1.2.2. Riprap-Lined Channels

Soil, climatic, agronomic, or hydraulic conditions may make it difficult or impossible for vegetation to control channel erosion. For such cases, riprap can be used to protect the channel against erosion if suitable rock or manufactured bed protection is available.

5.1.2.2.1. Highway Research Board Procedure

The riprap must be of adequate size to resist the forces that are acting on the individual stones. Anderson, Paintal, and Davenport (1970) have developed an equation for stable riprap sizes. In US customary units, the equation is

$$\tau_c = 4d_{50} \tag{5.1}$$

where τ_c is the tractive force in pounds per square foot and d_{50} is the particle size in feet of which 50 per cent is finer. In SI units, where τ_c is in newtons per square metre Pascals and d_{50} is in metres, equation (5.1) becomes

$$\tau_c = 192\, d_{50}. \tag{5.2}$$

The tractive force may be computed from the equation

$$\tau = \gamma RS \tag{5.3}$$

where γ is the unit weight of water, R is the hydraulic radius, and S is the slope of the bed. The system of units for equation (5.3) must, of course, be consistent with the units of equations (5.1) or (5.2).

My experience using these equations, although limited to laboratory model studies, is completely satisfactory. Anderson (1973) also reports satisfactory results of limited field experiences using the tentative design procedure as follows:

> Two of the four completed channels have been subjected to discharges that approached the design discharges and hence provided reasonably definitive tests. Both channels appeared to be stable and in good condition after the floods. Although these results are somewhat sparse, it appears that drainage channels designed according to the proposed procedures will convey design discharges without significant erosion.

Plate 27 shows the excellent condition of a highway ditch that had experienced a flow which amply tested the design procedure.

5.1.2.2.2. Soil Conservation Service Procedure

Using the results presented by Anderson, Paintal, and Davenport (1970), the Soil Conservation Service (SCS) of the United States Department of Agriculture has developed a procedure for designing riprap gradient control structures. As shown in Figure 5.1, the SCS structure consists of a riprap prismatic channel with riprap transitions at both upstream and downstream ends (See POSTSCRIPT, page 355).

The SCS structures' essential feature is that the specific energy of the flow at the

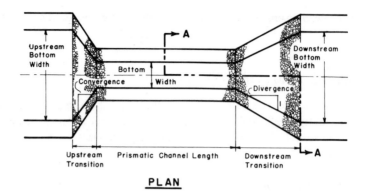

PLAN

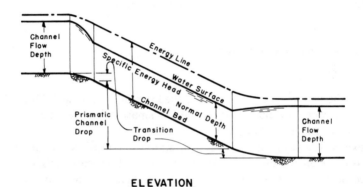

ELEVATION

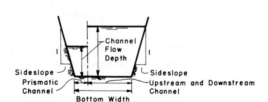

SECTION A-A

Figure 5.1 Soil Conservation Service riprap gradient control structure (After Goon, 1975)

design discharge is constant throughout the structure and is equal to the specific energy of the flow in the channel immediately upstream and downstream of the structure. At the design discharge the flow in the channel is at uniform depth and the specific head is constant throughout the structure. For the design discharge, energy is dissipated at the same rate as the energy gain due to the gradient of the riprap structure. The structure, which is steeper and narrower than the upstream and downstream channels, maximizes energy dissipation.

Goon (1975) has described the SCS design procedure and the computer program prepared for designing the prismatic channel and the upstream and downstream transitions. Brevard (1976) presents graphical procedures for designing riprap gradient control structures having prismatic channel side slopes of 1 on 2 and 1 on 3. And, finally, Goon (1978) presents a procedure and computer program for investigating the capacity and stability of the riprap structure for discharges and tailwater conditions other than the design values. The computer program determines the water surface profile, maximum tractive stresses, and other hydraulic parameters.

5.1.2.3. Chutes

Two types of chutes are useful for erosion control: A plain chute to transport the flow at high velocity with minimum energy dissipation, and a baffled chute to dissipate the energy as fast as it is generated by the chute slope. The emergency spillway used with soil and water conservation and farm pond dams is another form of chute. These three types of chutes will be described.

5.1.2.3.1. Emergency Spillways

As noted under the headings 'HIGH DROP STRUCTURES, Island Structures' and 'CLOSED CONDUIT STRUCTURES, Hydraulic Design', emergency or auxiliary spillways are required at many water and erosion control structures. Emergency spillways frequently are made of earth protected by vegetation or other material. Vegetated emergency spillways can be designed using the methods cited in 'LINED CHANNEL STRUCTURES, Vegetation-Lined Channels'. Renfro (1969), Culp and Gregory (1956) and Gregory (1957) discuss the procedures and considerations involved in the design of earth spillways. Figure 5.2 shows the basic features of earth spillways. In climatic (e.g. arid) areas where vegetation cannot be depended upon to protect the emergency spillway from erosion damage, soil cement or other surface protection can be used. Plate 28 shows an emergency spillway in New Mexico, USA, paved with soil cement.

5.1.2.3.2. Plain Chute

Water in a steeply sloping plain chute will flow at supercritical velocity. Because supercritical velocities are difficult to handle in trapezoidal channels, the preferred channel cross-section is rectangular. A suitable entrance is required at the upstream end of a chute. The entrance should admit the flow smoothly and eliminate the possibility of surface cross waves that might overtop the chute sidewalls. A suitable exit or stilling basin is also required to dissipate the large amount of energy in the flow at the chute exit and thereby eliminate or control the scour at the chute exit. Figure 5.3 shows the elements that comprise a chute spillway.

Doubt (1955) gives the most complete information on the design of chutes and their entrances and exits. Blaisdell and Moratz (1961) present information on two

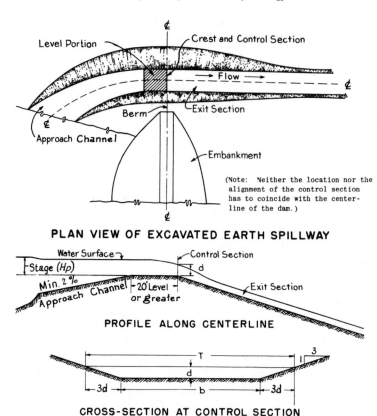

PLAN VIEW OF EXCAVATED EARTH SPILLWAY

PROFILE ALONG CENTERLINE

CROSS-SECTION AT CONTROL SECTION

Figure 5.2 Basic features of excavated earth spillway (Renfro, 1969)

types of chute entrance and one type of chute stilling basin, and refer to other stilling basin designs. Beauchamp (1969) shows small chutes useful for small, intermittent flows and low drops such as may exist on farms. Hoffman (1973) gives information on chutes for larger dams than those considered in the other references
cited.

5.1.2.3.3. Baffled Apron Chute

The baffled apron chute has multiple rows of blocks or baffles equally spaced along
the chute. A typical arrangement is shown in Figure 5.4. The flow passes over,
around, and between the baffles. Like the riprap lined channels, energy is dissipated
along the chute as fast as energy is generated by the chute gradient. The baffled
chute provides its own energy dissipation, so no stilling basin is required at the
chute exit. The chute exit and the bottom one or more rows of baffles are extended
below the bottom of the downstream channel. If the downstream channel degrades,

Plate 26 Vegetation-lined channel being tested at the Water Conservation Structures Laboratory, Stillwater, Oklahoma, USA. The cover is Bermuda grass except for a 1.22 m wide strip of Reed's Canary grass down the centre. The Bermuda grass stems averaged 89 mm in length, had a density of 1722 stems per square metre, and was dormant and brittle. The Reed's Canary grass averaged 267 mm in length, had 603 stems per square metre, and was green. The channel cross-section is V-shaped with 1 on 20 sideslopes. The channel slope is 6 per cent, the discharge is 3.50 m^3 sec^{-1}, the velocity is 2.50 m sec^{-1}, the Manning n is 0.033, and the retardance coefficient class is D

Plate 27 Riprap-lined channel of 25 mm mean size rock with grass growing through the riprap approximately three years after construction and after a flow for which the discharge equalled or exceeded the design discharge of $0.17 \text{ m}^3 \text{ sec}^{-1}$. An observer wrote, 'the ditch is in surprisingly good condition' (Anderson, 1973)

(a)

(b)

Plate 28 Emergency spillway 158 m wide in New Mexico, USA. It has a capacity of 188 m^3 sec^{-1} at a flow depth of 0.82 m. The 17 per cent downstream slope was paved with 200 mm of soil cement in 1970. The pictures were taken in 1976. (A) The spillway crest and upstream end of the soil cement paving. (B) Closeup view of the soil cement paving. Small cracks and weathering are evident; however, the spillway is still in excellent condition

Plate 29 Riprap, baffle plate low-drop spillway stilling basin being developed at the US Department of Agriculture Sedimentation Laboratory, Oxford Mississippi.

(a)

(b)

Plate 30 Baffle wall low-drop spillway stilling basin being developed at the US Department of Agriculture Water Conservation Structures Laboratory, Stillwater, Oklahoma. (A) The crest of low-drop shows beneath the horizontal baffle wall. (B) Flow with the horizontal baffle wall submerged

Plate 31 Stepped channel trash rack. This two-way drop inlet is 0.91 m wide, 2.74 m long and 12.2 m high. The steel channels are 229 mm wide, overlap 76 mm, and have a 314 mm clear horizontal spacing. The gate is for municipal water supply

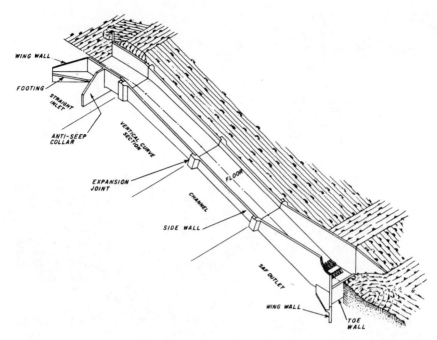

Figure 5.3 Plain chute spillway with a straight inlet and a SAF stilling basic exit (Beauchamp, 1969)

successive rows of baffles are exposed, so the baffled chute should terminate below the anticipated degraded downstream channel bed. If downstream channel aggradation occurs, successive rows of baffles are buried, but the performance of the baffled chute is not changed. In contrast to some other types of chute exits, the baffled chute performance does not change as the tailwater level varies with aggradation or degradation of the downstream channel.

The total flow capacity of the baffled chute can be varied by varying the chute width and the discharge per unit width. Hayes (1974) shows recommended discharges as low as 0.5 m^3 sec^{-1} m^{-1}. At the other extreme, Rhone (1977) reports model tests with unit discharges as high as 28 m^3 sec^{-1} m^{-1}. Peterka (1964) shows many examples of the application of the baffled chute and, as does Rhone (1977), reports the successful operation of these chutes for many years, sometimes at twice the design discharge for short periods. Hayes (1974) also presents information on and examples of the design of baffled chute drops, and cautions that trash may become lodged on the baffles.

5.1.3. LOW DROP STRUCTURES

A low drop is defined by W. Campbell Little of the US Department of Agriculture Sedimentation Laboratory as a drop in which the bed drop height is equal to or less

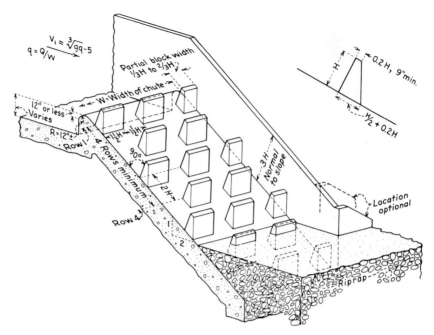

Figure 5.4 Basic proportions of the baffled chute spillway. Rhone (1977) shows other arrangements of the baffles at the chute entrance (Bureau of Reclamation, 1973)

than the upstream specific head; that is, less than the upstream flow depth plus the velocity head. For this condition and for the design flow, Little considers the drop as a large roughness element that causes a standing wave or undular hydraulic jump downstream of the drop crest.

Energy dissipation is small in the undular jump (a maximum of 5 per cent), so special measures are required to dissipate the destructive energy in the flow. Also, for water surface drops that are less than one-third the upstream specific head, the jet may float on the downstream water surface and perhaps meander to attack the downstream channel banks. Low drop structures may not be required to control the channel grade at high flows where the available energy grade line slope equals the average bed slope. However, for lesser flows the bed slope may exceed the required energy slope and the excess energy will degrade the channel.

I know of no publications describing low drop erosion control structures. I do know of two current studies and I will describe the structures.

5.1.3.1. Oxford, Mississippi, Study

At the United States Department of Agriculture Sedimentation Laboratory in Oxford, Mississippi, W. C. Little is developing a trapezoidal, riprap, low-drop, grade control structure. Several structures have been modelled and built. However, the laboratory study giving general design criteria is incomplete, so no publication describing the structure is yet available. As can be seen in Plate 29, Little's low drop has a crest comprised of a steel sheet pile cutoff wall across a trapezoidal channel. Riprap protects the stream bed upstream of the crest. Downstream of the crest, riprap slopes downward into a 'scour hole' shaped stilling basin lined with riprap. In the basin, there is either a baffle pier or baffle plate transverse to the flow. The baffle pier is constructed by driving sheet piles to form a rectangular box which is filled with rock and capped with concrete. The baffle plate is constructed by driving two or three H-piles on which either wood planks or interlocked sheet piles are fastened horizontally. The pier or plate are positioned with respect to the location of the standing wave in the stilling basin. Little prefers the plate to the pier because sediment accumulates downstream of the pier but is swept out by flow under the plate. Little anticipates that vegetation on the sediment deposit downstream of the pier might stabilize the deposit and impair the structure's performance.

Limited experience with this low-drop structure has been generally satisfactory. Difficulties that have been encountered are ascribed to improper riprap placement by a contractor.

5.1.3.2. Stillwater, Oklahoma, Study

At the US Department of Agriculture Water Conservation Structures Laboratory in Stillwater, Oklahoma, W. R. Gwinn is developing a different type of low-drop grade control structure. Gwinn's low-drop structure has a rectangular weir crest and a vertical drop into a stilling basin (Plate 30). The stilling basin has vertical sidewalls. At the discharge and tailwater elevation at which the nappe would tend to float on the tailwater surface, a horizontal baffle wall spanning between the stilling basin sidewalls is positioned to intercept the upper surface of the nappe and deflect the nappe down into the stilling basin. At the high flow shown in Plate 30b the baffle wall might be overtopped, but it still effectively prevents nappe flotation and ensures satisfactory energy dissipation.

This drop structure would be built of concrete or other structural material. Riprap would be used, if necessary, only in the approach to and beyond the exit of the structure. The study is incomplete so no publication is available. No structures of this type have been built.

5.1.4. HIGH DROP STRUCTURES

High drop structures are those in which the upstream water levels are normally unaffected by downstream conditions, such as high tailwater levels that submerge the spillway crest. There is a range of conditions where the high drop may perform as a low drop, but generally there is a definite, sudden drop in the water surface through a high drop, whereas the water level drop through a low drop is small or nil.

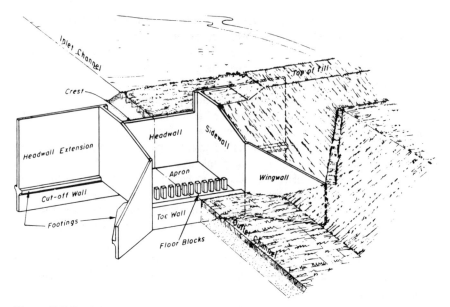

Figure 5.5 Straight drop spillway and its stilling basin (Beauchamp, 1969)

Three types of high drop structures will be described: (1) the straight drop spillway; (2) the dissipation bar drop structure; and (3) the box inlet drop spillway.

5.1.4.1. Straight Drop Spillway

The straight drop spillway is a rectangular overfall weir. Water flowing over the spillway falls onto a horizontal apron. The nappe is broken up by floor blocks, which also prevent damaging scour of the downstream channel banks. Scour of the downstream channel bed near the stilling basin exit is prevented by an end sill. Flaring wingwalls, triangular in elevation, prevent erosion of the dam fill by reverse currents downstream of the basin exit. These reverse currents cause a controlled local widening of the downstream channel near the basin exit in which energy dissipation, begun in the stilling basin, is completed. The components of the straight drop spillway and its stilling basin are shown in Figure 5.5. The straight drop spillway is used as an erosion control structure in gullies, as a grade control and tile outlet structure in drainage ditches, as a grade control structure in rivers, as an irrigation check and drop structure, and as a dam spillway.

The stilling basin can be used for a range of discharges from those encountered on farm drainageways to those encountered in rivers, and for a similar wide range of heads on the crest, lengths of crest, heights of drop, and downstream tailwater levels. The rules for designing the straight drop spillway stilling basin can be found in the papers by Donnelly and Blaisdell (1965) and by Blaisdell and Moratz (1961).

Field experience with the straight drop spillway stilling basin has been uniformly good. Since the introduction of this stilling basin design in 1953, the sizes of straight drop spillway basins have been gradually increased over the years. One recent example is a structure designed for a bank-full, 5-year frequency storm of 317 m^3 sec^{-1}. This spillway has passed several capacity flows with performance as anticipated from the small scale model studies. (At this location, auxiliary spillways handle larger floods.) Donnelly and Blaisdell (1966) and Blaisdell, Anderson, and Hebaus (1976) show several pictorial examples of field applications. Murley (1970) mentions six drop structures used on a 28 m^3 sec^{-1} diversion canal where the Donnelly and Blaisdell (1965) structures were used alternatively with Katsaitis (1966) structures. Murley (1972) compares the performance of these structures and concludes: 'Performance to date indicates that both types are very efficient for large drops and big discharges. The Donnelly and Blaisdell type is favored because of the smaller teeth and the end sill and because the apron is shorter giving a slightly more economical structure.' The smaller teeth are preferred by Murley because of the lesser tendency to collect and be damaged by trash. Murley (1972) states that the expected collection of debris on the teeth did not occur, but logs did break teeth in Australia. I know that floating ice has broken teeth in Minnesota. The possibility of damage by trash must, therefore, be anticipated.

5.1.4.2. Dissipation Bar Drop Structure

Katsaitis (1966) has reported tests and design criteria for a vertical drop like the straight drop spillway but with two rows of taller baffle piers and no end sill. (Figure 5.6). The uses of the dissipation bar drop structure are similar to those of the straight drop spillway. Murley (1970; 1972) compares these two stilling basins, showing the straight drop spillway to be shorter and commenting as quoted above on his preference.

5.1.4.3. Box Inlet Drop Spillway

The box inlet drop spillway is a rectangular box open at the top and at the downstream end. Runoff is directed to the box by dikes and headwalls, enters the box over the upstream end and two sides, and leaves the box through the open downstream end. An outlet structure is attached to the downstream end of the box. The outlet and one form of the box inlet crest is shown in Figure 5.7.

The long crest of the box inlet permits large flows to pass over the crest with relatively low heads, yet the width of the spillway need be no greater than that of the exit channel. The low heads also minimize upstream flooding and permit the use of low dikes, which is especially desirable if the dikes are long.

The uses of the box inlet drop spillway are similar to those of the straight drop spillway. I prefer the straight drop spillway where it can be used because the stilling basin operation causes less scour in the downstream channel. Blaisdell (1973) has

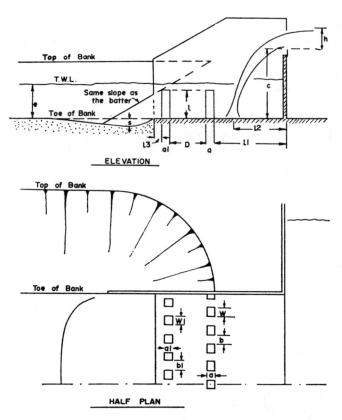

Figure 5.6 Dissipation bar drop structure (Katsaitis, 1966)

shown the extent of the downstream channel scour for one box inlet drop spillway model. However, I know of no field structures that have scour as severe as that shown in the cited paper, and I have observed no scour at field structures sufficient to cause any concern regarding the safety of the box inlet drop spillway.

Blaisdell and Donnelly (1956; 1966) and Blaisdell and Moratz (1961) are sources of information for the design of the box inlet drop spillway and its outlet structure.

5.1.4.4. Island Structures

Economy can sometimes be achieved by using an island type of design. In the island design, the principal spillway is a straight, box inlet, or similar drop structure. The principal spillway capacity is designed to fill the downstream channel bank-full. Larger flows are then passed over structural or earth auxiliary spillways located beyond either or both ends of the principal spillway. The flow over the auxiliary spillway(s)—generally overland type flow—cannot drop into the downstream channel and erode the banks because flow from the principal spillway already has

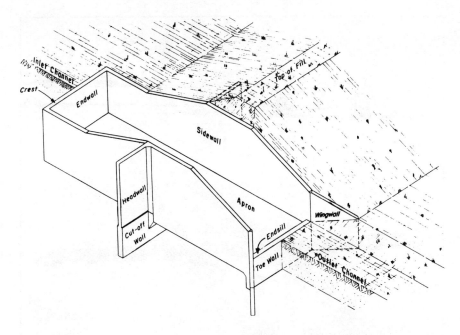

Figure 5.7 Box inlet drop spillway and its outlet (Beauchamp, 1969)

filled the downstream channel bank-full. The basic principals of island spillways are shown in Figure 5.8. The name 'island design' derives from the fact that the principal spillway is partly or completely surrounded by water when flow is passing over the auxiliary spillway(s). Blaisdell and Moratz (1961) describe how the island design is made.

5.1.5. CLOSED CONDUIT STRUCTURES

This type of spillway is made of concrete or tile pipe cradled in concrete, corrugated- or plain-metal pipe, or concrete poured in place to form a box culvert. It is used ordinarily as the principal spillway for earth dams in connection with farm ponds, gully and erosion control, floodwater retardation, permanent storage, and similar soil- and water-conservation structures. Highway culverts are one form of closed conduit spillway. The conduit sizes may range from small pipes to the large morning glory spillways used for major flood-control, water-power, and irrigation reservoirs. The fall through the spillway will vary from 2 m to 15 m for the ordinary farm and soil-conservation installation. The slope of the barrel portion of the conduit may range from horizontal to as steep as it is practical to install. A hood inlet such as might be typically used for farm ponds is shown in Figure 5.9. The components and some criteria for a two-stage, two-way drop inlet for a closed conduit

Figure 5.8 Island spillway (Beauchamp, 1969)

spillway is shown in Figure 5.10. This latter inlet is used at large US Soil Conservation Service and other upstream water control structures.

5.1.5.1. Hydraulic Design

The head-discharge relationships of the closed conduit spillway have special characteristics that must be considered by the designer. Before the structure flows full, the head-discharge relationship is controlled by the inlet and is given by a weir equation. During weir flow control, the discharge increases rapidly as the head increases, thus preserving reservoir storage for later use in reducing the flood outflow peak. After the spillway fills completely, the head-discharge relationship is given by a full pipe equation. During full pipe control, the discharge increases slowly with the head and excess reservoir inflow is stored. Flow in the downstream channel can be controlled by sizing the closed conduit spillway to give, for example, bank-full flow. The excess inflow to the reservoir is then temporarily stored for later release. The reservoir storage volume must be sufficient to store the volume under the hydrograph that does not pass through the spillway during the period of excess reservoir inflow. Because the capacity of closed conduit spillways is limited, prudence requires that an emergency spillway be provided in addition to the principal spillway.

Figure 5.9 Hood inlet spillway (Beauchamp, 1969)

Blaisdell and Moratz (1961) briefly describe how to design closed conduit spillways. Details of the hydraulic design of closed conduit spillways and the performances of the many forms of the spillway and its appurtenances are given in the following papers. Blaisdell (1958a; 1975) presents the theory of flow in closed conduit spillways and siphon spillways. Blaisdell (1958b) describes the hydraulic performance and presents discharge coefficients for five forms of the closed conduit spillway, and discusses vortices and their effect on the spillway capacity. Blaisdell (1958c) reports the results of miscellaneous laboratory tests, tests on models of field structures, and tests of field structures Blaisdell and Donnelly (1958; 1975) describe the hood inlet for closed conduit spillways, report the results of tests, and give criteria for the design of hood pipe inlets. The hood inlet is formed by cutting the pipe entrance at an angle so that the crown overhangs the invert by three-fourths of a pipe diameter. The form of the hood inlet ensures that pipes with a slope steeper than the hydraulic grade line slope will flow full and induce siphonic action. The hood inlet is now a standard structure recommended by the US Soil Conservation Service for use as farm pond principal spillways. The hood inlet has been used for pipe spillways ranging from 0.3 m to 2 m in diameter.

Because air was used for conducting tests on closed conduit spillways , the use of air is discussed by Blaisdell and Hebaus (1966). Donnelly, Hebaus, and Blaisdell (1974) and Blaisdell, Donnelly, Yalamanchili, and Hebaus (1975) present the per-

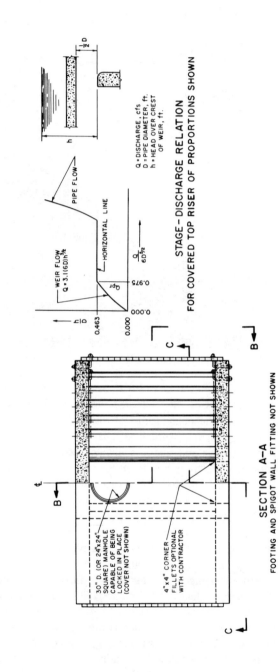

Q = DISCHARGE, cfs
D = PIPE DIAMETER, ft.
h = HEAD OVER CREST
OF WEIR, ft.

STAGE–DISCHARGE RELATION
FOR COVERED TOP RISER OF PROPORTIONS SHOWN

WEIR FLOW
$Q = 3.1(6D)h^{3/2}$

PIPE FLOW

HORIZONTAL LINE

$\frac{Q}{6D^{5/2}}$

$\frac{h}{D}$

PIPE FLOW

$\frac{1}{2}D$

h

SECTION A–A
FOOTING AND SPIGOT WALL FITTING NOT SHOWN

30" D. (OR 24"x24"
SQUARE) MANHOLE
CAPABLE OF BEING
LOCKED IN PLACE
(COVER NOT SHOWN)

4"x4" CORNER
FILLETS OPTIONAL
WITH CONTRACTOR

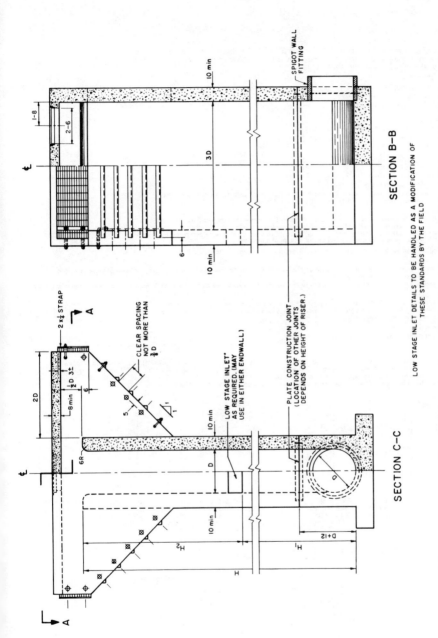

Figure 5.10 Covered top, two-stage, two-way drop inlet for a closed conduit spillway. Dimensions without units are in inches (1 in = 2.54 cm). (US Soil Conservation Service Drawing ES-150)

SCOPE

1. The covered top riser is a standard for two-stage risers, and also for single-stage risers in multi-purpose sites if the maximum sediment elevation is set at least (2D+12") below the crest.

2. Height Ranges of Riser:

   High stage, $H_2 = (2D+6")$ to 20 feet
   Low stage, $H_1 = 0$ to 30 feet
   Sum, $H = H_2 + H_1$, ≤ 40 feet.

CRITERIA

1. Pipe Diameters and Associated Discharges:

D	$Q_{pr} = 0.975(6D^{5/2})$	$Q_{max} = \frac{30}{4}\pi D^2$
24	33	94
30	58	148
36	92	212
42	135	288
48	188	376

   Note:
   Maximum allowable nominal velocity in pipe = 30 fps.

2. Hydraulic Losses (pipe flow):
   Head loss between pool water surface and the projected hydraulic grade line at the pipe entrance = 1.0 times the velocity head in the pipe.

3. Trashracks:
   Required net area for National Standard Detailed Drawings–to be computed from Q_{max} as listed in Criteria (1), and an allowable average velocity of 2.0 fps.
   All bolts, nuts, pipe sleeves, and grating, to be galvanized or otherwise protected by corrosion resistant coating except when made of aluminum.

4. Cover slab live load = 100 psf plus weight of any equipment on the slab.

5. Flotation:
   When riser is in reservoir— the ratio of the weight of riser to the weight of the volume of water displaced by the riser shall not be less than 1.5
   When riser is in the embankment— add to the weight of the riser, the buoyant weight of submerged fill over footing projections.

6. Dry Dams:
   Where sediment is not a problem – set crest of low stage inlet at required elevation.
   Where sediment is a problem – use a series of slotted openings up the longitudinal sides. Trashracks are not required for these openings.

7. Materials:
   Concrete: Class B, $f'_c = 4000$ psi, $f_c = 1600$ psi.   Reinforcing Steel: Intermediate grade.
   Trashrack: Structural steel or structural aluminum.

NOTES

1. Riser Analyses: Standards to be developed for risers located in the embankment (at berm) and for risers located in reservoir area. 2. Round Bottom: May be obtained by use of a pipe cut longitudinally along a diameter, or may be formed by removable semi-circular forms acceptable to the engineer. 3. Drainage of Pool: Provision of means of draining pool to be handled as a modification of these standards by the Field.

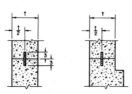

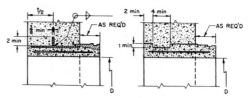

$\frac{1}{4}$ MIN x 6 STEEL PLATE TO BE CONTINUOUS AROUND RISER. TO BE EITHER BUTT WELDED, LAPPED AND FILLET WELDED, OR LAPPED AND BOLTED AT ALL JOINTS (SPLICES)

PLATE CONSTRUCTION JOINT DETAIL

SPIGOT WALL FITTING FOR PIPE AS DESIGNED AND MANUFACTURED UNDER A.W.W.A. SPECFICATION C-300, C-301, AND C-302, AND A.S.T.M. DESIGNATION C-361.

SPIGOT WALL FITTING FOR PIPE AS DESIGNED AND MANIFACTURED UNDER A.W.W.A. SPECIFICATION C-302 AND A.S.T.M. DESIGNATION C-361.

SPIGOT WALL FITTING DETAIL

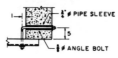

$\frac{3}{4}"\phi$ PIPE SLEEVE

$\frac{5}{8}"\phi$ ANGLE BOLT

BOLT DETAIL

BEARING BAR $1\frac{1}{4}" \times \frac{3}{16}"$

SPACING AND SIZE GIVEN ARE MINIMUMS

GRATING DETAIL– STEEL OR ALUMINUM

formance characteristics and information for the design of two-way drop inlets for closed conduit spillways. See Figure 5.10 for typical features and dimensions. This is the primary type of drop inlet currently used by the US Soil Conservation Service for upstream watershed protection, flood prevention, water conservation, and other single or multiple use purposes for structures of medium to large size with an anticipated extended length of life.

The two-way drop inlet gets its name because water flows over the two long sides of the drop inlet. End walls support a solid antivortex plate that covers and extends beyond the two long sides of the drop inlet. A vertical grating at the outer edges of the antivortex plate extends to below the spillway crest to exclude floating trash and admit water and air. A trash rack slopes from the bottom of the grating downward and back to the drop inlet. At low flows, the two sides of the crest act as weirs. As the flow increases, the pool level rises until it touches the antivortex plate. There is little further rise in the pool level until the spillway is completely full. Within certain limits, the antivortex plate height can be set to control the pool level, a desirable feature for recreation reservoirs. However, present Soil Conservation Service practice for drop inlets over about 13 m tall, drop inlets placed in a berm or where sediment may accumulate to near crest level and thus encroach on the area below the spillway crest normally reserved for trash racks, and present US Army Corps of Engineers practice is to place the antivortex plate slightly above the head pool elevation at which the flow control changes from weir to pipe.

The hood drop inlet has special application to ensure full conduit flow for drop inlet heights between 0 and 5 pipe diameters, a range of heights for which Blaisdell (1958b, Part V) shows square-edged barrel entrances ordinarily cause part-full flow. The drop inlet crest length can also be set to ensure a controlled reservoir rise at which the pipe fills. Yalamanchili and Blaisdell (1975) and Blaisdell and Yalamanchili (1975) present the performance characteristics and design criteria for the hood drop inlet.

Antivortex devices are required to eliminate the harmful effects of vortices on the spillway performance and capacity. Donnelly and Blaisdell (1976) report the results of tests on antivortex walls for hood drop inlets, rectangular drop inlets, square drop inlets, and circular drop inlets.

Low-stage inlets are used many times in the two-way drop inlet to establish the level of a sediment detention pool or permanent pool. The reservoir storage between the low stage and principal spillway crests is then used to temporarily store the frequent small floods. The stored water is discharged at a low rate through the low stage inlet. The results of tests on low-stage inlets for two-way drop inlets presented by Donnelly and Blaisdell (1976) show how to size and locate low-stage inlets to ensure satisfactory performance.

The US Soil Conservation Service is building some two-way drop inlets so high that cavitation and cavitation damage are possible. Anderson (1979) reports results of tests on double-circular and elliptical elbows for use between the two-way drop inlet and the crown of the barrel. For the semicylindrical bottom drop inlet used with circular barrels, transitions to change the crown cross section from semisquare to semicircular are also described. These elbows and transitions have low cavitation potential. Pressure coefficients are presented to permit computations to evaluate the cavitation potential.

When the barrel is circular in cross-section, the US Soil Conservation Service now uses a semicylindrical bottom two-way drop inlet. This reduces the potential for trash to collect at the barrel entrance. Design criteria for the two-way drop inlet with a semicylindrical bottom are presented by Yalamanchili and Blaisdell (1979).

5.1.5.2. Trash Racks

Some type of rack is required to prevent trash from plugging closed conduit spillways. The ideal trash rack will pass harmless, nonplugging trash—trash small in comparison to the spillway size—and intercept any trash that might plug the spillway. An additional requirement is that the area of the trash rack must be sufficient so that the velocity through the rack will be less than about 0.5 m sec^{-1} to 0.7 m sec^{-1} to prevent packing the trash and effectively sealing the inlet.

Hebaus and Gwinn (1975) report tests on various types of trash racks used by the US Soil Conservation Service. Racks located on the drop inlet crest performed poorly; racks on the drop inlet crest provide insufficient area, so the trash-free velocity through the rack is high and when trash collects it packs tightly and little flow can pass through. Racks supported away from the drop inlet crest—where the velocity through the rack was low—were satisfactory.

The best design (Figure 5.10) use a solid plate supported above and extending well beyond the drop inlet crest. Ventilated skirts or screens extending downward from the edges of the solid plate to below the spillway crest level excluded floating trash. Flow is under the skirts, upward under the solid plate, and horizontally over the spillway crest. This type of trash rack is restricted to drop inlets located in pools deep enough so flow can enter from below the skirted, solid plate.

Gwinn (1976) reports tests on the stepped baffled trash rack conceived by M. M. Culp of the US Soil Conservation Service and currently used by that Service on two-way drop inlets. Gwinn (1976) describes the rationale and the form as follows:

Earlier experiments on trash racks for spillway crests showed that in the weir flow range open racks permitted a drawdown of the surface of the approaching flow, attracting floating sticks and debris to the rack. When a solid wall or skirt, extending below the crest level and above the water surface, was placed around the drop inlet entrance at a reasonable distance from the crest, the drawdown outside the skirt was eliminated, and flow near the surface and trash accumulation on the rack were reduced. The skirt, therefore, is a desirable feature for a trash rack. However, for a reservoir expected to fill to the crest with sediment, the area below the skirt could become plugged with trash and sediment. To overcome the possibility of zero flow and yet retain the advantage of the skirt, Culp designed the baffle rack. In this rack, rack members themselves become segments of a skirt . . . by arranging structural steel channels in stair-step fashion outward from the entrance so that each vertical web becomes a small skirt which could exclude floating trash. Pre-cast, concrete rectangular beams could also be used as rack members.

Plate 31 shows a stepped channel trash rack in Pennsylvania, USA, before completion of the dam earthwork.

Another description of the stepped baffle trash rack is: A series of overlapping stair risers with the bottom of the lowest riser below the weir crest. The stair treads are omitted. The baffles are all covered by a flat plate that also covers the drop inlet. To enter the drop inlet the water flows under the upstream baffles, upward through the space between the baffles that otherwise would be occupied by a stair tread, and over the downstream baffles. Floating trash has no horizontal access to the spillway and is thereby excluded, except for the small amount that may be trapped between the baffles as the water level rises. Gwinn comments: 'Models of baffled, covered-top drop inlets . . . performed satisfactorily when subjected to trash-laden flows'.

5.1.6. OUTLETS

Outlets for most lined channel structures, low drop structures, the baffled apron chute, and high drop structures are themselves part of the structures, so no further comment on them is needed. However, plain chutes and closed conduit spillways require separate outlet structures. Although many types of outlet structure are available, we will limit our presentation to preformed plunge pools, the US Bureau of Reclamation impact basin, a sloping apron stilling basin, and the St. Anthony Falls (SAF) stilling basin. Other types of stilling basins are described by Peterka (1964), the Bureau of Reclamation (1973; 1974) and in some open channel flow text books.

5.1.6.1. Preformed Plunge Pool

No stilling basin may be required for small closed conduit spillways. For these small spillways, the spillway exit may be supported on a pile bent and cantilevered beyond the bent. The flow is allowed to scour the channel bed and 'self form' its energy dissipation pool. For larger spillways it is preferable to pre-excavate and perhaps line with riprap a plunge pool in which the flow can dissipate its destructive energy by turbulence.

To supply a need of the US Soil Conservation Service for information on the design of armoured plunge pool energy dissipators, Culp (1969) prepared design criteria. These criteria have been used with reasonable success. However, in some cases the riprap near the downstream end of the basin has been eroded.

Unfortunately, I know of no presently available researched information for the design of preshaped, armoured, plunge pool energy dissipators. However, the Science and Education Administration–Agricultural Research is conducting research at the St. Anthony Falls Hydraulic Laboratory in Minneapolis, Minnesota, to determine the flow-formed shape of scour at cantilevered pipe outlets. This information will be used to develop criteria for designing the shape, size, and riprap protection required to build preformed plunge pool energy dissipators.

5.1.6.2. Impact Basin

The US Bureau of Reclamation has developed an impact basin in which the flow from a pipe strikes a wall placed across the width of the outlet structure (Figure 5.11). Water exits primarily under but partially over the wall. The energy of the incoming water jet is dissipated by striking the wall and by intensive turbulence. A feature of this basin is that no tailwater is required to ensure its proper operation. However, tailwater and riprap are required downstream of the basin to protect the bed and banks downstream of the basin exit from scouring.

The proportions and hydraulic design of the impact stilling basin are given by Young (1974), Peterka (1964), and Blaisdell and Moratz (1961).

5.1.6.3. Sloping Apron-Stilling Basin

The sloping apron-stilling basin described by Peterka (1964) (Figure 5.12) is suggested for use at plain chute exits where the tailwater level is high, or where the chute discharges into a gully or stream bed which has not eroded to a stable grade and where the tailwater level may change during the life of the structure. For such unstable conditions, the chute is extended below the stream bed to an elevation corresponding to the minimum tailwater elevation minus the tailwater depth required for the sloping apron stilling basin. As the bed erodes, the hydraulic jump moves down the chute until it ultimately reaches the design elevation. Blaisdell and Moratz (1961) also give the required tailwater depth.

5.1.6.4. SAF Stilling Basin

The SAF stilling basin (Figures 5.3 and 5.14) dissipates energy by a forced hydraulic jump. Chute blocks at the end of a sloping open channel, floor blocks in the stilling basin, and an end still to prevent bed scour near the stilling basin exit cause rapid dissipation of energy using a relatively short stilling basin. Flaring wingwalls, triangular in elevation, are used at the stilling basin exit to prevent scour of the dam by reverse currents generated in the downstream channel near the basin exit by flow leaving the basin. The reverse currents cause a local widening of the downstream channel in which energy dissipation, begun within the basin, is completed. This controlled completion of energy dissipation downstream of the basin exit is the reason the SAF stilling basin is so much smaller than other stilling basins.

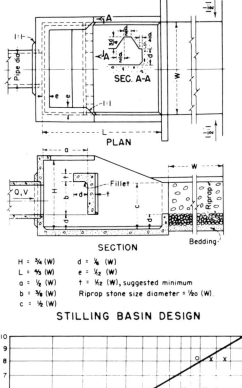

H = ¾ (W)   d = ⅙ (W)

L = ⁴⁄₃ (W)   e = ½ (W)

a = ½ (W)   t = ½ (W), suggested minimum

b = ⅜ (W)   Riprap stone size diameter = ¹⁄₂₀ (W).

c = ½ (W)

STILLING BASIN DESIGN

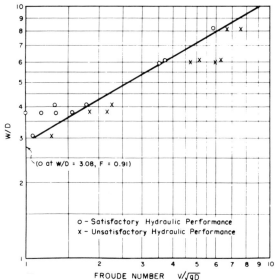

DESIGN WIDTH OF BASIN

"w" is the inside width of the basin.

"D" represents the depth of flow entering the basin
  and is the square root of the flow area.

"v" is the velocity of the incoming flow.

Figure 5.11 Dimensional criteria for the impact
stilling basin (Bureau of Reclamation, 1973)

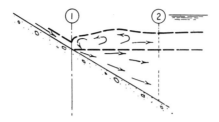

Figure 5.12 Sloping apron stilling basin (Peterka, 1964)

The SAF stilling basin has been widely and successfully used since its development in the 1940s. Every SAF stilling basin I have seen and know of that has been designed following the SAF design procedure has performed as anticipated. In some field structures, the tailwater level has been sufficiently lower than planned so that the jump has apparently washed out of the basin. In such cases, the scour downstream has lengthened but not deepened. This performance, which increases the scoured area but does not endanger the structure, is as predicted by the original model tests.

The design criteria for the SAF stilling basin may be found in many texts. The development of the SAF design is described by Blaisdell (1948). A design handbook is Blaisdell (1959). A third source is Blaisdell and Moratz (1961).

5.1.7. TRANSITIONS

Transitions are frequently required at the entrance to stilling basins. They are used to change the flow cross section from circular to the rectangular form required by most stilling basins, from trapezoidal to rectangular, and to increase the Froude number by widening and decreasing the flow depth to achieve greater economy and increased energy dissipation. I will mention four transitions that have been used for these purposes.

5.1.7.1. Abrupt Circular Pipe to Rectangular Channel Transition

It is possible to abruptly change the flow cross-section from circular to rectangular if the transition is built as shown in Figure 5.13. However, the width of the rectangular channel must exactly equal the circular jet diameter. Greater widths allow waves, originating where the circular jet hits the flat floor, to climb past the jet and over the sidewalls. When the jet and transition widths are identical, the climbing wave is suppressed by the falling jet. Blaisdell, Donnelly, and Yalamanchili (1969) give criteria for the design of abrupt transitions.

After the cross-section becomes rectangular, the flaring transition described below can be used to increase the Froude number and thereby improve the economy and increase the energy dissipated.

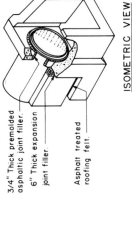

3/4" Thick premolded asphaltic joint filler.

6" Thick expansion joint filler.

Asphalt treated roofing felt.

ISOMETRIC VIEW

3/4" Premolded asphalt joint filler all around pipe.

45°

D+2t_p+2"

a

Cradle Cradle

a a

PLAN

6"Min

a

$\mathbb{C}$ Pipe a

3/4" $S=\tan\alpha$

Last Pipe Section

$S=\tan\alpha$

$t_p + 3/4"$

x_1 y_1 45°

Jet Section

x_2 y_2

P.T.

Parallel Section $S_v=\tan\alpha$

x_3 y_3

P.C.

Floor Curve $S_o = 1/2$

$\frac{1}{2}$

$\frac{1}{2}$

SECTION ON CENTERLINE

① Horizontal distance from invert at outlet end of pipe to P.C.
② Difference in elevation between these 2 points.
③ Normal height of chute sidewalls at P.C.

NOTE: THIS PROCEDURE APPLIES FOR VALUES

OF $\dfrac{Q}{D^{5/2}} \leq 20$

GENERAL FORMULAS	REF.
JET $y = (t_p + \frac{3''}{4})\sec\alpha$, slope length $= x_1 \sec\alpha$, $y_1 = y + x_1 \tan\alpha$ $x_1 = \sqrt{\dfrac{2 v_p^2 \cos^2\alpha \, y}{g}}$	1
PARALLEL SECTION slope length $= 0.1\dfrac{Q}{D^{3/2}}$ $y_2 = $ (slope length) $\sin\alpha$ $x_2 = $ (slope length) $\cos\alpha$	2
FLOOR CURVE $x_3 = 50\left(\dfrac{1}{z} - \tan\alpha\right)$ $y_3 = 25\left(\dfrac{1}{z^2} - \tan^2\alpha\right)$ Note: Coordinates of any point between P.T. and P.C. are given by equation: $y = 0.01x^2 + S_v x$	3
WIDTH OF CHUTE SIDEWALLS (inside) $= D$	2
MINIMUM HEIGHT OF CHUTE SIDEWALLS height $= D + 0.5'$	
REFERENCES (1) National Handbook Section 5- Hydraulics page 5.6-2 (2) See footnote *, page 1 (3) National Handbook Section 14 - Chute Spillways page 2.120, 2.141, 2.142, 2.143 and 2.144	

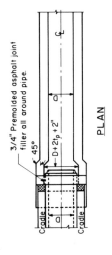

Figure 5.13 Abrupt transition from a circular pipe to a rectangular open channel (Blaisdell, Donnelly, and Yalamanchili, 1969)

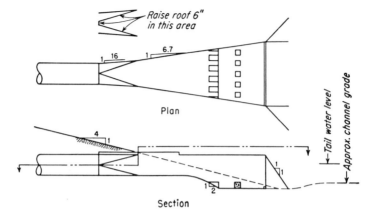

Figure 5.14 Shaped transition from a circular pipe to a rectangular open channel, followed by flaring sidewalls in supercritical flow and a SAF stilling basin (Blaisdell and Moratz, 1961). Copyright © 1961 McGraw-Hill Book Company, Inc. Used with permission of McGraw-Hill Book Company

5.1.7.2. Shaped Circular to Rectangular Channel Transition

Another form of transition from circular to rectangular is comprised of plane triangular sides, bottom and top, and oblique conical quadrants having a base diameter equal to that of the pipe (Figure 5.14). The sidewalls flare slightly and the top plane surface is raised to ensure that the flow does not cling to the top of the transition and cause uneven flow distribution at the transition exit. Blaisdell and Moratz (1961) describe the design and show a sketch of this transition.

The flaring sidewall transition described below can follow this transition used to change the flow cross section from circular to rectangular. The advantages of using the flaring transition are described above.

5.1.7.3. Trapezoidal to Rectangular Channel Transition

Stilling basins that are trapezoidal in cross-section usually perform poorly and are not recommended. This is because a channel change from trapezoidal to rectangular is conducive to the formation of cross waves and poor flow conditions when the velocity is supercritical. However, one way to change the cross section from trapezoidal to rectangular is shown in Figure 5.15 and is briefly described by Blaisdell and Moratz (1961).

The trapezoidal to rectangular transition is formed by extending the trapezoidal chute sloping sidewalls on the same plane until they intersect the horizontal floor of the stilling basin. The spacing of the vertical side walls of the stilling basin is slightly greater than the width of the water surface in the trapezoidal chute, and the

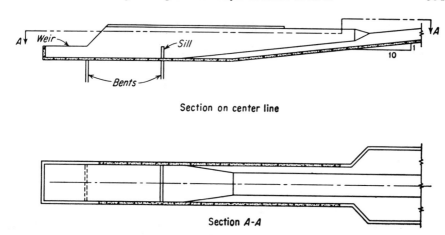

Section on center line

Section A-A

Figure 5.15 Trapezoidal to rectangular open channel transition. (Blaisdell and Moratz, 1961). Copyright © 1961 McGraw Hill Book Company, Inc. Used with permission of McGraw-Hill Book Company

vertical walls extend upstream to a chute elevation equal to the maximum water surface height in the stilling basin. This transition form prevents the supercritical velocity flow in the chute from impinging on the basin sidewalls, thereby creating cross waves and poor flow distribution in the stilling basin.

5.1.7.4. Flaring Sidewalls in Supercritical Flow

As has been hinted previously, the Froude number at the exit of a closed conduit spillway is frequently so low that it is economical to use a transition between the outlet and the stilling basin (e.g. Figure 5.14). However, there is less advantage in using a flaring transition at a chute exit where the Froude number is already relatively high. A flaring transition in supercritical flow will have little effect on the flow velocity but will decrease the flow depth, increase the Froude number, and increase the precentage of the energy that can be dissipated in a hydraulic jump or stilling basin. Another advantage is that a decrease in the flow depth at a stilling basin entrance will decrease the required flow depth at the stilling basin exit and thereby decrease the depth below the downstream bed level at which the basin floor must be placed to achieve the required basin tailwater depth.

Transition design for flows at supercritical velocities is entirely different from that for flow at subcritical velocities. If the flow is supercritical, any change in the flow boundaries may create surface waves that might overtop the channel sidewalls. Also, the waves may continue and require higher sidewalls for long distances in the downstream channel. However, Blaisdell and Moratz (1961) present design data for a simple, straight-sidewall transition in which the sidewalls flare 1 transverse in 3 times the Froude number longitudinally. As far as I know, transitions using this sidewall flare have performed satisfactorily.

5.1.8. ACKNOWLEDGEMENTS

I have by no means exhausted the list of engineering structures useful for erosion control. The structures I have listed are those that have been found particularly useful to the US Soil Conservation Service. Properly designed, they can be applied anywhere their use is indicated.

I acknowledge my indebtedness to my many associates in the US Soil Conservation Service whom I joined in 1935 near the beginning of the soil and water conservation movement. It has been my privilege to participate with Soil Conservation Service engineers in developing many of the structures in current use. Through the years I have maintained and benefited from this close association with Soil Conservation Service engineers as my group conducted research on hydraulic structures for, first, the Research Division of the Soil Conservation Service and, later, the succeeding agencies—the Agricultural Research Service and now the Science and Education Administration-Agricultural Research.

Finally, I acknowledge the significant contributions of my coresearchers, both at my own and at other locations. Most of their names may be found as authors and coauthors of the publications I have cited.

REFERENCES

Anderson, A. G., 1973, *Tentative Design Procedure for Riprap-lined Channels— Field Evaluation,* Project Report No. 146, St. Anthony Falls Hydraulic Laboratory, University of Minnesota, Minneapolis, Minnesota, 55414, 53 pp.

Anderson, C. L., 1979, *Hydraulics of Closed Conduit Spillways, Part XVI: Elbows and Transitions for the Two-way Drop Inlet,* AAT-NC-1, US Department of Agriculture, Science and Education Administration-Agricultural Research, Minneapolis, Minnesota, 55414, 44pp.

Anderson, A. G., Paintal, A. S., and Davenport, J. T., 1970, *Tentative Design Procedure for Riprap-lined Channels,* National Cooperative Highway Research Program Report 108, Highway Research Board, National Academy of Sciences, 2101 Constitution Avenue, Washington, D.C., 20418, 75 pp.

Beauchamp, K. H., 1969, Chapter 6: Structures, In *Engineering Field Manual for Conservation Practices,* US Department of Agriculture, Soil Conservation Service, Washington, D.C., 20013, pp. 6-1–6-91.

Blaisdell, F. W., 1948, Development and hydraulic design, Saint Anthony Falls stilling basin, *Trans. Amer. Soc. Civ. Engrs.,* 113, 483–520 (*Discussions,* 521–561).

Blaisdell, F. W., 1958a, *Hydraulics of Closed Conduit Spillways, Part I: Theory and its Application,* Technical Paper No. 12, Series B, St. Anthony Falls Hydraulic Laboratory, University of Minnesota, Minneapolis, Minnesota, 55414, 22 pp.

Blaisdell, F. W., 1958b, *Hydraulics of Closed Conduit Spillways, Parts II–VII: Results of Tests on Several Forms of the Spillway,* Technical Paper No. 18, Series B, St. Anthony Falls Hydraulic Laboratory, University of Minnesota, Minneapolis, Minnesota, 55414, 50 pp.

Blaisdell, F. W., 1958c, *Hydraulics of Closed Conduit Spillways, Part VIII: Miscellaneous Laboratory Tests; Part IX: Field Tests,* Technical Paper No. 19, Series B, St. Anthony Falls Hydraulic Laboratory, University of Minnesota, Minneapolis, Minnesota, 55414, 54 pp.

Blaisdell, F. W., 1959, *The SAF Stilling Basin,* Agriculture Handbook No. 156, US Department of Agriculture, Agricultural Research Service, United States Government Printing Office, Washington, D.C., 20402, 16 pp.

Blaisdell, F. W., 1973, *Model Test of Box Inlet Drop Spillway and Stilling Basin Proposed for Tillatoba Creek, Tallahatchie County, Miss.,* ARS-NC-3, US Department of Agriculture, Agricultural Research Service, Minneapolis, Minnesota, 55414, 50 pp.

Blaisdell, F. W., 1975, Theory of flow in long siphons, In *Proceedings of the Symposium on Design and Operation of Siphons and Siphon Spillways,* BHRA Fluid Engineering, Cranfield, Bedford, MK43 0AJ, England, pp. C7-89–C7-98.

Blaisdell, F. W., Anderson, C. A., and Hebaus, G. G., 1980, *Soil and Water Conservation Structures, and Field Applications,* Hydraulic Models AAT-NC-6, US Department of Agriculture, Agricultural Research Service, Minneapolis, Minnesota, 55414, 33 pp.

Blaisdell, F. W., and Donnelly, C. A., 1956, The box inlet drop spillway and its outlet, *Trans. Amer. Soc. Civ. Engrs.,* 121, 955–986 (*Discussions,* 987–994).

Blaisdell, F. W., and Donnelly, C. A., 1958, *Hydraulics of Closed Conduit Spillways, Part X: The Hood Inlet,* Technical Paper No. 20, Series B, St. Anthony Falls Hydraulic Laboratory, University of Minnesota, Minneapolis, Minnesota, 55414, 41 pp.

Blaisdell, F. W., and Donnelly, C. A., 1966, *Hydraulic Design of the Box-inlet Drop Spillway,* Agriculture Handbook No. 301, US Department of Agriculture, Agricultural Research Service, United States Government Printing Office, Washington, D.C., 20402, 40 pp.

Blaisdell, F. W., and Donnelly, C. A., 1975, The hood inlet self-regulating siphon spillway, In *Proceedings of the Symposium on Design and Operation of Siphons and Siphon Spillways,* BHRA Fluid Engineering, Cranfield, Bedford, MK43 0AJ, England, pp. C11-137–C11-154.

Blaisdell, F. W., Donnelly, C. A., and Yalamanchili, K., 1969, *Abrupt Transition from a Circular Pipe to a Rectangular Open Channel,* Technical Paper No. 53, Series B, St. Anthony Falls Hydraulic Laboratory, University of Minnesota, Minneapolis, Minnesota, 55414, 67 pp.

Blaisdell, F. W., Donnelly, C. A., Yalamanchili, K., and Hebaus, G. G., 1975, The two-way drop inlet self-regulating siphon spillway, In *Proceedings of the Symposium on Design and Operation of Siphons and Siphon Spillways,* BHRA Fluid Engineering, Cranfield, Bedford, MK43 0AJ, England, pp. C4-31–C4-53.

Blaisdell, F. W., and Hebaus, G. G., 1966, *Hydraulics of Closed Conduit Spillways, Part XI: Tests Using Air,* Technical Paper No. 44, Series B, St. Anthony Falls Hydraulic Laboratory, University of Minnesota, Minneapolis, Minnesota, 55414, 53 pp.

Blaisdell, F. W., and Moratz, A. F., 1961, Chapter 41: Erosion-control Structures, In *Agricultural Engineers' Handbook,* (Eds. C. B. Richey, P. Jacobson, and C. W. Hall), pp. 426–491, McGraw-Hill, New York.

Blaisdell, F. W., and Yalamanchili, K., 1975, The hood drop inlet self-regulating siphon spillway, In *Proceedings of the Symposium on Design and Operation of Siphons and Siphon Spillways,* BHRA Fluid Engineering, Cranfield, Bedford, MK43 0AJ, England, pp. C6-69–C6-88.

Brevard, J. A., 1976, *Graphical Solution for the Hydraulic Design of Riprap Gradient Control Structures,* Technical Release No. 59, Supplement 1, Design Unit, Engineering Division, Soil Conservation Service, US Department of Agriculture, 10,000 Aerospace Road, Lanham, Maryland, 20801, 32 pp, 26 charts.

Bureau of Reclamation, 1973, *Design of Small Dams, Second Ed.,* Bureau of Reclamation, United States Department of the Interior, United States Government Printing Office, Washington, D.C., 20402, 816 pp.

Bureau of Reclamation, 1974, *Design of Small Canal Structures 1974*, Bureau of Reclamation, United States Department of the Interior, United States Government Printing Office, Washington, D.C., 20402, 435 pp.

Chow, V. T., 1959, *Open-channel Hydraulics*, McGraw-Hill, New York, 680 pp.

Coyle, J. J., 1969, Chapter 7, Grassed waterways and outlets, In *Engineering Field Manual for Conservation Structures*, US Department of Agriculture, Soil Conservation Service, Washington, D.C., 20013, pp. 7-1–7-43.

Culp, M. M., 1969, *Armored Scour Hole for Cantilever Outlet*, Design Note No. 6, Design Branch, Engineering Division, Soil Conservation Service, US Department of Agriculture, Washington, D.C., 20013, 3 pp, 7 charts.

Culp, M. M., and Gregory, A. R., 1956, *Earth Spillways*, Technical Release No. 2, Design Unit, Engineering Division, Soil Conservation Service, US Department of Agriculture, 10,000 Aerospace Road, Lanham, Maryland, 20801, 4 pp, 5 charts.

Donnelly, C. A., and Blaisdell, F. W., 1965, Straight drop spillway stilling basin, *J. Hydraulics Div. Proc. Amer. Soc. Civ. Engrs.*, **91**, HY3, 101–131.

Donnelly, C. A., and Blaisdell, F. W., 1966, Closure to: Straight drop spillway stilling basin, *J. Hydraulics Div. Proc. Amer. Soc. Civ. Engrs.*, **92**, HY4, 140–145.

Donnelly, C. A., and Blaisdell, F. W., 1976, *Hydraulics of Closed Conduit Spillways, Part XIV: Antivortex Walls for Drop Inlets; Part XV: Low-stage Inlet for the Two-way Drop Inlet*, ARS-NC-33, US Department of Agriculture, Agricultural Research Service, Minneapolis, Minnesota, 55414, 37 pp.

Donnelly, C. A., Hebaus, G. G., and Blaisdell, F. W., 1974, *Hydraulics of Closed Conduit Spillways, Part XII: The Two-way Drop Inlet with a Flat Bottom*, ARS-NC-14, US Department of Agriculture, Agricultural Research Service, Minneapolis, Minnesota, 55414, 66 pp.

Doubt, P. D., 1955, Section 14: Chute spillways, Of *National Engineering Handbook*, US Department of Agriculture, Soil Conservation Service, Washington, D.C., 20013, 249 pp.

Goon, H. J., 1975, *Hydraulic Design of Riprap Gradient Control Structures*, Technical Release No. 59, Design Unit, Engineering Division, Soil Conservation Service, US Department of Agriculture, 10,000 Aerospace Road, Lanham, Maryland, 20801, 47 pp.

Goon, H. J., 1978, *Water Surface Profiles and Tractive Stresses for Riprap Gradient Control Structures*, Technical Release No. 59, Supplement 2, Design Unit, Engineering Division, Soil Conservation Service, US Department of Agriculture, 10,000 Aerospace Road, Lanham, Maryland, 20801, 34 pp.

Gregory, A. R., 1957, *Earth Spillways*, Technical Release No. 2, Supplement A, Design Unit, Engineering Division, Soil Conservation Service, US Department of Agriculture, 10,000 Aerospace Road, Lanham, Maryland, 20801, 3 pp, 7 charts.

Gwinn, W. R., 1976, Stepped baffled trash rack for drop inlets, *Trans Amer. Soc. Agric. Engrs.*, **19**, 1, pp. 97, 98, 99, 100, 101, 102, 103, 104, and 107.

Hayes, R. B., 1974, Chapter VI, Energy dissipators, B. Baffled apron drops, In *Design of Small Canal Structures*, Bureau of Reclamation, US Department of the Interior, United States Government Printing Office, Washington, D.C., 20403, pp. 299–308.

Hebaus, G. G., and Gwinn, W. R., 1975, *A Laboratory Evaluation of Trash Racks for Drop Inlets*, Technical Bulletin No. 1506, United States Department of Agriculture, Agricultural Research Service, Stillwater, Oklahoma, 74074, 70 pp.

Hoffman, C. J., 1973, Chapter IX: Spillways, In *Design of Small Dams, Second Ed.*,

Bureau of Reclamation, US Department of the Interior, United States Government Printing Office, Washington, D.C., 20402, pp. 345–447.

Jacobson, P., 1961, Chapter 40: Waterways for erosion control, In *Agricultural Engineers' Handbook* (Eds. C. B. Richey, P. Jacobson and C. W. Hall), pp. 419–425, McGraw-Hill, New York.

Katsaitis, G. D., 1966, The use of dissipation bars in channel drop structures, *J. Inst. Engrs., Australia,* **38**, 1–2, 1966, 9–18.

Murley, K. A., 1970, Irrigation channel structures, Victoria, Australia, *J. Irrig. and Drainage Div., Proc. Amer. Soc. Civ. Engrs.,* **96**, IR2, 131–150.

Murley, K. A., 1972, Closure to Irrigation channel structures, Victoria, Australia, *J. Irrig. and Drainage Div., Proc. Amer. Soc. Civ. Engrs.,* **98**, IR1, 138–143.

Palmer, V. J., Law, W. P., and Ree, W. O., 1954, *Handbook of Channel Design for Soil and Water Conservation,* SCS-TP-61, Soil Conservation Service, US Department of Agriculture, Washington, D.C., 20013, 34 pp.

Palmer, V. J., Law, W. P., and Ree, W. O., 1966, *Handbook of Channel Design for Soil and Water Conservation (Metric System),* SCS-TP-61, Soil Conservation Service, US Department of Agriculture, Washington, D.C., 20013, 22 pp, 18 figures.

Peterka, A. J., 1964, *Hydraulic Design of Stilling Basins and Energy Dissipators,* Engineering Monograph No. 25, US Department of the Interior, Bureau of Reclamation, United States Government Printing Office, Washington, DC., 20403, 222 pp.

Ree, W. O., and Palmer, V. J., 1949, *Flow of Water in Channels Protected by Vegetative Linings,* USDA Technical Bulletin No. 967, Soil Conservation Service, US Department of Agriculture, Washington, D.C., 20013, 115 pp.

Renfro, G. M., 1969, Chapter 11: Ponds and Reservoirs, In *Engineering Field Manual for Conservation Structures,* US Department of Agriculture, Soil Conservation Service, Washington, D.C., 20013, pp. 11-1–11-60.

Rhone T. J., 1977, Baffled apron as spillway energy dissipator, *J. Hydraulics Div., Proc. Amer. Soc. Civ. Engrs.,* **103**, HY12, 1391–1401.

Yalamanchili, K., and Blaisdell, F. W., 1975, *Hydraulics of Closed Conduit Spillways, Part XIII: The Hood Drop Inlet,* ARS-NC-23, US Department of Agriculture, Agricultural Research Service, Minneapolis, Minnesota, 55414, 78 pp.

Yalamanchili, K., and Blaisdell, F. W., 1979, *Hydraulics of Closed Conduit Spillways, Part XVII: The Two-way Drop Inlet with a Semicylindrical Bottom,* AAT-NC-2, US Department of Agriculture, Science and Education Administration-Agricultural Research, Minneapolis, Minnesota, 55414, 22 pp.

Young, R. B., 1974, Chapter VI: Energy dissipators, C. Baffled outlets, In *Design of Small Canal Structures,* Bureau of Reclamation, US Department of the Interior, United States Government Printing Office, Washington, D.C., 20402, pp. 308–322.

POSTSCRIPT

Since Chapter 5.1 was written, recent (December 1980) observations of four SCS riprap lined channels have shown that the channels and upstream transitions perform satisfactorily. However, in one down-stream transition some movement of the riprap was evident and downstream of two structures scour holes upto 4.6 m deep and perhaps a hundred or more meters long had formed. It appears that the downstream transition between the riprap channel and the downstream channel needs improvement.

Tropical Agricultural Hydrology
Edited by R. Lal and E. W. Russell
© 1981, John Wiley & Sons Ltd.

5.2

The Need for Soil Conservation Structures for Steep Cultivated Slopes in the Humid Tropics

T. C. Sheng

5.2.1. INTRODUCTION

It is always a challenge to a watershed conservationist to decide what kind of land treatment will be best suited to an area. Professional bias may overshadow the actual needs. Preference and prior experience may also introduce improper measures from one region to another. Examples are not lacking in which the use of inappropriate conservation measures has caused serious waste.

There is a tendency in many developing countries to welcome any soil conservation measure which is easy to apply or is of low cost without looking into its effectiveness. Many such works appear superficially good but in fact produce very limited real benefits. On the other hand, in many rich countries there has been a trend to construct unnecessarily massive structures. Therefore the conservationist needs to find out, between the most effective and the least expensive treatments, the optimum one. This is, of course, his real challenge.

Heavy and frequent rains as well as excessive runoff and steep slopes limit the application of purely agronomic or biological conservation measures in hilly watersheds in the humid tropics. The erosion-prone conditions of these areas have been discussed in detail (Sheng, 1979). It is therefore necessary to use conservation structures to protect slopes against erosion and for runoff disposal. This paper will describe these structural measures in soil conservation, their needs, types, effectiveness, economics, and application based on the author's experience in the watersheds of four tropical countries: Jamaica, El Salvador, Taiwan, and recently Northern Thailand.

5.2.2. THE NEED FOR STRUCTURES

Of the four major factors that affect soil erosion, namely rainfall, soil, slope, and land use, it is often only feasible to change slope in the developing countries. To change rainfall and major soil characteristics is impossible, while to change land use, for instance, from cultivated crops to grass or forest in order to minimize erosion may also be impractical, especially in the hilly watersheds populated with small farmers. In these areas, farmers who cultivate an average of one hectare each just cannot accept such changes for the simple reason of survival. Changing land use can only become practical if the government has effective programmes of land adjustment, consolidation, reform, or settlement. When these are lacking, hilly lands in developing countries must be protected against erosion whilst remaining in extensive cultivation.

The use of agronomic measures such as strip cropping, contour cultivation, close planting, grass barriers, etc., is not sufficient for erosion control under torrential rains and on steep terrain (Gil, 1976; Sheng, 1979). They can only be used on gentle slopes (below 7° or 12 per cent) or as supplementary measures to structures on steep slopes.

Therefore, the most suitable method of reducing soil erosion from hills in the humid tropics is to bring about some changes in the slope of the land by structural measures. There are two ways to change a slope for erosion control and they are corresponding to two slope factors which cause erosion, namely steepness and slope length:

(1) To change a steep slope to many continuous flat strips running along the contour across a hillside.

(2) To change a long slope to a series of shorter slopes by using a discontinuous type of structure.

Both of the above measures can be designed to divert runoff to flow along the contour at a non-erosive velocity towards protected drainage channels or waterways. Although structures such as walls and trenches are used in some hilly tropical countries, the best form, or at least the most widely used form, is benching or terracing. The flat benches not only control erosion but also, if properly arranged, can be used to facilitate irrigation, drainage, and transportation as well as for their primary purpose of cultivation.

Terracing steep slopes is not a new practice in many parts of the world, especially in Southeast Asia. Irrigated rice, or so called paddy rice, for example, has been grown on terraces or paddy fields for over a thousand years. However, most of these old systems have many disadvantages such as uneven width of the terraces with poor accessibility, and so are unsuited to the use of power driven machineries. There is a need for improvement through better and more scientific design.

5.2.3. TYPES OF SOIL CONSERVATION STRUCTURES

5.2.3.1. For Land Treatment

The types of soil conservation structure for land treatment which are suited for hilly regions in the humid tropics, and which are described below, are all of reversed-slope benches, continuous, or discontinuous, differing in width to suit different crops and slopes. Their advantages are as follows:

(1) They are structures which provide the drainage that is necessary in the humid tropics. Any runoff will first concentrate at the toe or the cut-side of the bench and then drain safely on a controlled horizontal gradient to a protected waterway.

(2) Discontinuous types of structures are inexpensive, flexible, and can be built over a period of years. They break a long slope into many shorter ones and reduce erosion quite significantly especially when supplemented by agronomic conservation measures.

(3) They can be applied safely on slopes up to 30° (58 per cent) and therefore increase the area which can be safely used for agriculture.

(4) They include types of structures which can be used for transportation, irrigation or for future mechanization.

(5) They are suited to annual crops, semi-permanent crops and also mixed crops.

(6) They are suitable for both small and medium-sized farms.

This paper will not discuss irrigated rice terraces because they are well-known old practices. In Southeast Asia, for example, most farmers know how to build them.

Eight types of conservation structure including recently developed ones, are briefly described here, and their cross-sectional views and specifications are shown in Figure 5.16 and Table 5.1 respectively. Handbooks and volume tables for most of these structures are available in English, Spanish, and Chinese. Sample sheet of tables can be seen in FAO Conservation Guide No. 1 pp. 176–177.

5.2.3.1.1. Bench Terraces

A series of level or nearly level strips running across the slope supported by steep risers. The reverse slope of a bench is 5 per cent and the slope along the contour or horizontal is up to 1 per cent. They can be built and cultivated by manual labour, animal drawn implements or by machines. They are very effective in erosion control (See Section 5.2.4), and have many other advantages including facilitating irrigation, mechanization, transportation, cultivation, and increased production. They can be used on slopes up to 25° (47 per cent) and are mainly used for upland crops.

5.2.3.1.2. Hillside Ditches

They are a discontinuous type of narrow (1.8 m or 2 m), reverse-slope benches built across the hill slope in order to break long slopes into many shorter ones. In addition to draining the runoff they can also be used for roads. The width of the cultivable strips between two ditches is determined by the slope of the lands. They should be supplemented with agronomic conservation measures. The ditches can also be used for cultivation if needed by small farmers. This treatment can be applied to slopes up to 25° (47 per cent).

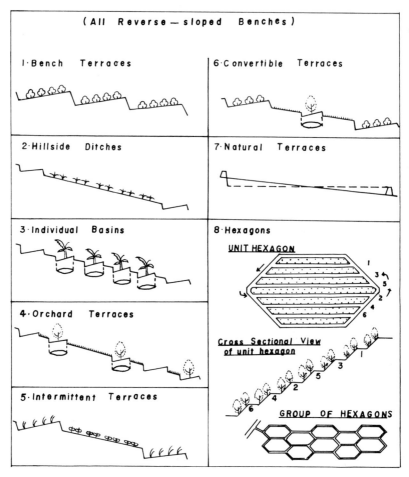

Figure 5.16 Cross-sectional view of eight type structures for land treatments

5.2.3.1.3. Individual Basins

Individual basins are small round benches for planting individual plants. They are particularly useful for establishing semi-permanent or permanent tree crops to control erosion. They also prevent fertilizers being washed away by rain, retain soil moisture if mulched, and make weeding easier. They can be applied to dissected terrain and shallow soils. However, they should normally be supplemented by hillside ditching, orchard terracing, and cover cropping.

5.2.3.1.4. Orchard Terraces

They are a discontinuous type of narrow terrace (1.75 m) which are applicable on steep slopes up to 30° (58 per cent). Their spacing is determined by the planting

Table 5.1. Specifications and applications of eight type land treatment structures

Kind	Specification						Applications	
	Width of flat bench	Length	Horizontal grade	Reverse grade	Riser slope	Land slope	VI[a] or spacing	Auxiliary treatments
1. Bench terraces (a) Hand made	2.5–5.0 m	<100 m	up to 1%	5%	0.75:1	7°–25°	$\dfrac{S \times W_b}{100 - S \times 0.75}$[c]	—
(b) Machine built	3.5–8.0 m	<100 m	1%	5%	1:1	7°–20°	$\dfrac{S \times W_b}{100 - S \times 1}$ $\dfrac{S \pm 4}{10}$ or $\dfrac{S+6}{10}$	Agronomic con. measures[b]
2. Hillside ditches	1.8–2.0 m	<100 m	1%	10%	0.75:1	<25°		Hillside ditches Orchard terraces and agro. con. measures
3. Individual basins	1.5 m (Round)	—	—	10%	0.75:1	<30°	Distance of crop	
4. Orchard terraces	1.75 m	<100 m	1%	10%	0.75:1	25°–30°	11–13 m along slope	Agro. measures Individual basins
5. Intermittent terraces	2.5–5.0 m	<100 m	1%	5%	0.75:1	7°–25°	3 times BT	Agro. measures Individual basins
6. Convertible terraces	2.5–5.0 m	<100 m	1%	5%	0.75:1	7°–25°	3 times BT	Individual basins Agro. measures
7. Natural terraces	8–20 m	—	—	—	1:1	<7°	1 m VI	Agro. measures
8. Hexagons (a) Terraces and operation routes	3.5 m	<100 m	1%	5%	1:1	7°–20°	8–13 m along slope	Individual basins Agro. measures Grass or marling
(b) Peripheral road	3.5 m		<7° (12%)	5%	1:1	7°–20°	—	Cross drains

[a]VI is vertical interval between two succeeding terraces, which determines spacing.
[b]To be applied mostly between the terraces (or on the individual basins) such as contour planting, close planting, cover cropping, mulching, etc.
[c]S: Slope as percentage W_b: Width of bench

distance of the tree crop. As they are usually used on steep slopes, the spaces between two terraces should normally be kept under permanent grass or legume cover. The tree crops are preferably planted on individual basins in the spaces between terraces. One terrace will serve two lines of trees; one up-slope and the other down-slope.

5.2.3.1.5. Intermittent Terraces

Fundamentally they are bench terraces which are to be built over a period of several years. Normal bench terraces are staked out but to begin with only one out of three is built. Both the spaces between the terraces and the benches can be used for cultivation of either the same or different types of crops. The benches will intercept runoff from the slopes above and reduce erosion.

5.2.3.1.6. Convertible Terraces

This type, when staked out and built, is similar to intermittent terraces, but, when they are used, the spaces between two terraces are planted to semi-permanent or tree crops. Should a farmer later on wish to cultivate more food crops he can convert the spaces to more benches. On the other hand he can convert the whole area into tree crops when he so wishes, if age, labour shortage, or economic circumstances should make this desirable.

5.2.3.1.7. Natural Terraces

They are constructed initially with contour bunds 50 cm high on slopes not over 7° or 12 per cent and on soils with good infiltration rates. They are designed and constructed in such a way that the top of the lower bund is levelled with the mid-slope between two bunds so that a natural terrace will be formed after a few years of cultivation.

5.2.3.1.8. Hexagons

A unit hexagon is a special arrangement of a farm road that surrounds or envelops a piece of slope treated with discontinuous terraces which are thus accessible to four-wheel tractors. The circumferential or peripheral road is in the form of a hexagon connecting each terrace or operation route and enables them to be entered at an obtuse angle. A group of hexagons forms a honeycomb on uniform terrain. This treatment is primarily for mechanization of orchards on larger blocks of land, but it can also be used for orchards on steep slopes (up to 20°) or on small farms (quarter to half hectare). In the latter case, the operation route (terraces) can be cultivated to produce cash crops until the tree crops in the spaces between the terraces grow up.

5.2.3.2. For Runoff Disposal

Excess runoff is inevitable on cultivated slopes in the humid tropics and when it is concentrated, a safe disposal system or a protected waterway is needed to drain it safely to the down slope. However, this is often neglected and when it is lacking a gully may be formed in the middle of a terraced field. Usually natural depressions are not sufficient to carry additional runoff and should only be used where the amount of runoff is small and with extreme care. In many cases, they need to be reshaped and protected with grass or structures.

A waterway therefore is an integral part of land treatment in the humid tropics. On steep slopes when the flow velocity exceeds 1.8 m to 2 m per second, grass waterways may not be adequate to prevent erosion. Experience in Taiwan and Jamaica has indicated that on slopes over 11° (20 per cent) structures should normally be used. Figure 5.17 and Table 5.2 show seven major types of waterways and structures, together with their uses and limits.

Waterway structures can be expensive. The selection of a suitable type depends on the climatic and physical environment of the area, its purpose, and the experience and judgement of the conservation engineer. However, in order to avoid overbuilding and excessive expense the following principles should be observed:

(1) Divide rather than concentrate runoff from a field if possible.
(2) Use adjacent grass land or forest area as a sump for runoff if it is available. Otherwise, a good protected waterway should be built to take care of all the run-off.
(3) Use as much locally available material as possible.
(4) Plan and select suitable waterway sites carefully to reduce the cost of structures.

5.2.4. EFFECTIVENESS OF THE STRUCTURES

Soil loss and runoff plot studies in Jamaica, El Salvador, and Taiwan have demonstrated that conservation structures can significantly reduce soil erosion. A four year study (1969-1973) of soil losses from 294 runoff producing rains on a clay loam soil with a 17° (30 per cent) slope cropped to yellow yams (*Dioscorea Cayennensis*) showed an average annual soil loss per hectare from the check plots with no conservation treatment was 133 metric tons (hereafter referred to as tons) while from the bench terrace plots it was 17 tons. On plots treated with hillside ditches supplemented with continuous mounds the soil loss was 26 tons and from hillside ditches with individual mounds 38 tons. Bench terraces thus reduced soil loss by 87 per cent and hillside ditches by 71 per cent to 80 per cent. Figure 5.18 shows the cumulative soil loss of the plots over a four year period (Sheng and Michaelsen, 1973). A further study in bananas under different treatment showed that from 1973 to 1974, loss from check plots was on average 183 tons of soil per hectare per year against results of 17 tons from bench terraces and 21 tons when hillside

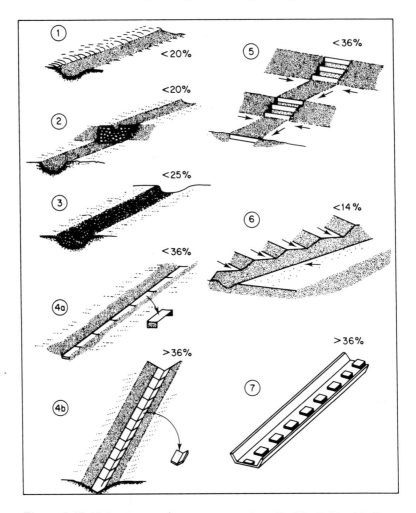

Figure 5.17 Major types of waterways as described in Table 5.2 (Sectional views: stilling basins not shown)

ditches with individual basins and cover crops were used. If the check plot is given an index figure of 100, that for the bench plot is only 9 and for hillside ditch plot 12.

A similar plot study in maize followed by beans on 17° (30 per cent) slope with loam to clay loam soils in Northern El Salvador showed average soil losses for 2 years (1975-1976) as follows: check plots 100 t ha^{-1} yr^{-1} and bench terrace plots 30 t ha^{-1} yr^{-1}. However, the grass barrier plots yielded an average 65 t ha^{-1} yr^{-1} which was quite high (Michaelsen and Heymans, 1976).

Table 5.2. Major types of protected waterways: their uses and limits [a]

Type		Shape	Channel protection	Velocity limit	Slope limit	Uses
1.	Grassed waterway	Parabolic	By grass	1.8 m sec⁻¹	< 11 ° (20%)	For new waterway or depression
2.	Grassed waterway with drops	Parabolic	By grass and concrete or masonry	1.8 m sec⁻¹	Between two structures: 3%, overall slope < 11 ° (20%) < 15 ° (26%)	For discontinuous type of channel
3.	Ballasted waterway	Parabolic	By stones or stones in wire mesh	3 m sec⁻¹		Where stones are available
4.	Prefabricated (a) Parabolic waterway	Parabolic	By concrete structures and grass	—	< 20 ° (36%)	A stilling basin is usually needed and where rainfalls are frequent and flows are constant
	(b) V-notch chute	90 ° V-notch	By concrete structures and grass	—	> 20 ° (36%)	Same as above and on very steep slopes
5.	Stepped waterway	Parabolic and rectangular	By grass and concrete or masonry drops	On grass part: 1,8 m sec⁻¹	Overall slope < 20 ° (36%) < 8 ° (14%)	For 4 wheel tractors and in the middle of bench terraces
6.	Waterway and road ditch	Parabolic	By grass and stone ballasting	3 m sec⁻¹		For 4 wheel tractors mechanisation
7.	Foot-path and chute complex	Trapezoid or rectangular	By concrete or masonry structures	—	> 20 ° (36%)	For paths on small farms and on very steep slopes

[a] These limits are approximations for general reference. In practice, the volume and velocity of runoff and site conditions should all be taken into consideration.

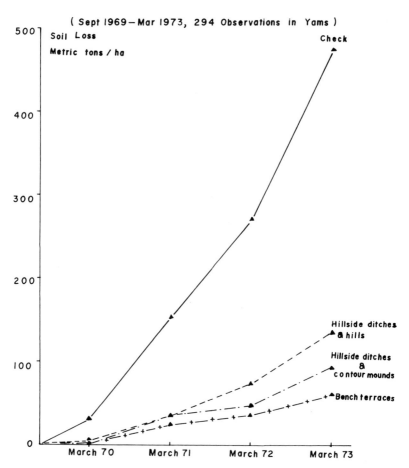

Figure 5.18 Accumulaion of soil loss from runoff plots in Jamaica (September 1969–March 1973, 294 observations in yams)

In Taiwan, many such plot studies have been carried out since the 1950s, using various crops and treatments. Wu (1967) after reviewing the results of over a dozen studies conducted by research institutes since 1951 concluded that bench terraces could reduce erosion by 95 per cent. Figure 5.19 gives a selection of some recently published results. The effects on erosion control of discontinuous types of structure such as hillside ditches and individual basins have unfortunately not been studied much in Taiwan. However, a recent research on the use of grass barriers or strips planted 11 m apart in a loamy soil on a $18°$ (32 per cent) slope, with the land between the strips in various closely planted crops has shown that these measures were not effective in erosion control. The average total loss of soils from 5 treatments and four replications ranged from 273 t ha^{-1} to 417 t ha^{-1} in two years (Hsu *et al.*, 1977).

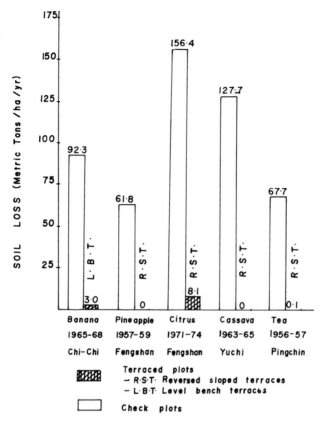

Figure 5.19 Some results of runoff plot studies in Taiwan
(Soil losses of check plots *vs.* bench terraced plots)

There is still much argument on whether the various types of terraces reduce runoff as well as erosion. Results from the yellow yams experiment in Jamaica already quoted showed no significant differences between the plots in the rate of runoff. Once the soil are saturated or nearly saturated any additional heavy rain will become runoff. In addition, risers of bench terraces which occupied 20 per cent to 25 per cent of the plot area contribute considerable runoff as they have a steep slope and are compacted or grassed. In the new runoff plot studies in Northern Thailand, it seems that the bench terraced plots have not yielded less runoff than the check plots. The structures may reduce peak flows but this could not be shown in small plots. However, this subject needs more studies.

The effectiveness of protected waterways is quite apparent, for without them, gullies may readily form.

5.2.5. ECONOMICS OF THE STRUCTURES

5.2.5.1. Cost

The cost of various conservation structures per unit area depends on slope, soil, type of terraces, width of bench, presence of rocks or tree stumps, and the tools to be used for building them. A set of volume tables and specifications for cutting and filling for several types of terraces has been worked out and used satisfactorily in Jamaica, El Salvador, and Northern Thailand. For example, one hectare of 4.6 m bench terraces on 15° slope may need 425 man-days for their construction. If a D-6 bulldozer is used it may need 47 hours. A man with hand tools can move, on average, 4 m³ in 8 hours. A D-6 bulldozer can move some 40 m³ an hour. From the daily wage rates or rate per tractor hour, the cost per hectare can thus be computed.

Before detailed costs can be estimated the correct specification for the system such as width of benches and the type of structures required, must be agreed. For general budgeting purposes, however, some criteria can be established using specification tables in conjunction with local experience. In Northern Thailand, for instance, some round figures for land treatment cost per hectare have been established in manual labour. The present daily wage is US$1.25 which may be representative for the rural area of the developing countries. The following figures include 25 per cent leeway as the work efficiency there is around 75 per cent:

Type of terraces	Man days per hectare	Labour costs per hectare
Bench terraces 4 m wide on slopes less than 20° with top soil preserved	500 man days ha⁻¹	12,500 Baht (US$625) ha⁻¹
Hillside ditches 2 m wide on slopes less than 25°	100 man days ha⁻¹	2,500 Baht (US$125) ha⁻¹
Intermittent or convertible terraces 4 m wide on slopes less than 20°	167 man days ha⁻¹	4,175 Baht (US$208) ha⁻¹
Orchard terraces 1.75 m wide on slopes of 20°–27°	112 man days ha⁻¹	2,800 Baht (US$140) ha⁻¹
Individual basins 1.5 m in diameter, 200 basins per ha	12 man days ha⁻¹	300 Baht (US$15) ha⁻¹
Natural terraces with initially building contour bunds < 7° slope	100 man days ha⁻¹	2,500 Baht (US$125) ha⁻¹

Waterway cost varies with slope, drainage area, and quantity and velocity of run-off. The length of waterway needed per hectare depends on local topography, but a figure of 100 m should normally be sufficient. While the peak runoff can be estimated from local climatic data, the type of waterway needed depends mainly on the steepness of the slope. Some reference figures of cost per hectare at 1976 wage level in El Salvador (US$1.2 per man day) are as follows (Sheng, 1977a):

Grassed waterway	Used on slopes below 11 ° (20%)	18.3 man days	55 ¢ (US$22)
Prefabricated waterway	Used on slopes above 11 ° (20%)	32 man days plus material	399 ¢ (US$160)
Stepped waterway	Used for mechanization below 20 ° (36%)	51 man days plus material	630 ¢ (US$252)

The structures for land treatments and runoff disposal are quite permanent for they can have economic life of 25 to 30 years. Their annual cost per hectare, however, vary a great deal. For instance, it ranges from less than US$20 for simple structures, based on a preliminary study in Northern Thailand (Mae Sa Integrated Watershed Project, 1978) to US$200 for bench terraces with top soil preservation, and with stepped waterway structures and access roads for mechanization in Jamaica (Sheng, 1977b).

Since these structures can be expensive, a cost sharing or subsidy scheme should be introduced by the governments of developing countries who must also determine what portion or percentage of the cost should be borne by the farmers, and what by the nation or from other sources. This is because much of the indirect or off-site benefits from soil conservation measures accrue to people other than the farmers directly involved and to society as a whole.

5.2.5.2. Benefit

5.2.5.2.1. Direct Benefit

These are, in the main, increases in production and employment. In many instances, crop yields have been more than doubled due to the combined effects of fertility improvement, moisture conservation and erosion control; and in general an increase of 20 to 30 per cent can easily be attained. Usually, the farmers who adopt conservation structures are the same ones who initiate other improvements such as better seed, use of fertilizers, etc.

In Jamaica, the net return per hectare of yams, after deducting the annual cost of maintaining the terraces, waterways and road, was around US$1870 (Powell, 1974). In Northern Thailand, a preliminary study of cost and benefit showed that taro (*Colocasia* sp.) planted on intermittent terraces yielded a net profit of 1,330 Baht per rai (1,600 m^2) or US$416 per ha.

The benefit to employment is quite apparent for the construction and maintenance of most of the conservation structures are labour intensive; so rural employment will almost certainly be increased if a government has a national policy of giving incentives to farmers who build conservation structures such as providing technical assistance, food, materials or subsidies.

5.2.5.2.2. Indirect and Associated Benefits

The benefit of reducing soil erosion has been discussed at length under Section 5.2.4. Among other indirect and associated benefits, structures like bench terraces will make irrigation possible and simultaneously provide safe drainage on steep slopes. Also they can serve as access roads and facilitate mechanization with proper arrangement. In areas where shifting cultivation is practised, structures promote intensive land use, enable the cultivation of crops which previously could not be grown because of rain wash, and most important, they will encourage permanent farming.

5.2.6. APPLICATIONS OF THE STRUCTURES

5.2.6.1. For Small Farmers

For small farmers, there are two systems of conservation structures that can be applied in the hilly humid tropics: The ordinary system of intensive bench terracing and the simpler system for those farmers who at the moment cannot afford complete terracing or do not want to use land intensively:

A. Ordinary System (Continuous type of structures):
   (1) For irrigated crops ——————— Level terraces or paddy fields
   (2) For high value upland ——————— Reverse-sloped (or drainage
        crops mainly, up to   type) bench terraces
        25 ° (47 per cent) slope
B. Simple System (Discontinuous type of structures supplements with agronomic conservation measures):
   (3) For upland crops, particularly ——————— Hillside ditches
        semi-permanent crops, up to
        25 ° (47 per cent) slope
   (4) For upland crops or partly ——————— Intermittent
        irrigated crops where the   terraces
        structures are to be built
        over several years, up to
        20 ° or 25 ° (36 per cent or
        47 per cent) slope
   (5) For mixed farming, or for ——————— Convertible
        flexible land use in the   terraces
        future, up to 20 ° or 25 °
        (36 per cent or 47 per cent)
        slope
   (6) For semi-permanent crops ——————— Individual
        or tree crops on shallow   basins

soils and/or dissected terrains,
up to 25 ° or 30 ° (47 per cent
or 58 per cent) slope

(7)   For tree crops on steep   ————   Orchard terraces
slopes, from 25 ° to 30 ° (47
per cent to 58 per cent)

5.2.6.2. For Medium-sized Farms

The following additional systems can be used on many medium sized farms on gentle slopes where mechanization is either practical now or is likely to be in the near future:

(8)   For upland crops, or semi-   ————   Natural terraces
permanent crops on gentle   (initially with
slopes up to 7 ° (12 per cent)   contour bunds)

(9)   For mechanizing tree crops   ————   Hexagons
on slopes up to 20 ° (36 per cent)

The above recommendations on suitable types of structures for small or medium farmers are however not rigid. Usually the farmer must make his final decision after discussions with the officials concerned, and according to his particular interests, and the availability of finance and labour.

5.2.7. CONCLUSIONS

For cultivation of steep slopes in the humid tropics, soil conservation structures are necessary. Not only are they much more effective in erosion control than agronomic or biological measures but also they provide the basis for modern and permanent farming with the possibilities of introducing irrigation, improved drainage, access and even mechanization. In comparison with most agronomic conservation measures which are short-lived or end with the reaping of the crops, conservation structures are relatively permanent. Although they cost more their benefits are high, versatile, and long lasting. Governments in developing countries and international financial organizations should render sufficient assistance, technically and financially, to the farmers who are willing to apply them for the good of the whole society.

REFERENCES

Gil, N., 1976, The role of soil conservation in watershed management on agricultural lands, *Paper presented to FAO Expert Consultation on Soil Conservation and Management in Developing Countries.*

Hsu, S. C., *et al.*, 1977, Experiment on soil conservation practices for upland crops on slope-land (1st Report), in *J. Chinese Soil & Water Conserv.*, **8(2)**, Taichung, Taiwan.

Joint Commission on Rural Reconstruction, 1977, *Abstracts on Soil Conservation Research in Taiwan,* **1**, 1958–1976, Taipei, Taiwan.

Liao, M. C., 1976, Effects of bench terraces and improved hillside ditch, in *J. Chinese Soil & Water Conserv.,* **7(2)**, Taipei, Taiwan.

Mae Sa Integrated Watershed Project, 1978, *Results of Conservation Farming at Pong Khrai Demonstration Area (Year 1978),* Thai Government and FAO THA/76/001, Project, Chiang Mai, Thailand.

Michaelsen, T., and Heymans, L. E., 1976, Informe anual de las parcelas para la investigación del control de la erosion y escorrentia superficial en le distrito forestal de Metapán, *Direccion General de Recursos Naturales Renovable,* San Salvador.

Powell, W. I., 1974, Hillside agriculture in demonstration watersheds in Jamaica, *UNDP/FAO Jamaica 505 Project,* Technical Report 11.

Sheng. T. C., 1977a, Conservacion de suelos, *PNUD/FAO EIS/73/004 Documento de Trabajo No. 13,* FAO, Rome.

Sheng, T. C., 1977b, Protection of cultivated slopes, terracing steep slopes in humid regions, in *FAO Conservation Guide No. 1,* Guidelines for Watershed Management, 147–179.

Sheng, T. C., 1979, Erosion problems associated with cultivation in humid tropical hilly regions, *Paper prepared for the Symposium on Soil Erosion and Conservation in the Tropics,* ASA annual meetings, Colorado, August, 1979.

Sheng, T. C., and Michaelsen, T., 1973, *Runoff and Soil loss Studies in Yellow Yams,* UNDP/FAO Jamaica 505 Project working Document, Kingston.

Wu, C. M., 1967, A study of watershed sediment problem, in *Lecture Notes for Watershed Management Seminar* (in Chinese), Taiwan Provincial Government, Natou, Taiwan.

PART 6

Estimating Soil and Water Loss

Tropical Agricultural Hydrology
Edited by R. Lal and E. W. Russell
© 1981, John Wiley & Sons Ltd.

6.1

Simulation of Erosion and Sediment Yield from Field-sized Areas

G. R. FOSTER AND L. J. LANE

6.1.1. INTRODUCTION

A method is needed to evaluate sediment yield from field-sized areas under various management practices to control non-point-source pollution. In response to this need, we developed a reasonably simple simulation model that incorporates fundamental principles of erosion, deposition, and sediment transport mechanics.

Sediment load in overload flow and open channel flow is controlled by either transport capacity or sediment available for transport. If sediment load is less than transport capacity, detachment can occur, and deposition occurs when sediment load exceeds the transport capacity. The model provides comprehensive representation of a field by considering complex overland flow slopes, concentrated channel flow, and impoundments or ponds. The model estimates transport of sediment composed of primary particles (sand, silt, and clay) and large and small aggregates. In deposition, sediment sorting is calculated which can result in enrichment of the finer particles.

Sediment yield is a function of sediment production by erosion and the subsequent transport of the sediment. On a given field, either erosion or sediment transport capacity may limit sediment yield, depending on topography, soil characteristics, cover, and rainfall–runoff rates and amounts. The controlling mechanism can change from season to season, from storm to storm, and even within a storm. The relationships for erosion and transport are different, which prevents lumping them into a single equation. Since erosion and transport for each storm are best considered separately, lumped equations such as the Universal Soil Loss Equation, USLE, (Wischmeier and Smith, 1978), or Williams' (1975) modified USLE (a flow transport–sediment yield equation) cannot give the best results over a broad range of conditions on field-sized areas. Furthermore, the interrelation between erosion and transport is non-linear and interactive for each storm, which prevents using separate equations to linearly accumulate erosion or sediment transport capacity over several storms. Therefore, to simulate erosion and sediment yield on an individual storm basis and to satisfy the need for a continuous simulation model, we selected a more fundamental approach with separate equations used for erosion and sediment transport.

Several fundamentally based models (e.g. Beasley *et al.*, 1977; Li, 1977) compute erosion and transport at various times during the runoff event. Although these models are powerful, they require excessive use of computer time which practically prohibits simulating 20 to 30 years of record. The model described herein uses characteristic rainfall and runoff factors for a storm to compute erosion and sediment transport for that storm. In terms of computational time, this corresponds to a single time step for models which simulate over the entire runoff event.

The model is intended to be useful without calibration or collection of research data to determine parameter values. Therefore, established relationships, such as the USLE, were modified and used in the model.

6.1.2. BASIC STRUCTURE OF MODEL

The erosion-sediment yield model is a component of a more comprehensive model consisting of hydrologic, erosion, nutrient, and pesticide components (Knisel, 1978). Briefly, the erosion component receives as input from the hydrologic model rainfall and runoff data and provides input to the chemical transport components. This paper describes the erosion–sediment yield component of the more comprehensive model.

6.1.2.1. Basic Processes

The model was developed for quasi-steady state conditions by using characteristic measures for hydrologic inputs. Rainfall is described by volume and the product of storm energy and maximum 30-minute intensity. Volume and peak rate attenuated for travel time are used to describe runoff. These terms drive soil detachment and subsequent transport in overland and open channel flow.

The principal governing equation is the continuity equation expressed as:

$$\frac{dG}{dx} = D_L + D_F \tag{6.1}$$

where G is sediment load, x is distance, D_L is the rate of lateral inflow of sediment, and D_F is the rate of sediment removal (deposition) or addition (detachment). Equation (6.1) applies to overland flow and flow in channels. The flow path is divided into segments and equation (6.1) is applied sequentially to each segment.

The minimum potential sediment load at the lower end of a segment is the sum of incoming sediment at the upper end of the segment and that added by lateral inflow within the segment. This potential load is compared with sediment transport capacity. If sediment transport capacity exceeds the potential load, the potential exists for detachment by flow. The detachment rate will be the lesser of the detachment rate to satisfy transport capacity or the detachment capacity of the flow. When soil is detached by flow, it adds particles whose distribution is specified in the input data. On overland flow areas, lateral inflow of sediment is from inter-rill erosion and has the input distribution. Lateral inflow of sediment into the channels is from overland flow or other channels. This sediment has the distribution from the sediment yield calculations.

If the potential sediment load exceeds transport capacity, deposition occurs at the rate of:

$$D = \alpha(T_c - G) \tag{6.2}$$

where D is deposition rate (mass area^{-1} time^{-1}), α is a first order reaction coefficient (length^{-1}), and T_c is transport capacity (mass width^{-1} time^{-1}). The coefficient α is estimated from:

$$\alpha = E V_s/q \tag{6.3}$$

where E is 0.5 for overland flow (Davis, 1978), and 1.0 for channel flow (Einstein, 1968), V_s is particle fall velocity, and q is water discharge per unit width, Since α is large for coarse and heavy particles, they deposit rapidly, leaving the sediment relatively enriched in smaller and lighter particles. Also, the transport capacity equation considers the transportability of particles with various sizes and densities.

These equations are solved in the same way for both overland and channel flow. However, detachment by flow is computed differently for overland and channel flow.

6.1.2.2. Overland Flow

To describe sediment detachment by raindrop impact and inter-rill and rill flow, a modification of the Universal Soil Loss Equation is used for individual storm events. Inter-rill detachment (D_{IR}) in the overland flow element is expressed as

$$D_{IR} = 4.57 \, EI(S + 0.014)KCP(q_p/Q) \tag{6.4}$$

where EI is storm rainfall energy times maximum 30 minute intensity, S is overland flow slope, q_p is peak rate of runoff, Q is runoff volume, K is the soil erodibility factor, C is the cover-management factor, and P is the supporting practice factor. Notice that D_{IR} is equivalent to D_L in equation (6.1). The rill detachment process is described by

$$D_R = (6.86 \times 10^6)n_x Q q_p^{1/3}(x/22.1)^n x^{-1} S^2 KCP(q_p/Q) \tag{6.5}$$

where D_R is the rill detachment rate, n_x is the slope length exponent, x is the distance downslope, and the other variables are as described above. Only the USLE contouring part of the P factor is used. The model is structured to directly account for other highly variable USLE P-factors such as strip cropping and deposition in terrace channels.

Sediment transport capacity is calculated using the Yalin sediment transport equation (Yalin, 1963). Sediment transport capacity, W_s, in units of mass time^{-1} flow-width^{-1} is calculated using

$$W_s = 0.635 \; \delta \; V_* s \rho_w d [1 - \frac{1}{\sigma} \log(1 + \sigma)] \tag{6.6}$$

where:

$$\sigma = A\delta \tag{6.7}$$

$$A = 2.45 s^{-0.4}(Y_{cr})^{0.5} \tag{6.8}$$

$$\delta = \begin{cases} 0 & Y < Y_{cr} \\ \dfrac{Y}{Y_{cr}} - 1 & Y \geqslant Y_{cr} \end{cases} \tag{6.9}$$

$$Y = \frac{V_*^{\,2}}{(s - 1.0)gd}, \text{ and} \tag{6.10}$$

$$V_* = (gRS_f)^{1/2} = (\tau/\rho_w)^{1/2} \tag{6.11}$$

With this notation, V_* is the shear velocity, τ is the shear stress, g is acceleration of gravity, R is hydraulic radius defined as the cross-sectional area divided by the wetted perimeter, S_f is the friction slope, s is particle specific gravity, d, is particle diameter, Y_{cr} is the critical lift force from the Shields' diagram extended to low particle Reynolds numbers, and ρ_w is the mass density of the fluid. The constant 0.635 and the Shields' diagram were empirically derived. Shear stress required for the Yalin equation is computed using the Manning equation.

The sediment load may have fewer particles of a given type than the flow's transport capacity for that type. At the same time, the sediment load of other particle types may exceed the flow's transport capacity for those types. The excess transport capacity for the deficit types is assumed to be available to increase the transport capacity for the types where available sediment exceeds transport capacity.

The Yalin equation was modified to shift excess transport capacity. For large sediment loads (sediment loads for each particle type clearly in excess of the respective transport capacity for each particle type), or for small loads (sediment loads for each particle type clearly less than the respective transport capacity for each particle type), the flow's transport capacity is distributed among the available particle types based on particle size and density and flow characteristics.

6.1.2.3. Concentrated Flow

The concentrated flow or channel element of the erosion model assumes that peak rate of runoff is the characteristic discharge for the channel, and detachment–deposition is based on that discharge. Detachment can occur when the shear stress developed by the characteristic discharge is greater than the critical shear stress for the channel. Bare channels, grassed waterways, and combinations of bare and grass channels can be considered by the model with as many as 10 channel segments. Discharge is assumed to be steady state, but spatially varied, increasing downstream with lateral inflow. Friction slope and shear stress are estimated from solution of the spatially varied flow equations. The solutions consider draw-down or back-water effects in the channel as a result of channel outlet control.

The concentrated flow relationships discussed here are limited to upland areas typical of farm-field situations and include: (1) erosional channel development in areas of fields where flow is concentrated such as stream headwaters, terrace channels, etc., (2) small channels as permanent features of the landscape which are normally 'tilled over' during cultivation, and (3) temporary channels developing when rows or terraces overtop and flow proceeds cross-contour to the field edge. Specifically excluded are permanent stream channels and active gully systems of a scale larger than described above. Such large-scale features are more in the range of a basin-scale model than the field-scale model. Finally, gully systems are beyond the scope of the current field-scale modelling effort due to our lack of understanding of gully dynamics. An exception to the channel size limitations is in development of a final or equilibrium channel width. As discussed below, relationships developed for field-sized channels also apply to larger channels.

Hydrologic inputs to the channel system consist of overland flow hydrographs or volume and peak rate of runoff and a duration of runoff. In the latter case, it is necessary to choose a characteristic discharge and a time distribution with the speci-fied peak, volume, and duration. In this analysis, we assumed that the peak rate is the characteristic discharge, and that the temporal distribution of shear stress in the channel is triangular.

Given the hydrologic input to the channel system, the next step was to apply equations (6.1)–(6.2) with the same logic as used for the overland flow areas. Lateral inflow of sediment is from overland flow or contributing channels and D_f is erosion or deposition in the channel.

An estimate of shear stress along the channel was required. The routing equations were simplified by using the peak or characteristic discharge with the assumption of steady flow. This eliminated the unsteady flow equations, leaving steady but spatially varied flow.

As described by Chow (1959), the dynamic equation for spatially varied flow with increasing discharge is

$$\frac{dy}{dx} = \frac{S_0 - S_f - 2\alpha Q q_* / gA^2}{1 - \alpha Q^2 / gA^2 D} \tag{6.12}$$

where

$\dfrac{dy}{dx}$ = slope of water surface,

S_0 = bed slope,

S_f = friction slope,
α = energy coefficient,
Q = discharge at point of interest,
q_* = lateral inflow per unit length of channel,
A = cross-sectional area,
D = hydraulic depth, and
g = acceleration due to gravity.

To avoid solving equation (6.12) for each runoff event, it was solved under a variety of conditions, and regression equations were used to approximate the solutions (Foster *et al.*, 1980). Given a representative channel and flow conditions, we derived regression equations for the friction slope as a function of position along the channel. Given the friction slope $S_f(x)$, the average shear stress at distance x downstream is then

$$\tau(x) = \gamma R(x) S_f(x) \tag{6.13}$$

where:

$\tau(x)$ = shear stress, force per unit area,
γ = specific gravity of water, and
$R(x)$ = hydraulic radius, length.

In many field situations, outlet control for a channel has a significant influence on sediment detachment and transport capacity. Consequently, back-water or drawdown at the outlet can significantly affect sediment yield. If the outlet rating is known (critical depth or a rating table), then the friction slope at the outlet is

$$S_f = \frac{n^2 Q^2}{A^2 R^{4/3}} \tag{6.14}$$

where n is Manning's resistance coefficient. Subsequent values of friction slope at positions above the outlet are obtained from the spatially varied flow equation (equation (6.14)) or the approximating regression equations. The procedure is similar to computation of backwater profiles except that spatially varied flow is considered.

Solutions for equations (6.12)–(6.14) are used to derive S_f and which are required to solve the detachment capacity equation, the Yalin equation, and equations (6.1)–(6.3). The following discussion emphasizes the detachment capacity equations developed for and used in the model.

The detachment equations are based on a simplified channel morphology–erosion– sediment yield model. Objectives of this simplified model included developing equations which would: (1) be relatively simple with a minimum number of parameters, (2) incorporate what is known of channel hydraulics, (3) reproduce observed relationships between sediment yield and time for developing channel systems. Limiting assumptions were: (1) steady-state discharge, (2) erosion occurs at potential rate (no depositional-erosional cycles), and (3) the shear stress distribution around the wetted perimeter can be approximated using data from rectangular channels. Moreover, the quasi-steady state morphological relationships could be tested using existing channels, but the dynamic relationships require data from developing, dynamic channel systems.

The model simulates channel development in homogeneous-erodible material and in material with an erosion-resistant or 'non-erodible' boundary. Input to the model consists of a flow rate Q, a channel slope S, hydraulic resistance parameter n, soil erodibility factor K_{ch}, a critical shear stress τ_{cr}, and parameters for the shear stress distribution around the channel cross-section. If a non-erodible boundary is present, then the depth to this boundary, d_{side}, is also required.

Detachment rate was assumed to be given by:

$$D_p = K_{ch}(\tau - \tau_{cr})^c \tag{6.15}$$

where D_p is detachment rate at a point along the wetted perimeter, K_{ch} is a soil erodibility factor for channel erosion, τ is shear stress at a point along the wetted perimeter, τ_{cr} is a critical shear stress, and c is an exponent. Under a continuous steady discharge, the channel reaches an equilibrium width, W_{ac}, that moves downward at the rate that the middle of the channel erodes. Although the actual cross-sections are irregular and dynamic in an eroding channel, as a first approximation we assume a rectangular cross-section with specified width. This width is given by:

$$W_{ac} = \left[\frac{nQ}{S^{1/2}} \right]^{3/8} \frac{W_*}{R_*^{5/8}} \tag{6.16}$$

where W_* and R_* are geometric properties that depend on the shear stress distribution τ_{cr}, Q, n, and S. The functions for W_* and R_* were numerically derived. The corresponding erosion rate E_0 is

$$E_0 = W_{ac}K_{ch}(1.35\bar{\tau} - \tau_{cr})^c \tag{6.17}$$

where $\bar{\tau}$ is the average shear stress in the cross section, 1.35 is the ratio of the maximum to the average shear stress in small rectangular channels, and c is an exponent with a value of 1.05 to 1.10.

When the channel reaches the non-erodible layer, downward movement of the channel ceases and the channel widens. As it widens, erosion rate decreases. The channel continues to widen until the shear stress at the non-erodible layer equals the critical shear stress. The final width at which erosion ceases is:

$$W_f = \left[\frac{nQ}{S^{1/2}} \right]^{3/8} \left[\frac{1 - 2x_*}{x_*^{5/8}} \right]^{3/8} \tag{6.18}$$

where x_* is the normalized distance from the water surface to where τ equals τ_{cr} divided by the wetted perimeter. The final width W_f is a function of the distribution of τ, τ_{cr}, Q, n, and S.

The erosion rate immediately after the channel reaches the non-erodible boundary is:

$$E_i = 2K_{ch}(\tau_b - \tau_{cr})^c d_{soil} \tag{6.19}$$

where τ_b is the shear stress at the non-erodible boundary and d_{soil} is the depth of the soil above the non-erodible layer.

Once the non-erodible boundary is reached but before the final eroded width is reached, erosion rates decrease exponentially with time as

$$E(t) = E_i \exp(-\alpha t_*) \tag{6.20}$$

where $E(t)$ is erosion rate with time, α is a decay constant, and t_* is the normalized time. This normalized time t_* is computed from

$$t_* = tE_i / [(W_f - W_{ac})d_{soil}\rho_{soil}] \tag{6.21}$$

where t is time since the non-erodible boundary is reached, E_i is the initial erosion rate from equation (6.20), W_f is the final eroded width from equation (6.18), W_{ac} is the equilibrium width from equation (6.16), and ρ_{soil} is the apparent mass density of the soil.

Equations (6.16)–(6.21) provide a means of computing widths and associated erosion rates for eroding channels in homogeneous soil and under circumstances where a non-erodible boundary is present.

6.1.2.4. Impoundment Component

Impoundments often occur in field situations, either where a channel flows through a restriction (for instance a fence line or a road culvert) or in an impoundment-type terrace. Any such restriction reduces the flow velocity giving coarse-grain sediments and aggregates an opportunity to settle out of the flow. Deposition in impoundments is a function of the fall velocity of the particles and travel time through the impoundment. The fraction of particles, FP of a given size, i, is given by the exponential relation

$$FP_i = A_i e^{B_i d_i} \tag{6.22}$$

where d_i is the equivalent sand-grain diameter and A and B are coefficients.

6.1.2.5. Enrichment

Besides calculating the sediment transport fraction for each of the five particle size classes, the model also computes the sediment enrichment ratio based on the specific surface area of the sediment and organic matter and specific surface area for the residual soil. That is, the enrichment ratio, ER, is

$$ER = \frac{SSA_{sed}}{SSA_{soil}} \tag{6.23}$$

where SSA is specific surface area and the subscripts 'sed' and 'soil' refer to the sediment and residual soil, respectively. As deposition of sediment occurs in transport, the organic matter, clay, and silt are the principal particles transported. This results in high enrichment ratios, important in adsorbed chemical transport.

Table 6.1. Possible elements and their calling sequence
used to represent field-sized are

Sequence Number	Elements and their Sequence
1	Overland
2	Overland–Pond
3	Overland–Channel
4	Overland–Channel–Channel
5	Overland–Channel-Pond
6	Overland–Channel–Channel–Pond

6.1.2.6. Watershed Elements

Every model is a representation and a simplification of the prototype. Various techniques, including planes and channels (Li, 1977), square grids (Beasley *et al.*, 1977), converging sections (Smith, 1977), and stream tubes (Onstad and Foster, 1975) have been used. Most erosion–sediment yield models have adequate degrees of freedom to fit observed data. Some models, depending on their representation scheme, distort parameter values more than others do. Distortion of parameter values greatly reduces the transferability of parameter values from one area to another (Lane *et al.*, 1975). An objective in this model development was to represent the field in a way that minimizes parameter distortion.

Overland flow, channel flow, and impoundment (pond) elements are used to represent major features of a field. The user selects the best combination of elements to represent the field and enters the appropriate sequence according to Table 6.1. The model (computer program) calls the elements in the proper sequence. Typical systems that the model can represent are illustrated in Figure 6.1.

Computations begin in the uppermost element, which is always the overland flow element, and proceed downstream. Sediment concentration (for each particle type) is the output from each element and becomes the input to the next element in the sequence.

6.1.3. APPLICATIONS

6.1.3.1. Overland Flow

The overland flow component has been tested using data from several agricultural areas of the United States. The erosion relationships in the overland flow element gave good results for a watershed at Treynor, Iowa. Estimates were considerably better than those from the USLE using storm EI (Foster *et al.*, 1977) and better than those obtained from a procedure using runoff volume and peak discharge alone as an erosivity factor (Onstad *et al.*, 1976) in the USLE. Both rainfall and runoff seem to be important for estimating detachment on overland flow areas.

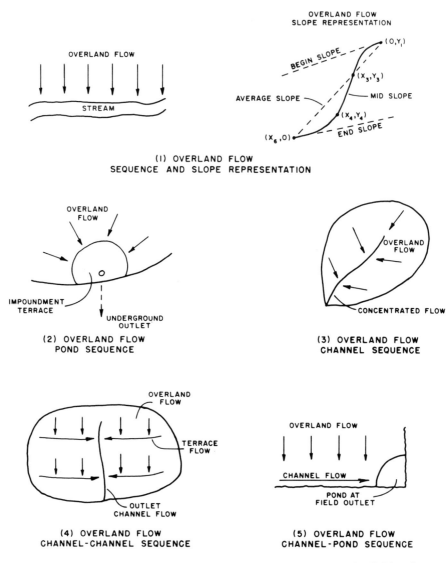

Figure 6.1 Schematic representation of typical field systems in the field-scale ero-
sion/sediment model

More comprehensive models like ARM (Donigian and Crawford, 1976) or ANSWERS
(Beasley *et al.*, 1977) use modifications of the USLE or require data for calibration
or both. Our model preserves the USLE form when simulated over a range of
slopes, lengths, and storms. For long-term simulation, our model produces results
comparable to those of the USLE. Information to select overland flow erosion
parameters is as readily available for this model as it is for the USLE.

The validity of the model has been partially assessed by comparing output with observed data and by comparing this model with other models. Output from the model has been compared with measured sediment yield from concave field plots under simulated rainfall, single terrace watersheds, small watersheds with impoundment terraces, and a small watershed with conservation tillage. The simulations were made using measured rainfall and runoff values. Parameter values were selected from the *User's Manual* (Foster *et al.*, 1980) without calibration, except as noted.

Three concave plots 10.7 m long were carefully shaped in a soil where soil properties were uniform with depth. Slope along the plots continuously decreased from 18 per cent at the upper end to 0 per cent at the lower end. Simulated rainfall at 64 mm hr^{-1} was applied to one of the plots, and deposition was observed to begin 7 m from the upper end. Plot ends were installed at 7.0 and 8.8 m on the other two plots. The measured particle size distribution was used, and the soil erodibility factor and Manning's n were adjusted in the model to give the observed soil loss and particle size distribution for the 7.0 m plot. The estimated sediment yield for the 8.8 m plot was 3.9 g m^{-1} sec^{-1} compared with 2.5 g m^{-1} sec^{-1} observed. For the 10.7 m plot, the estimated and observed values were 1.7 and 1.4 g m^{-1} sec^{-1}. Calculated and observed particle size distributions are shown in Table 6.2.

6.1.3.2. More Complex Watersheds

Soil loss was simulated for 8 years of data from small, single-terrace watersheds at Guthrie, Oklahoma (Daniel *et al.*, 1943). The simulations were made without calibration, and represented trends in the observed data quite well (Table 6.3).

Soil loss was simulated for six selected storms representing a range of rainfall and runoff characteristics for the Eldora, Charles City, and Guthrie Center, Iowa locations from an impoundment terrace study (Laflen *et al.*, 1978). The model was run using the *User Manual* instructions without calibration. Observed and computed sediment yield data for the three impoundment terraces are shown in Table 6.4. In the Julian data column, the first two digits represent the data year, and the last three digits represent the date in days since the first of the year.

Simulations were run without calibrating for about two-and-a-half years of data from the 1.3-ha P2 watershed at Watkinsville, Georgia in a conservation tillage system for corn (Smith *et al.*, 1978). Deposition in the backwater from the flume at the watershed outlet was modelled. Deposition measured in the flume backwaters was about equal to the measured sediment yield on a similar nearby watershed (Langdale *et al.*, 1979). The computed total sediment yield for the period of record was 1.47 kg m^{-2}, while the measured value was 1.85 kg m^{-2}.

6.1.3.3. Channel Morphology Model

The procedures described earlier were applied to data from an experimental rill erosion study at Lafayette, Indiana. The procedure was to measure channel (rill)

Table 6.2. Calculated and observed particle size distribution, in per cent, for transport of soil aggregates on concave field plots under simulated rainfall

Plot		Aggregate Size (μm)								
Length	Slope at end	< 2		2-210		210-500		500-1000		>1000
		Ob	Cal	Ob	Cal	Ob	Cal	Ob	Cal	Ob Cal
8.8 m	3%	7	8	53	58	7	24	12	9	21 1
10.7 m	0%	10	19	79	80	8	1	1	0	2 0

Table 6.3. Comparison of simulated sediment yield from single terrace watersheds with measured values

		Sediment Yield	
Terrace	Grade	Simulated	Observed
		(kg m^{-2})	(kg m^{-2})
2B	Variable, 0.0033 at outlet to 0.0 at upper end	6.4	12.2
3B	Variable, 0.005 at outlet to 0.0 at upper end	11.9	13.8
3C	Constant, 0.005	10.6	12.1
5C	Constant, 0.0017	4.6	4.8

cross sections, flow variables, and sediment yield under controlled conditions. A non-erodible boundary at the depth of disking was present below the soil surface.

Comparisons of observed and computed sediment yields with time showed a good fit using the simplified model. Total sediment yields over the seven replicated runs produced a relation between observed sediment yields Q_s, and computed sediment yield $\hat{Q}_s$ as:

$$\hat{Q}_s = -11.0 + 0.93\, Q_s \tag{6.24}$$

with an $R^2 = 0.97$. Therefore, the simplified model reproduces observed sediment yields within measurement accuracy.

In the rill erosion studies, discharge, slope, and Manning's n values were measured. However, to apply the model to selected discharge-width data from the literature, it was necessary to estimate the n values (Barnes, 1967). Given these estimates, the model was used to compute final widths, W_f, and these were compared with measured

Table 6.4. Summary of observed and simulated sediment yield from impoundment terraces in Iowa

Watershed	Area (ha)	Julian date	Observed sediment field (kg)	Computed sediment field (kg)
Charles City	1.9	70147	542	24
		70152	33	6
		70244	2	72
		70323	26	2
		71151	127	133
		71157	95	72
Eldora	0.73	68198	128	68
		68220	26	25
		69187	479	251
		69232	56	103
		71163	152	63
Guthrie Center	0.57	69207	116	124
		69249	10	40
		70144	55	29
		70162	90	56
		70167	10	13
		70229	5	24

values. Osterkamp (1977) selected several streams in the mountains and high plains of the United States and related channel width to a characteristic discharge. Observed and computed data for Osterkamp's 32 streams and the 7 rills are shown in Figure 6.2. Widths and discharges were then related by regression of the form

$$W = aQ^b \tag{6.25}$$

following the procedure outlined by Leopold and Miller (1956). The regression results for the observed and computed channel widths are shown in Figure 6.2. For the data from the experimental rill study, the coefficients and exponents in equation (6.25) are quite similar. Again, these results are for small rills under controlled experimental conditions. For natural streams, the exponent b was larger for the computed than for the observed widths. In these wide, natural streams, the distribution of shear stress around the channel cross section may be more uniform and nearer to the average shear stress over the wetted perimeter than is assumed in the small rills. Nonetheless, as shown in Figure 6.2, the model produced a reasonable approximation to the observed width-discharge relationship.

The concept of a quasi-steady state channel developed in a homogeneous soil due to a constant discharge is an oversimplification of processes occurring in natural channels. However, the simplified model described here does seem to explain width-

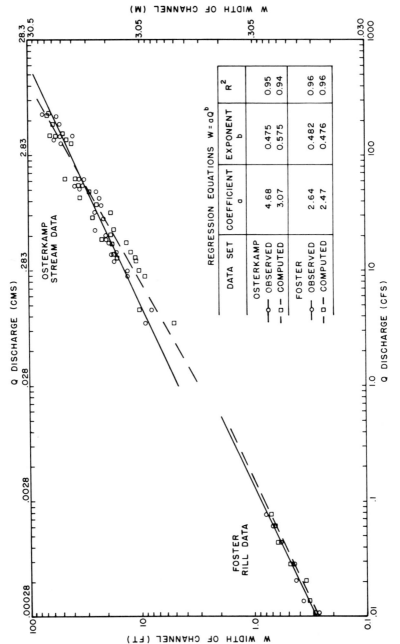

Figure 6.2 Relation between channel discharge and width for natural streams and experimental rill systems

discharge relationships that have been empirically derived over the past few decades (e.g. Leopold and Miller, 1956; Osterkamp, 1977). Moreover, width-discharge equations such as equations (6.16) and (6.18) seem to be an improvement over multiple linear regression equations used to predict channel width using discharge and other 'independent' variables. There is a hydraulic basis for the functional form in equations (6.16) and (6.18), whereas multiple linear regression equations can result from spurious correlations.

6.1.3.4. Selection of Best Management Practices

The model may be used to evaluate sediment yield from field-sized areas under various management practices to control nonpoint-source pollution. Given basic inputs that represent a specific field and the rainfall, the model is run to evaluate the various practices by using parameter values that characterize each specific practice. Results for such simulation runs are shown in Table 6.5. The field selected for this analysis had a typical slope length of 50 m on a uniform 6 per cent slope with a moderately erodible soil in continuous corn. The soil was assumed to be quite sandy. The analysis considered only 14 storms occurring over a two-month period around seedbed time. Several cropping years would be considered in a more complete analysis.

Practice 1 in Table 6.5 is a baseline with its uniform slope and clean tillage with no conservation practices. Practice 2 represents the effect of deposition on the toe of the concave portion of a highly convex–concave profile having an average 6 per cent slope. This is really not a management practice, since topography generally cannot be radically changed. Practice 3 shows that deposition in two grass strips along a uniform slope, one in the middle and one at the toe, can significantly reduce sediment yield. Difficulties in uniformly constructing and maintaining the strip may in practice greatly reduce their effectiveness. Practices 4, 5, and 6 represent the effect of concentrated flow in a field. The difference between Practices 1 and 4 is due to erosion by concentrated flow. A well constructed grass waterway eliminates that erosion and traps sediment coming from overland flow areas. Ponding at the field outlet can also reduce sediment yield, as Practice 5 shows.

Conventional terraces effectively control erosion and sediment yield when properly installed. If their grades are too steep, as in Practice 7, they erode. On a flat grade as in Practice 10, they do not erode, but trap significant amounts of sediment. The delivery ratio of terraces is not constant, as frequently assumed. Impoundment terraces very effectively control sediment yield in many circumstances, as illustrated by Practice 11.

Practices 12 to 15 are cultural practices frequently referred to as conservation tillage. Their effectiveness mainly depends on surface cover of residue from the previous year's crop. Finally, Practices 16 and 17 show the influence of combining terraces with conservation tillage practices. The relative effectiveness of such structures varies with the tillage practice.

Table 6.5. Typical best management that can be analysed with the model and typical sediment yield estimates

Practice	All 14 storms		Small storm $EI = 6.1$ (kJ ha^{-1}) · (mm hr^{-1}) Runoff = 2.8 mm		Large storm $EI = 77.3$ (kJ ha^{-1}) · (mm hr^{-1}) Runoff = 44.2 mm	
	Sediment yield (t ha^{-1})	Computed delivery ratio	Sediment yield (t ha^{-1})	Computed delivery ratio	Sediment yield (t ha^{-1})	Computed delivery ratio
1 Conventional	29.5	1.00[a]	0.5	1.00[a]	23.8	1.00[a]
2 Conventional, complex slope w/concave at toe	5.1	0.17[a]	0.0	0.05[a]	4.8	0.20[a]
3 Stripcropping, grass buffer strip	1.7	0.06[a]	0.0	0.00[a]	1.6	0.07[a]
4 Conventional, concentrated flow	36.7	1.24[a]	0.5	1.00[a]	28.4	1.19[a]
5 Conventional, concentrated flow, restricted outlet	27.8	0.94[a]	0.3	0.67[a]	22.5	0.94[a]
6 Conventional, grass waterway	12.1	0.41[a]	0.1	0.24[a]	10.5	0.44[a]
7 Conventional, 12.5 m terr. int., 1% grade	22.1	0.75[b]	0.2	0.48[b]	19.9	0.84[b]

8 Conventional, 12.5 m terr. int., 0.8% grade	15.7	0.53[b]	0.2	0.48[b]	13.7	0.57[b]
9 Conventional, 12.5 m terr. int., 0.5% grade	10.4	0.35[b]	0.2	0.48[b]	8.5	0.36[b]
10 Conventional, 12.5 m terr. int., 0.25% grade	6.4	0.22[b]	0.1	0.24[b]	5.3	0.22[b]
11 Conventional, impoundment terr.	0.4	0.02[b]	0.0	0.03[b]	0.3	0.01[b]
12 Chisel, 5000 kg ha^{-1}, 50% cover	5.2	—	0.2	—	4.8	—
13 Chisel, 2000 kg ha^{-1}, 20% cover	13.2	—	0.2	—	11.9	—
14 No till, 5000 kg ha^{-1}, 80% cover	2.1	—	0.0	—	1.9	—
15 No till in killed sod	0.3	—	0.0	—	0.3	—
16 Chisel, 2000 kg ha^{-1}, 20% cover, 12.5 m terr., 0.5% grade	5.4	0.41[b]	0.0	0.09[b]	4.9	0.41[b]
17 No till, 5000 kg ha^{-1}, 80% cover, 12.5 m terr., 0.5% grade	2.9	1.41[b]	0.0	0.08[b]	2.7	1.47[b]

[a] Ratio of sediment yield at outlet to sediment yield from uniform slope, conventional management.
[b] Ratio of sediment yield at terrace outlet to sediment yield from uniform slope with no terraces. Slope length and steepness = 50 m and 6%, respectively. Corn at seedbed time.

If 7 t ha^{-1} is an allowable sediment yield, eight of the 17 practices in Table 6.5 (excluding the topographic effect of the concave slope of Practice 2) adequately control sediment yield, so one can choose a practice from these eight which would best fit the particular farming operation. The relative, and certainly the absolute results, are unlikely to be the same under different farming conditions at other locations. Nonetheless, this simulation study illustrates the intended use of the model described herein, for it provides a means of evaluating alternative management practices with respect to their relative influence on sediment yield for a given location with specified characteristics of climate, topography, soils, etc.

6.1.4. SUMMARY

An erosion–sediment yield model for field-sized areas is developed for use on a storm-by-storm basis. Our overall objective is to develop a model, incorporating fundamental erosion–sediment transport relationships, to evaluate best management practices. Although the procedure does not consider changes in parameter values within individual storms, it does allow these parameters to change from storm to storm throughout the season. However, parameters of the model allow for distribution of field characteristics along overland flow slopes and waterways. Many of the model parameters are selected using tested methods developed for the well known Universal Soil Loss Equation. For this reason, we feel that the model has immediate applications without extensive calibration.

Limited testing has shown that the procedures developed here give improved estimates over the USLE and modified USLE procedures. We tested specific components of the model using experimental data from overland flow, erodible channel, and impoundment studies. Testing and sensitivity analyses are described elsewhere. Initial results suggested that the model produces reasonable results and is a powerful tool for analyzing the influence of alternative management practices.

A simple model has been developed to compute channel widths and associated erosion rates or sediment yields with time for eroding channels. The model reproduced observed sediment yield data from an experimental rill erosion study. The quasi-steady state width-discharge relationships predicted by the simple model compared well with observed data from natural channels.

6.1.5. ACKNOWLEDGEMENTS

The authors wish to thank the USDA-SEA-AR, Lafayette, Indiana, and Southwest Watershed Research Center, Tucson, Arizona for their support.

REFERENCES

Barnes, H. H., 1967, Roughness characteristics of natural channels, *US Geological Survey Water Supply Paper 1849,* 213 pp.

Beasley, D. B., Monke, E. J., and Huggins, L. F., 1977, The ANSWERS model: A planning tool for watershed research, *Paper No. 77-2532,* American Society of Agricultural Engineers, St. Joseph, Michigan.

Chow, V. T., 1969, *Open-Channel Hydraulics,* McGraw-Hill Book Company, Inc., New York, NY, 680 pp.

Daniel, H. A., Elwell, H. M., and Cox, M. B., 1943, Investigations in erosion control and reclamation of eroded land at the Red Plains Conservation Experiment Station, Guthrie, Oklahoma, 1930–1940, *UDSA Technical Bulletin No. 837,* 34 pp.

Davis, S. S., 1978, Deposition of nonuniform sediment by overland flow on concave slopes, *M.S. Thesis,* Purdue University, West Lafayette, Indiana.

Donigian, A. S., Jr., and Crawford, N. H., 1976, Modelling nonpoint source pollution from the land surface, *US Environmental Protection Agency,* EPA-600/376-083, 279 pp.

Einstein, H. A., 1968, Deposition of suspended particles in a gravel bed, *J. Hyd. Div., Proc. Amer. Soc. Civ. Eng.,* **94(HY5),** 1197–1205.

Foster, G. R., Meyer, L. D., and Onstad, C. A., 1977, A runoff erosivity factor and variable slope length exponents for soil loss estimates, *Trans. Amer. Soc. Agric. Engrs.,* **20(4),** 683–687.

Foster, G. R., Lane, L. J., and Nowlin, J. D., 1980, A model to estimate sediment yield from field sized areas: Selection of Parameter values—A field scale model for chemicals, runoff, and erosion from agricultural management systems, Conservation Research Report No 26, USDA Science and Education Administration, Vol. II, Users Manual, Chapter 2, pp. 193–281.

Knisel, W. G., 1978, A system of models for evaluating nonpoint source pollution— An overview, *International Institute for Applied Systems Analysis,* A-2361, Laxenburg, Austria, Pub. CP-78-11, 17 pp.

Laflen, J. M., Johnson, H. P., and Hartwig, R. O., 1978, Sedimentation modelling of impoundment terraces, *Trans. Amer. Soc. Agric. Engrs.,* **21(6),** 1131–1135.

Lane, L. J., Woolhiser, D. A., and Yevjevich, V., 1975, Influence of simplification in watershed geometry in simulation of surface runoff, *Hydrology Paper No. 81,* Colorado State University, Fort Collins, Colorado, 50 pp.

Langdale, G. W., Barnett, A. P., Leonard, R. A., and Fleming, W. G., 1979, Reduction of soil erosion by no-till systems in the southern Piedmont, *Trans. Amer. Soc. Agric. Engrs.,* **22(1),** 82–86, 92.

Leopold, L. B., and Miller, J. P., 1956, Ephemeral streams—hydraulic factors and their relation to the drainage net, *US Geological Survey Professional Paper 282-A,* 32 pp.

Li, R. M., 1977, Water and sediment routing from watersheds, *Proceedings of River Mechanics Institute,* Colorado State University, Fort Collins, Colorado, Chapter 9.

Onstad, C. A., and Foster, G. R., 1975, Erosion modelling on a watershed, *Trans. Amer. Soc. Agric. Engrs.,* **18(2),** 288–292.

Onstad, C. A., Piest, R. F., and Saxton, K. E., 1976, Watershed erosion model validation for southwest Iowa, in *Proceedings of the Third Federal Interagency Sedimentation Conference,* Water Resources Council, Washington, DC, Chapter 1, pp. 22–24.

Osterkamp, W. R., 1977, Effect of channel sediment on width-discharge relations, with emphasis on streams in Kansas, *Bulletin No. 21,* Kansas Water Resources Board, 25 pp.

Smith, R. E., 1977, Field test of a distributed watershed erosion/sedimentation model, in *Soil Erosion: Prediction and Control,* Special Publication No. 21, Soil Conservation Society of America, Ankeny, Iowa, pp. 201–209.

Smith, C. N., Leonard, R. A., Langdale, G. W., and Bailey, G. W., 1978, Transport of agricultural chemicals from small upland Piedmont watersheds, *US Environmental Protection Agency* EPA-600/3-78-056, 364 pp.

Williams, J. R., 1975, Sediment-yield prediction with Universal Equation using runoff energy factor, in *Present and Prospective Technology for Predicting Sediment Yields and Sources,* ARS-S-40, USDA-Science and Education Administration, pp. 244–252.

Wischmeier, W. H., and Smith, D. D., 1978, Predicting rainfall erosion losses, *Agricultural Handbook, No. 537,* USDA-Science and Education Administration, 58 pp.

Yalin, Y. S., 1963, An expression for bedload transportation, *J. Hydr. Div. Proc. Amer. Soc. Civil Engr.,* **89(HY3)**, 221–250.

Tropical Agricultural Hydrology
Edited by R. Lal and E. W. Russell
© 1981, John Wiley & Sons Ltd.

6.2

Runoff, Erosion and Conservation in a Representative Catchment in Machakos District, Kenya

D. B. Thomas, K. A. Edwards, R. G. Barber, and I. G. G. Hogg

6.2.1. INTRODUCTION

Machakos District which lies to the east and south-east of Nairobi occupies an area of 14,500 km^2 at an elevation ranging from 1,000 to 1,800 m above sea level. Approximately 80 per cent of the district lies within the arid or semi-arid zones as defined by Pratt and Gwynne (1977) with a mean annual rainfall ranging from 500 to 800 mm. More humid conditions are found in the central hill mass which is characterized by steep slopes and intensive cultivation.

The problem of soil erosion and the struggle to control it have been described by many writers including de Wilde (1967) and Moore (1979). Much has been done to conserve the soil, particularly on cultivated land, but degradation is still continuing in many places.

Shortage of water is a major constraint to development and the potential for groundwater use is limited by the low waterholding capacity of the Basement Complex schists and gneisses. The need to conserve water has led to the construction of numerous earth dams but rates of sedimentation have been high, as in other parts of East Africa (Rapp et al., 1972).

Planning of water development has been handicapped by lack of data on stream flow, on sediment yield, and on the factors which influence runoff and soil loss such as precipitation, evaporation, transpiration, infiltration, soil type, slope, and management. To overcome this deficiency a joint project was initiated in 1976 by the Ministry of Water Development of the Kenya Government and the UK Ministry of Overseas Development. The project involves the monitoring of hydrological variables in four representative catchments, two of which are in Machakos District. One of these, at Iiuni in Kalama location 20 km south-east of Machakos town, is the subject of this paper.

To complement the data on runoff and sediment yield collected at the outfall of the catchment, a rainfall simulator was used to measure runoff and soil loss from small plots representing different types of land use within the basin. These trials were carried out jointly by the Departments of Agricultural Engineering and Soil Science of the University of Nairobi.

The paper describes the methods used to assess the rates of soil loss and runoff, firstly, from the catchment as a whole and, secondly, from small plots under simulated rainfall. In the discussion which follows, consideration is given to the comparative rates of erosion, the sources of eroded material and the way in which land use influences runoff and erosion. The final section deals with the significance of the findings for soil conservation and for the development of water supplies.

6.2.2. PHYSICAL BACKGROUND

6.2.2.1. Topography

The Iiuni catchment covers an area of 11.29 km² in the hills of Kalama Location and ranges in elevation from 1554-1932 m. These hills are erosion residuals rising from the peneplains and pediplains of younger erosion cycles and formed mainly of granitoid gneisses, quartzo-feldspathic gneisses and schists of the Pre-Cambrian Basement Complex.

Three main landforms were identified by Leslie and Mitchell (1979), using the system of Scott *et al.* (1970), as follows:

1. Uplands with slopes up to 47 per cent
2. Colluvial slopes ranging from 5 to 18 per cent
3. Valley sides ranging from 18 to 27 per cent.

Colluvial slopes are found at the foot of the uplands and represent an earlier phase of geologic erosion. At a later stage, lowering of the base level outside the catchment caused rejuvenation and the river has cut into the colluvium creating a dendritic drainage pattern and steep v-shaped valleys that merge with the uplands. As a consequence, no significant flood plains occur.

The frequency distribution of slopes of different gradient is given in Table 6.6. The slope divisions correspond to the limits of cultivation with or without conservation terraces as specified by the Basic Land Use Rules of the 1965 Agricultural Act (Laws of Kenya). It can be seen that the majority of slopes lie between 12 and 35 per cent, the category which requires conservation terraces to be built. In practice a considerable amount of land over 55 per cent slope is under cultivation and a proposal has been made to raise the legal limit to 35 per cent. The slope categories given in Table 6.6 correspond broadly to the land form components. Slopes in the 0–12 per cent category belong to the colluvial foot slopes and broad ridge tops. Slopes between 12 and 35 per cent fall mainly on the less steep valley sides. Slopes above 35 per cent are the steep, dissected slopes mostly unsuitable for

Table 6.6. Distribution of slope categories in Iiuni catchment

Category (per cent)	Legal classification	Percentage of catchment
0–12	no terracing required (i.e. not mandatory)	20.1
12–35	terracing required	57.8
over 35	no cultivation	22.1

agriculture and slopes greater than 55 per cent are confined to the cliffs and steep rocky slopes.

Figure 6.3 shows the catchment shape, the distribution of land forms and the location of the simulator experiments. The map is adapted from Leslie and Mitchell (1979).

6.2.2.2. Climate

The Machakos hills form a topographic barrier to the moisture-laden, easterly winds blowing off the Indian Ocean. The most important rainfall mechanism, therefore, is forced convection during periods of instability such as when the monsoonal circulation produces enhanced convergence over the eastern parts of Kenya.

Under normal circumstances, the air mass is sufficiently unstable for the convection mechanism to be initiated during two periods of the year. This gives rise to a bi-modal rainfall pattern with one rainy season during March to May, from the south-east monsoon, and the other during November and December, from the north-east monsoon. These two rainy seasons are of approximately equal magnitude in Machakos District with a tendency for the November–December rains to be more reliable and heavy towards the eastern part of the District and the March–May rains to be more dominant in the west.

Mean annual rainfall for the Iiuni catchment is of the order of 900 mm per annum, interpolating from the 1957–1972 isohyetal map (Figure 6.4). During 1978 when the experimental measurements were made, the areal mean rainfall from eight gauges in the catchment was 1148 mm, indicating that the total for the year was significantly higher than average. The annual total was inflated by the high recorded amounts in March and April (Table 6.7). Rainfall during the months of November and December totalled 333.5 mm; a figure much nearer the long term average for this season.

Detailed measurements of rainfall, streamflow and sediment yields were started in September 1978 following the completion of the river gauging station. The analyses of the first observations of sediment are preliminary in nature but, because they coincided with the rainfall simulator trials in the catchment, an attempt has been made to interpret them in terms of the probable long-term sediment yield.

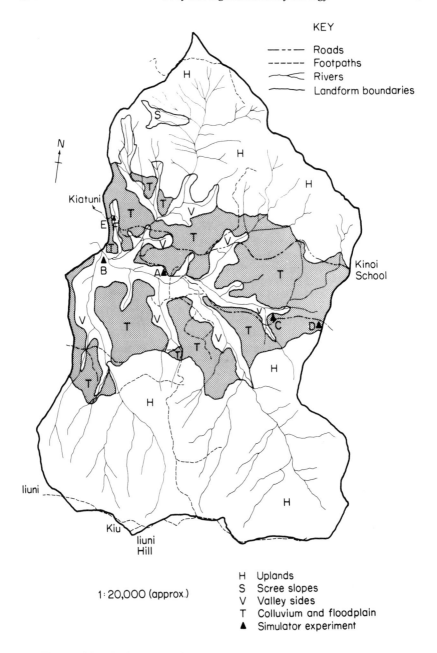

Figure 6.3 Distribution of land forms and location of the simulator experiments

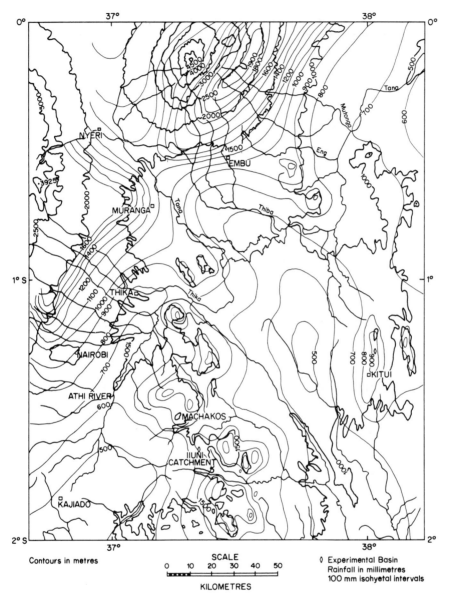

Figure 6.4 Mean annual rainfall (1957–77) isohyetal map

Table 6.7. Monthly rainfall at Iiuni 1978 (mm)

J	F	M	A	M	J	J	A	S	O	N	D	Total
73.9	74.3	196.7	384.5	15.9	0.1	0	1.0	0	67.8	160.5	173.0	1147.7

Table 6.8. Frequency of daily rainfall at Iiuni (mm)

Rainfall (mm)	(Based on data for November and December 1978)					
	0–10	10.1–20.0	20.1–30.0	30.1–40.0	40.1–50.0	>50
No. of events	19	7	3	2	0	0

6.2.2.3. Rainfall Intensity

The rainfall during November and December 1978 has been analysed in terms of the frequency of falls of different amounts for different periods of time. In terms of daily rainfall, Table 6.8 shows that of the 31 days with rain, only five had falls greater than 20 mm.

The frequency of maximum fifteen-minute rainfall during each storm followed the same pattern (Table 6.9) with only five storms out of thirty recording intensities greater than 25 mm hr^{-1} for the fifteen minute duration.

The indications are that the catchment is located in an area of low erosivity and this is borne out by the estimates of mean annual kinetic energy of rainfall greater than 25 mm hr^{-1} calculated by Moore (1978) for three surrounding stations (Table 6.10).

Using the five-minute rainfall totals recorded by the automatic weather station at Iiuni, two kinetic energy parameters have been calculated for the period during which sediment measurements were made and for the whole year. $KE_5 > 25$ is the kinetic energy of rainfall for intensities greater than 25 mm hr^{-1} and duration of five minutes; EI_{30} is the product of the total kinetic energy of a storm and its maximum intensity in mm hr^{-1} over a 30 minute period (Hudson, 1971). These values are shown in Table 6.11 together with the ratio of annual kinetic energy to the kinetic energy of storms in November and December.

6.2.2.4. Streamflow

Streamflow in the catchment is measured by means of a compound, rectangular, sharp-crested weir equipped with a digital water-level recorder. The flow duration curve during the period 18th November to 22nd December shown in Figure 6.5 was constructed from hourly mean discharge data. Hourly flow varied from 0.052 m^3 sec^{-1} to 2.690 m^3 sec^{-1} with a mean value of 0.102 m^3 sec^{-1}. The short duration of high flows and preponderance of very low flows is characteristic of eroded catchments with poor infiltration capacities. The total streamflow of 26 mm (293,000 m^3) from 287 mm of rainfall represents an overall runoff coefficient of 9 per cent. Runoff coefficients for individual storms varied from 3 per cent to 10 per cent.

6.2.2.5. Soils

The soils in the Iiuni catchment have been described by Leslie and Mitchell (1979) from about 100 auger hole observations. Although no soil map was produced, a

Table 6.9. Frequency of $I_{15} > 6$ mm hr^{-1} at Iiuni

(Based on data for November and December 1978)

6.1-10	10.1-15.0	15.1-20.0	20.1-25.0	25.1-30.0	30.1-35.0	35.1-40	40.1-45	>45 mm hr^{-1}
				No. of events				
17	2	4	2	1	2	1	1	0

Table 6.10. Mean annual kinetic energy for intensities greater than 25 mm hr^{-1} (After Moore, 1978)

	$KE_{15} > 25$ (kJ m^{-2} annum^{-1})
Makindu	4.45
Nairobi Airport	3.86
Voi	5.51

Table 6.11. Kinetic energy of storms during water year 1978-1979 at Iiuni

		$KE_5 > 25$ (kJ m^{-2})	EI_{30} (kJ m^{-2} mm hr^{-1})
(A)	Water year	4.78	263
(B)	November–December 1978	1.35	49
	Ratio A/B	3.5	5.4

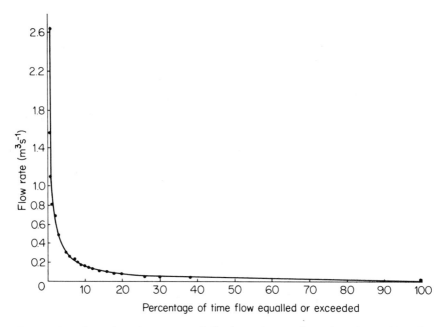

Figure 6.5 Flow duration curve of Iiuni catchment from hourly readings between 18 November and 22 December, 1978

land form map at a scale of 1:11,200 was produced giving the main features of the soils occurring in each land form. Most of the following soil descriptions are based on the data collected by Leslie and Mitchell. Detailed profile descriptions and soil analyses from 6 pits representing different land use types in the catchment have also been given by Thomas and Barber (1978) and Barber and Thomas (1979). Barber and Thomas found two of the pits to be luvisols, three to be ferralsols and one pit on the old colluvial footslopes to be a fluvisol.

The upland soils have developed mainly from Pre-Cambrian basement rocks consisting of granitoid gneisses, gneisses, and schists (Baker, 1954). On the old colluvial footslopes and flood plain areas the soils have developed from eroded sediments of variable composition. The upland soils are well-drained, red-to-brown in colour and with sandy clay loam to sandy clay top soils becoming finer textured with depth. On the gentler slopes (< 27 per cent), the soils are generally two metres or more in depth, becoming slightly shallower (1-2 m deep) on steeper land (27-47 per cent) and very shallow (< 30 cm) on slopes of more than 58 per cent. Very many of these soils contain stone or gravel layers which are usually between 45 and 170 cm from the surface. In the old colluvial areas the soils are generally red with very variable textures ranging from loamy sands to clays. The soils are frequently more than two metres deep and often show signs of impeded drainage.

Many of the soils at Iiuni are characterized by weak soil structures with a strong tendency to surface sealing. On the old colluvial footslopes the surface soils have

Table 6.12. Main land use categories

(1) Cultivated land with good conservation (well-constructed terraces, protected terrace banks and a good crop cover).	36%
(2) Cultivated land with poor conservation (steep slopes, poorly protected terrace banks or poor crop cover and evidence of erosion).	7%
Total Cultivated Land	43%
(3) Good grazing land (good grass cover and little or no evidence of erosion).	10%
(4) Degraded grazing land (sparse grass cover or bare with evidence of sheet, rill, or gulley erosion).	37%
(5) Bush and (6) Woodland (little or no evidence of erosion).	10%
	100%

extremely low organic matter contents due to the rapid accumulation of eroded sediment. In some of the small areas of alluvial deposits, former top soils are now buried by up to a metre of eroded soil.

The analytical data obtained for the six profiles show the soils to be generally of low fertility and moderately acid. The cation exchange capacity values are low, often less than 10 meq $(100 \text{ g})^{-1}$ and the soils are deficient in phosphorus, nitrogen, calcium, and, sometimes, in magnesium.

6.2.2.6. Land Use

Six land use types have heen recognized in the catchment and mapped at a scale of 1:12,500 by Moore (1979) from field observations and air photo interpretation. The proportions of the main land use types are given in Table 6.12.

The areas under different crop types are given by Leslie and Mitchell (1979) who gave a slightly lower estimate of total cultivated land, i.e. cropland, of 35 per cent based on field sampling. Maize and beans are the dominant crops and are planted each season. A very small amount of cotton is grown and a variety of food crops such as pigeon peas, bulrush millet, and cassava are also found. The predominance of annual crops and planting twice in the year means that there is a serious risk of erosion during the early part of each season when the most erosive rains can be expected (Fisher, 1978). The lack of perennial crops emphasises the need for physical conservation measures such as terraces.

Approximately 47 per cent of the land is available for grazing but about 37 per cent is now degraded with a sparse grass cover and showing signs of erosion. Livestock numbers are available for Kalama Location for 1974 (Table 6.13). Pereira *et al.* (1961) at Makateve in Kalama showed that, with very good management and fodder conservation, grazing land could be stocked at approximately 1.6 hectares per livestock unit without damage. Estimates of the stocking density in the area are very much higher than this (0.43 ha LU^{-1}).

Table 6.13. Livestock in Kalama location (1974)

	(Total land area 176 km^2)
Zebu cattle	15,000 approx
Grade cattle	8
Sheep	9,008
Goats	27,482

The high stocking rate coupled with a low standard of grazing management explains the serious denudation and erosion that has occurred, particularly during the droughts of the late sixties and early 1970s.

The natural vegetation has been greatly altered since the land was settled. In the upper part of the catchment there are some relics of original forest with large trees of *Albizia gummifera* which have been preserved as traditional shrines. The forest has been replaced mainly by bushes such as *Dodonea viscosa* and *Euclea divinorum*. Lower in the catchment a dry type of woodland may have existed in the past with *Euphorbia candelabra*, and species of *Acacia, Combretum,* and *Terminalia.* Now most of the large trees have gone and *Acacia hockii* predominates.

Grasses found in the catchment include *Dicanthium insculptum, Eragrostis superba, Cenchrus ciliaris, Hyparrhenia spp.* and particularly on overgrazed land, *Heteropogon contortus* and *Aristida spp.*

6.2.2.7. Erosion and Conservation

Soil erosion has been a major problem in Machakos District for about seventy years. Peberdy (1960) states that livestock were allowed to graze outside the district in 1911 and 1914 in an effort to alleviate overgrazing and quotes the District Commissioner's report of 1927 which states that:

> Since 1917, the reserve has become dessicated beyond all knowledge. Large areas which were good pasture land, and in some case thick bush, are now only tracts of bare soil.

Early travellers who passed through this area such as Lugard in 1890 and Gregory in 1896 make no mention of erosion or gullying. In the early part of the century, according to the older people, the land was well vegetated; what are now deep gullies were simply stock track; and valley bottoms with incised watercourses, which are now filled with sand, were used for growing sugar cane (Baker, 1954).

The oldest inhabitants in the Kalama location state that their grandparents came from the adjacent Mbooni hills which suggests that the settlement has occurred mainly within the last one hundred and thirty years. The extent of erosion indicates the rapidity with which land can deteriorate as a result of human activities.

The period of most rapid erosion appears to have started in 1929 when locusts and famine were recorded. Rainfall, which was low then, continued below average during the thirties and early forties.

Visible signs of erosion include bare ground, rilling, gullying, streambank erosion, exposure of subsoil, widespread surface accumulation of quartz especially on steeper slopes, exposure of grass roots, the sedimentation of cutoff ditches, and deterioration of terrace banks. Gullying can be seen on aerial photos. A comparison of air photos taken in 1948 and 1972 in an adjacent part of Kalama location (Thomas, 1974) showed that most of the gullies observable on the 1972 photography were already present in 1948. The main change in gully erosion since then appears to be due to widening and deepening rather than headward extension.

Soil conservation has been directed mainly towards cultivated land and has involved the digging of cutoff ditches, and the construction of terraces by digging a trench on the contour and throwing the soil uphill to form a bank. These terraces have steep backslopes, above which eroded sediment accumulates. Over a period of time, bench-type terraces develop which are more often forward sloping than level. On some farms terracing has been carried out so effectively that losses of soil or water from cultivated land have been almost eliminated.

6.2.3. MEASUREMENTS OF RUNOFF AND SEDIMENT FROM THE WHOLE BASIN

The sediment transported in rivers can be classified broadly into suspended load, that is the material being carried in the body of the flow by turbulence, and bed load, which consists mainly of the rolling and sliding sediments of larger size. Between the two groups is the saltating or jumping load which may remain in suspension for some time but returns frequently to a rolling and sliding motion on the bed.

Of these three components, it is easier to measure suspended load and, hence, national monitoring networks usually deal with only suspended sediment on a routine basis. In areas where coarse sediments predominate in the material being transported, as in Machakos District, the fine material in suspension represents a small part of the total load. Measuring only the suspended sediment therefore, will lead to a gross underestimation of the sediment being transported out of a catchment.

One method of measuring the total load is to increase the velocity in a section of a river channel so that more of the load is carried in suspension and can be sampled by conventional means. In practice, because of the damaging nature of a high velocity flow carrying large particles, it is not possible without very sophisticated equipment to sample the range of flows. In addition, the short duration of peak flow, and short time of concentration makes it difficult for observers to take manual samples. The compromise adopted in the Iiuni catchment was to build a weir for flow measurement. Suspended load passing over the weir is sampled automatically, at 15-minute intervals once a chosen stage is reached, by means of an automatic sediment sampler. Bed load and saltating load is deposited upstream in the approach channel where velocities are reduced. This is removed periodically by an excavator and lorry to keep the approach channel free of sediment, and its volume is determined by levelling the deposits with respect to the approach channel floor. Dry bulk density samples give the required conversion to a weight basis.

The river, in fact is adjusting its profile to a new and higher base level represented by the datum of the weir. Because of the velocity reduction in the approach channel, it is in a depositional phase and the observed proportion of suspended load to bed load is diminished as compared with conditions in natural channels. Nevertheless the total sediment load entering the reach within which measurements are made must be the same as that under natural conditions.

During the initial period of measurement, the relationship between suspended sediment as defined above (S) and discharge (q) was found to be:

$$S = 5.9q^{1.75} \tag{6.26}$$

where S is in t hr^{-1} and q is in m^3 sec^{-1}. This equation agrees with previous sediment rating curves derived by Dunne (1974) from 97 Kenyan Stations where the mean exponent was 1.7. Equation (6.26) was combined with the flow duration curve (Figure 6.5) to calculate the total weight of suspended sediment transported during the study period. This was found to be 213 t, of which 94 per cent was transported in ten storms and 77 per cent in only four storms. From this preliminary analysis, it is clear that storms with rainfall greater than 20 mm are the prime movers of sediment (see Tables 6.8 and 6.9).

In order to estimate the annual suspended sediment load from the measurements derived so far, attempts were made to relate suspended sediment to one or more of the intensity parameters (EI_{30}, $KE_5 > 25$, $KE_{30}. > 25$) and to the mean intensity or weighted mean intensity of the rainstorms which produced the floods in the river. No simple relationship was apparent, and lacking a much more precise predictive model of the flood hydrographs, it has not proved possible to correlate sediment load at the weir with rainfall incident upon the catchment. Since sediment load is related to the volume and velocity of flow, however, an order of magnitude approximation can be made by assuming that the total volume of sediment is closely related to the total volume of rainfall; the relationship between volume of rainfall and volume of runoff being much better understood.

If one assumes that the rainfall during the period represents approximately one-third of the long-term mean annual rainfall, the annual yield of suspended sediment is likely to be of the order of 640 t yr^{-1} or 57 t km^{-2} yr^{-1}. This is a very small quantity compared with regional estimates given by Edwards (1977). Using the relationship established there between annual sediment load and catchment area, one would expect a figure of 580 t km^{-2} yr^{-1} for a catchment this size (11.3 km^2).

During the same period, a sediment of 1800 t were deposited upstream of the weir, This suggests that the annual bed load is in the region of 5400 t and the total load 6040 t or 535 t km^{-2} yr^{-1}, a figure in keeping with the regional estimates above. Of the total load almost 90 per cent is bed load and, while this may be an overestimate, it is certain that most of the sediment transport is in the form of bed load and saltating load. The bulk of this material is medium and coarse sand (0.2–

2.0 mm) with larger particles up to small cobbles (64–128 mm) carried at the highest flows.

6.2.4. MEASUREMENTS OF RUNOFF AND SOIL LOSS USING SIMULATED RAINFALL ON DIFFERENT LAND USE TYPES

A rainfall simulator provides useful comparative data on runoff and soil loss from different forms of land use. In this study, three sites were selected within the Iiuni catchment to represent different conditions of grazing land. Data had already been obtained on soil losses and runoff from similar soils on cultivated land at Katumani, and have been reported by Barber *et al.* (1979).

6.2.4.1. Sites

The following sites were selected on grazing land:

Site A— bare land, representing degraded grazing land which was devoid of vegetation with an exposed subsoil and sealed at the surface by a very thin clay (?) layer.

Site B — new pasture, representing good grazing land that was ploughed and reseeded two years previously on formerly severely eroded land. At the time of the trials, the pasture was composed mainly of *Eragrostis superba, Harpachne schimperi*, and *Dicanthium insculptum*. The grass was protected from grazing and had an average basal cover of 20 per cent.

Site C — old pasture, representing good grazing land that had been in existence for possibly 20 years or more. The sward had been heavily grazed and trampled and consisted mainly of *Eragrostis superba, Dicanthium insculptum, Chloris pycnothrix*, and *Heteropogon contortus*. The average basal cover was 57 per cent and there were no signs of erosion.

In the previous study referred to above, a site had been selected on freshly ploughed land on a $6°$ slope at Katumani Research Station, some 10 km to the north-west. Selected properties of the surface soils are given in Table 6.14 and site characteristics, i.e. slope, initial moisture content and percentage of grass basal cover in Table 6.15.

6.2.4.2. Methods

Five plots, measuring 1 m wide by 1.5 m long, were laid out at each site. The plots were surrounded on three sides by a 15 cm deep metal sheet with a collecting trough at the lower end.

Simulated rain was applied to each plot in turn by a portable Morin rotating disc simulator. A simulated storm of 69 mm in one hour having an erosivity (R) value of 56, was applied to sites A, B, and C at Iiuni. At Katumani, a storm of

Table 6.14. Selected properties of the top soils

Location	Soil classification[a]	Depth (cm)	Organic carbon (%)	pH(H$_2$O) 1:1	CEC (me (100 g)$^{-1}$)	Erodibility K factor[c] (SI-units)	Dry unit weight (KN m^{-3})
Site A Iiuni	Chromic Luvisol[b]	0–10	1.16	5.7	15.9	0.25	11.97
Site B Iiuni	Orthic Ferralsol[b]	0–10	1.03	6.4	8.1	0.29	15.70
Site C Iiuni	Orthic Ferralsol[b]	0–10	2.07	6.4	11.3	0.26	15.21
Katumani	Ferral–chromic Luvisol	0–15	0.99	6.0	13.0	0.26	12.16

	Mechanical Analysis (%)						
	Coarse sand 2000–600 μm	Medium sand 600–200 μm	Fine sand 200–60 μm	Coarse silt 60–20 μm	Medium silt 20–6 μm	Fine silt 6–2 μm	Clay < 2 μm
Site A Iiuni	1.4	11.7	29.6	4.7	2.1	2.1	48.4
Site B Iiuni	2.2	14.6	41.0	6.8	2.1	2.1	31.2
Site C Iiuni	3.4	29.2	29.0	8.9	2.1	2.1	25.3
Katumani	2.6	10.9	31.0	12.5	0.0	8.2	34.8

[a] Based on the FAO–UNESCO legend of *Soil Map of the World*.
[b] Provisional classification
[c] Calculated from the nomograph in Arnoldus, 1977.

Table 6.15. Characteristics of the sites and mean values for runoff and soil loss

Site	Slope (%)	Basal grass Cover (%)	Initial soil moisture content (w/w; %)	Erosivity R value of simulated rainstorm	Runoff (mm)	Runoff as % of applied rain	Soil loss (g m^{-2})
A—bare land	12.4 (1.7)	0	10.1 (nd)	56	43.6 (3.4)	63.0 (4.2)	1234.2 (162.5)
B—new pasture	19.0 (2.5)	19.8 (7.9)	5.7 (1.3)	56	26.6 (6.5)	39.0 (9.8)	150.0 (100.7)
C—old pasture	18.5 (1.9)	57.3 (10.1)	17.3 (2.6)	56	44.0 (4.0)	63.8 (5.8)	60.2 (13.0)
Katumani—culitivated	10.5 (nd)	0	5.0 (nd)	59	10.3 (3.9)	20.7 (nd)	302.7 (182.5)

Standard deviations given in parentheses.

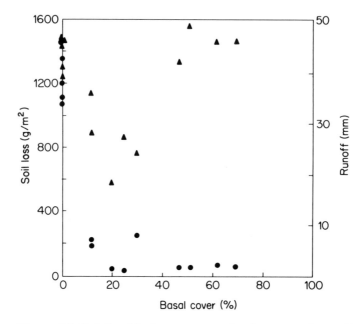

Figure 6.6 Relationship between grass basal cover and runoff and soil loss from simulated rainstorms of 69 mm/h intensity and 1 h duration. (From Moore *et al.*, 1979) ● soil loss; ▲ runoff

100 mm hr^{-1} was applied for 30 minutes, giving an erosivity value of 59. Storms of these intensities could be expected to occur once in 30 years and once in 45 years, respectively. Further details of the simulator are given in Barber *et al.* (1979).

Grass basal cover was measured on each plot at Iiuni using a point quadrat frame. Soil samples were taken for initial moisture content on sites adjacent to the plots.

6.2.4.3. Results

The mean amounts of runoff and total soil loss from the four sites are given in Table 6.15. The percentage runoff and soil losses were high from the bare land (site A gave means of 63 per cent and 1234 g m^{-2} respectively), but much lower from the new pasture (site B gave 39 per cent and 150 g m^{-2}) and from the culti-vated land at Katumani (21 per cent and 303 g m^{-2}). On the old pasture, site C, runoff was high (64 per cent) but soil losses were low (50 g m^{-2}). The percentage runoff and soil losses from all replicates at the three grazing land sites are plotted against grass basal cover in Figure 6.6. The grass cover clearly has little influence on volume of runoff. On bare land and old pasture, the runoff volumes are very similar despite the differences in initial moisture content, slope and grass cover. This is probably due to the greater influence of sealing on the bare surface and of the compact surface horizon on the old pasture. The lower runoff value (39 per cent) at the recently reseeded and largely ungrazed pasture at site B is probably because of the absence of surface compaction from grazing and the persistance of microtopography from the ploughing. A similar effect from residual microtopography could also account for the low runoff from the cultivated land at Katumani.

Soil erosion rates, on the other hand, are clearly influenced by grass cover as shown in Figure 6.6. Erosion losses were not related to the volume of runoff, or factors such as slope or moisture content which affect runoff, since the old pasture which gave the lowest erosion losses had a runoff volume as high as the bare land. Erosion losses were also unrelated to inherent soil erodibility differences between the three grazing sites, since the soil erodibility (K) factors were similar (Table 6.14). The soil losses from the cultivated land were greater than from the good grazing land but much lower than from the bare grazing land.

The relationship between erosion losses and grass basal cover in Figure 6.6 suggests that a critical value of 15 to 20 per cent is important. At values less than this, erosion is intense, whereas at values above 15 to 20 per cent there is little further reduction in soil loss. A similar relationship has been reported by Dunne (1977) in the rangeland areas of Kajiado district, where soil losses were greatly reduced with an increase in grass basal cover from 0 to 20 per cent. At higher basal covers there was little further reduction in soil loss. The influence of grass cover in reducing soil loss can be explained by the reduction in raindrop detachment of soil particles and the reduction in the velocity of runoff as the basal cover increases.

6.2.5. DISCUSSION

From the data obtained so far, it is only possible to draw preliminary conclusions but the seriousness of the erosion problem in Machakos District justifies such an attempt because the conclusions have a bearing on the way in which conservation measures should be implemented.

6.2.5.1. Erosion Potential

The potential for accelerated erosion within the Iiuni catchment is very high on account of the steep slopes, the erodible soils and the intensive land use. Approximately half the land is under cultivation and, although 80 per cent of this is described as terraced, the condition of terraces varies greatly. Only a very small proportion of the land is occupied by perennial crops and much of the cultivated land is exposed to raindrop impact twice a year, during the early part of each rainy season.

Rainfall simulator studies on the cultivated land have shown that very high rates of erosion can occur. Good terracing, however, can do much to prevent soil moving more than a short distance and a great deal of effort has gone into such conservation measures on cropland.

In contrast, much less attention has been paid to grazing land and yet the simulator trials have shown the high potential for accelerated erosion on such land and the very great importance of cover in controlling the amount of soil loss.

In the past, communal use of pasture land has been responsible for overgrazing and erosion. The recent demarcation and registration of individual holdings is partly responsible, no doubt, for the large variations in present conditions. Whereas some land has been fenced and reseeded other areas are bare and eroding rapidly.

The proper management of grazing land is made more difficult by large variations in rainfall and forage production from year to year. Overgrazing is almost inevitable when several years of below average rainfall follow in succession. Such a situation occurred between 1968 and 1976 and by the end of this period the exposure of bare soil became very noticeable. The combination of overgrazing, drought and termite activity can lead to the death of perennial grasses and to a reduction in basal cover. If heavy rain follows such a period the rate of erosion can be expected to be high.

6.2.5.2. The Influence of Land Use on Rates of Erosion

Simulated rainstorms of approximately the same erosivity have given the relative erosion rates and runoff values for the main land use types in the catchment, i.e., degraded grazing land (site A), good grazing land which is long established and with more than 19-20 per cent basal cover (site C) and cultivated land (Katumani). These three land use types constitute almost 90 per cent of the catchment area. The good grazing land which has been recently established (site B) forms only a small proportion of the total catchment area.

It is not possible to estimate the total annual soil losses for the whole of the catchment directly from the rainfall simulator data, because of the large number of unknown factors and unjustified assumptions that would have to be made. Nevertheless, the results do permit conclusions to be drawn as to the relative importance of these three land use types in contributing to the total soil losses and runoff within the whole catchment.

The soil loss values (Table 6.15) from sites A, C, and Katumani can be more validly compared when calculated for standard conditions, e.g. for an annual erosivity (R factor) of 189 (the estimated value for Machakos from data in Wenner, 1977, and Moore, 1978), and for a slope (S factor) of 20 per cent which is typical of much of the catchment. These calculations are based on the Universal Soil Loss Equation (USLE), (Smith and Wischmeier, 1962), and assume that the values and relationships between the R and S factors and soil losses are valid for these conditions. Table 6.16 (column 2) shows the depths of soil lost from the simulated rainstorms and is derived from the soil loss data in Table 6.15 and the dry unit weights of Table 6.14. These depths have been adjusted to give an estimate of the annual soil loss based on the erosivity value of 189 and slope of 20 per cent (column 3, Table 6.16). The last column in Table 6.16 shows the ratio of soil loss on degraded grazing and cultivated land to that from old pasture with a good cover. The ratio of estimated soil loss from degraded grazing land to cultivated land (i.e. 50:15) is likely to become further increased because of the steeper slopes, lack of effective

Table 6.16. Relative soil loss from the different sites

Land use type (site)	Soil loss from simulated rain storm ($R = 56$) (mm)	Estimated annual soil loss ($R = 189$ $S = 20\%$) (mm)	Ratio of soil loss per unit area from degraded grazing and from cultivated land to that from old pasture
Degraded grazing land (Site A) slope 12.4%	1.010	7.47	50
Old pasture (Site C) slope 18.4%	0.039	0.15	1
Cultivated land (Katumani) slope 10.5%	0.244	2.21	15

conservation measures, and longer slopes which are usually associated with the grazing land. Although estimates of potential soil loss from cultivated land have been seen to be high, about 84 per cent of the cultivated land is now terraced, whereas few effective conservation measures exist on the grazing land apart from occasional cut-off ditches and sisal hedges. Thus the movement of sediment into stream channels would be expected to be far greater from the degraded grazing land than from the cultivated land. Moreover, Thomas *et al.* (1980), have shown from a survey of farms in Iiuni and neighbouring areas that the most frequently occurring slope length in cultivated land is 5–10 m, whereas in grazing areas slope lengths are usually much greater.

The only estimates of absolute, annual soil loss from within the catchment have given values of 5 to 15 mm yr^{-1}, based on the depth of exposed *Cenchrus ciliaris* roots which died in the 1974 drought and the depth of exposed *Acacia spp.* roots within the degraded grazing areas (Moore, 1979).

6.2.5.3. The Influence of Land Use on Volumes of Runoff

The runoff percentages in Table 6.15 show markedly higher values from the grazing land, whether degraded or not, than from the cultivated land. When lower intensity storms of 25 mm hr^{-1} and of two-hour duration were applied to the cultivated land at Katumani there was no runoff (Barber *et al.*, 1979). When, however, the same storms were applied to another degraded grazing site and another old pasture site at Iiuni, the runoff percentages were 81 per cent and 24 per cent respectively (Thomas and Barber, 1978). Thus, the amounts of runoff are much greater from the grazing land than from the cultivated land for both high and low intensity storms. Such differences would probably be accentuated by the steeper and longer slopes and fewer conservation structures in the grazing areas. These conditions would encourage the concentration of runoff, leading to the initiation of gullies which Moore (1979) has observed to occur most frequently in the grazing land areas. Gulley development would also be accelerated by the geologically recent lowering of stream base levels (Leslie and Mitchell, 1979) which has led to the downcutting by numerous V-shaped valleys into the old colluvial footslopes, and the absence of any significant flood-plain areas along present water courses.

6.2.5.4. Relationship between the Simulator Data and Measurements of Sediment Loss

The fact that most of the sediment leaving the catchment is medium and coarse sand suggests that the source of this sediment is streambank erosion and gully erosion upstream. The lower proportion of the total load made up by suspended material further suggests that much of the soil lost from the cultivated land is being deposited within the catchment. Until more data is obtained, it is difficult to extrapolate results from the small runoff plots to erosion on the catchment scale. It is possible, however, to exploit the comparative values of rates of soil loss from different land use types to apportion the total sediment loss measured at the weir.

The estimated total loss of sediment from the catchment, given in Section 3, is 535 t km^{-2} yr^{-1}. Assuming the sediment delivery ratio for a catchment this size is 0.2 or less, (i.e. 80 per cent of the soil eroded is deposited within the catchment) the mean off-field soil loss is $535/0.2 = 26.8$ t ha^{-1} yr^{-1}. Using the ratios of soil loss from the land use types given in Table 6.16, minimum rates of soil loss can be calculated as:

Degraded grazing land	53.3 t ha^{-1} yr^{-1}
(37% of land area)	
Cultivated land	16.0 t ha^{-1} yr^{-1}
(43% of land area)	
Good grazing land, bush and woodland	1.1 t ha^{-1} yr^{-1}
(20% of land area)	

This represents, in the case of the degraded grazing land, a loss of soil of 4.4 mm per annum compared with 7.7 mm per annum based on simulator data. Bearing in mind that 4.4 mm is a minimum figure, it can be seen that it is in agreement with the mean annual soil loss estimates of Moore cited previously (5-15 mm yr^{-1}). This is clearly an unacceptably high rate of erosion and underlines the urgent need for conservation measures, not only to control the loss from the fields but also to reduce the rate of sedimentation of surface water reservoirs downstream.

6.2.6. SIGNIFICANCE OF RESULTS FOR CONSERVATION MEASURES

6.2.6.1. Cultivated Land

The potential for erosion from cultivated land is very great but measures for its control by cutoff drains and terracing have generally been accepted as necessary and desirable by the local farmers. The greatest problems arise on very steep slopes—land being cultivated on slopes greater than 35 per cent in some places. Regular maintenance of cutoff drains and terraces is very important since failure of either can lead to gullying (Thomas, 1978). The development of cropping practices or cultivation methods which reduce the risk of erosion has not yet taken place and is handicapped by the increasing population for whom maize and beans are the staple diet.

6.2.6.2. Grazing Land

Improving cover on grazing land should be a priority in any efforts to reduce erosion. The grazing land is important not only for supporting livestock now, but also as future cropland when population expands.

Where grazing land is situated above cropland, the cover will influence the amount of sediment deposited in cutoffs designed to protect the latter.

Where pasture is situated below cropland the cover will play an important role in retarding movement of soil from terraced lands or cutoff drains to the watercourses.

Where cover has deteriorated either through mismanagement or a succession of dry years, replanting or reseeding must be a priority and breaking the surface is needed to promote infiltration. This in itself will expose soil to erosive forces and rapid revegetation should be sought by exclusion from grazing. Diversion ditches may be of temporary value in controlling runoff during revegetation but are less satisfactory as permanent features because of rapid siltation and damage by livestock.

Every effort should be made to prevent deterioration of cover to the point at which reseeding is necessary. The trend to stall feeding and the increased use of fodder such as Bana grass (*Pennisetum purpureum* and *P. americanum*) may provide one way of reducing grazing pressure on natural pasture during critical periods, e.g. during the early part of the rainy season when cover is usually poor and the erosion risk is high. The need to relate stock numbers to the carrying capacity of the land is still not fully recognized or implemented. Inability to adjust stock numbers to fluctuations in rainfall and forage production means that conservative stocking rates are needed, and the inevitable overgrazing in drought years must be compensated for by undergrazing and regeneration in wet years.

REFERENCES

Arnoldus, H. M. J., 1977, Predicting soil losses due to sheet and rill erosion, in *Guidelines for Watershed Management*, FAO Conservation Guide No. 1, pages 99–124, FAO, Rome.

Baker, B. H., 1954, *Geology of the Southern Machakos District*, Report No. 27, Geological Survey of Kenya, Nairobi.

Barber, R. G., and Thomas, D. B., 1979, Measurements of soil loss and runoff from simulated rainstorms at Kabete, Katumani and Iiuni, *Paper presented to the Third Annual General Meeting of the Soil Science Society of E. Africa, Muguga, Kenya*, July 25–27, 1979.

Barber, R. G., Moore, T. R., and Thomas, D. B., 1979, The erodibility of two soils from Kenya, *J. Soil Sci.*, **30**, 579–591.

De Wilde, J. C., 1967, *Experiences with Agricultural Development in Africa*, Johns Hopkins Press, Baltimore.

Dunne, T., 1974, Suspended sediment data for the rivers of Kenya, *Report to Ministry of Water Development*, Unpublished, 122 pp.

Dunne, T., 1977, Intensity and controls of soil erosion in Kajiado District, *FAO/Government of Kenya Report*, Nairobi.

Edwards, K. A., 1979, Regional contrasts in rates of soil erosion and their significance with respect to agricultural development in Kenya, in *Soil Physical Properties and Crop Production in the Tropics*. Eds. Lal, R. and Greenland, D. J. J. Wiley & Sons, Chichester, U.K., 441–454.

Fisher, N. M., 1978, Cropping systems for soil conservation in Kenya, in *Soil and Water Conservation in Kenya*, Occasional paper No. 27, pages 25–39, Institute for Development Studies, University of Nairobi, Nairobi.

Hudson, N. W., 1971, *Soil Conservation*, B. T. Batsford Ltd., London.

Leslie, A., and Mitchell, A. J. B., 1979, Geomorphology, soils and land use of Utangwa, Iiuni, Kitui, and Kune catchment areas: volume 1, *Report to the Institute of Hydrology, Wallingford,* Land Resources Development Centre, Surbiton, Surrey.

Moore, T. R., 1978, An initial assessment of rainfall erosivity in East Africa, *Technical Communication No. 11,* Department of Soil Science, University of Nairobi, Nairobi.

Moore, T. R., 1979, Land use and soil erosion in the Machakos Hills, Kenya, *Ann. Ass. Am. Geogr.,* **69**, 419–431.

Moore, T. R., Thomas, D. B., and Barber, R. G., 1979, The influence of grass cover on runoff and soil erosion from soils in the Machakos area, Kenya, *Trop. Agric.,* **36**, 339–344.

Peberdy, J. R., 1960, *Machakos District Gazeteer,* Ministry of Agriculture, Nairobi.

Pereira, H. C., Hosegood, P. H., and Thomas, D. B., 1961, The productivity of tropical semi-arid thorn-scrub country under intensive management, *Emp. J. Exp. Agr.,* **29**, 269–286.

Pratt, D. J., and Gwynne, M. D., 1977, *Rangeland Management and Ecology in East Africa,* Hodder and Stoughton, London.

Rapp, A., Murray-Rust, D. H., Christiansson, C., and Berry, L., 1972, Soil erosion and sedimentation in four catchments near Dodoma, Tanzania, *Geograf. Ann.,* **54A**, 255–318.

Scott, R. M., Webster, R., and Lawrance, C. J., 1970, *A Land System Atlas of Western Kenya,* Military Vehicles and Engineering Establishment, Christchurch, Hampshire.

Smith, D. D., and Wischmeier, W. H., 1962, Rainfall erosion, *Adv. Agron.,* **14**, 109–148.

Thomas, D. B., 1974, Air photo analysis of trends in soil erosion and land use in part of Machakos District, Kenya, *M.Sc. Thesis,* University of Reading, Reading.

Thomas, D. B., 1978, Some observations on soil conservation in Machakos District, with special reference to terracing, in *Soil and Water Conservation in Kenya,* Occasional paper No. 27, Institute for Development Studies, University of Nairobi, Nairobi.

Thomas, D. B., and Barber, R. G., 1978, Report on rainfall simulator trials at Iiuni, Machakos, Kenya, *Mimeo. Department of Agricultural Engineering,* University of Nairobi, Nairobi.

Thomas, D. B., Barber, R. G., and Moore, T. R., 1980, Terracing of cropland in low rainfall areas of Machakos District, Kenya, *J. Ag. Eng. Res.,* **25**, 57–65.

Wenner, C. G., 1977, *Soil Conservation in Kenya,* Mimeo, Ministry of Agriculture, Nairobi.

PART 7

Watershed Modelling

Tropical Agricultural Hydrology
Edited by R. Lal and E. W. Russell
© 1981, John Wiley & Sons Ltd.

7.1

Models of Surface Water Flow

E. M. MORRIS

7.1.1. THE RANGE OF SURFACE WATER FLOW MODELS

Mathematical models used to model surface runoff range from detailed 'distributed' models, which are based on a full analyais of the physics of shallow water flow, to very much simpler 'black-box' models which describe mathematically the relation between precipitation and surface runoff, without describing the physical processes by which they are related. The choice of model to be used depends on the intended application and on how much information is available on the behaviour of the system to be modelled.

Each model consists of a set of equations which are used to calculate the output data (hydrographs of storm runoff) from the input data (effective rainfall hyetographs). In the equations there are certain constants, the *parameters* of the model, which describe fixed characteristics of the catchment to be modelled. Determination of appropriate values for these parameters is called calibration of the model for a given catchment.

Distributed, physics-based models are most useful when flow data for calibration are not available, for example if the catchment to be modelled is ungauged or if the effect of a hypothetical land use change is to be investigated. Values of the parameters of these models may be determined by separate laboratory or field work.

If some past flow records are available for a catchment they may be used to adjust the parameters of any model to obtain the best values for that particular catchment. This procedure is most useful with the so-called 'lumped, conceptual' models. These models are based on simplified descriptions of physical processes and their parameters describe spatially averaged rather than point values of catchment characteristics. The mathematical structure of the models is relatively simple so it is often not too expensive to determine the values of the parameters by optimization. Lumped, conceptual models will usually produce good predictions of flow for a given catchment, but they are not necessarily good models of the physical processes involved. They are particularly suitable for applications involving the extension of flow records.

For real time forecasting applications the simplest, 'black-box', forms of surface runoff model are used. The model parameters may be adjusted as new, up-to-date measurements become available. This type of model is often used in conjunction with a stochastic method of estimating the error in the flow forecast from known previous errors.

Surface flow models form a continuous spectrum of types, although there is a broad division into the three groups of distributed, physics-based models; lumped, conceptual models; and black-box models. In Section 7.1.2 a selection of models is described, in order of decreasing complexity. In this way the connections between various models and the simplifications inherent in any particular model should become clear.

7.1.2. SOME EXAMPLES OF SURFACE WATER FLOW MODELS

The most detailed descriptions of surface water flow used in hydrological modelling are based on the fundamental physical principles of conservation of mass and momentum. These lead to mathematical expressions which describe the spatial and temporal variation of the depth and velocity of the water layer. If the surfaces considered have a complex form, as is in the case for natural catchments, the equations will also be highly complex. Thus in practice it is necessary to simplify the geometry of the problem so that the equations become manageable. This may be achieved by dividing the surface water flow in a catchment into *overland* and *channel* flow. In the simplified geometry overland flow is represented by sheet flow of a thin layer of water over a set of rough planes. Channel flow is modelled as the flow of a deeper layer of water within a rough channel of some simplified form. The division between the two types of flow is determined by the level of detail in the model. A highly detailed model will treat flow in small hillslope streams as channel flow, with each stream specifically included in the model. A less detailed model will treat such flow as part of the general overland flow on the hillslope.

The problem of modelling surface flow on the various hillslopes and streams of the catchment is thus reduced to that of modelling the flow on a series of plane surfaces representing these hillslopes and streams. Relatively simple continuity and momentum equations, expressing the conservation of mass and momentum, may be written for flow on each of the plane surfaces. In most distributed physics-based models some further simplification of the equations is made. Some channel flow models do consider flux in two or three dimensions and a two dimensional overland flow model has been constructed by Chow and Ben-Zvi (1973). However, on the whole flow is modelled in one dimension only, with the assumption that water velocity does not vary significantly with depth in the thin layer. The appropriate equations are then the Saint-Venant equations for shallow surface water flow.

7.1.2.1. The Saint-Venant Equations

The Saint-Venant equations describe the flow of water over a plane surface. For one-dimensional flow the equation of continuity for unit width of the plane is

$$\frac{\partial h}{\partial t} + u\frac{\partial h}{\partial x} + h\frac{\partial u}{\partial x} = q \tag{7.1}$$

where x is the direction of flow, t is time, h the depth of the water layer, and u its velocity averaged over depth. The lateral inflow per unit area of the surface per unit time is q. The momentum equation is

$$\frac{\partial u}{\partial t} + u\frac{\partial u}{\partial x} + g\cos\theta\,\frac{\partial h}{\partial x} = g(\sin\theta - S_f) - q\frac{u}{h} \tag{7.2}$$

where g is the acceleration due to gravity, θ the constant angle of the slope, $\rho g h\, S_f$ the frictional retarding force per unit area exerted by the plane on the water, and ρ is the density of water. The velocity of the lateral inflow in the downstream direction is assumed to be negligible. Equations (7.1) and (7.2) may be used to describe overland flow on hillslopes or flow in fairly wide channels of constant width and with straight sided banks. For overland flow the lateral input, q, is the rainfall less any infiltration or plus any seepage from the ground. For a channel, q is the overland flow plus throughflow from, or seepage to, hillslopes on each side, per unit width of the channel. The equations can be rewritten for a non-rectangular channel of cross-sectional area A and surface width of flow B thus:

$$B\frac{\partial h}{\partial t} + u\frac{\partial A}{\partial x} + A\frac{\partial u}{\partial x} = Bq \tag{7.3}$$

$$\frac{\partial u}{\partial t} + u\frac{\partial u}{\partial x} + g\cos\theta\,\frac{\partial h}{\partial x} = g(\sin\theta - S_f) - \frac{B}{A}qu \tag{7.4}$$

There are various empirical expressions for the friction slope S_f all based on the assumption that resistance to unsteady flow is the same as that to steady flow at the same depth and velocity.

7.1.2.2. Empirical Expressions for the Friction Slope

For flow over natural surfaces a parameter describing an 'effective' friction slope, which takes account of factors such as large scale irregularities in the surface and non-uniform roughness or lateral inflow, must be defined. In practice the frictional slope is commonly defined either as $S_f = u^2 n^2/h^{4/3}$ where n is Manning's constant or as $S_f = u^2/C^2 h$, where C is Chezy's constant the former being used for channel flow studies and the latter for overland flow.

7.1.2.2.1. For River Channels

There are three main components of effective flow resistance in river channels. These are

   (a) the boundary resistance, which depends on the roughness of the bed

(b) the channel resistance, which arises because in practice the channel geometry will not be uniform

(c) free surface resistance, which arises from distortions of the water surface produced either by unstable flow (roll waves) or obstacles penetrating the surface.

The effect of size, shape and spacing of roughness elements on a channel bed on boundary and free surface resistance to flow has been studied in some detail in the laboratory. Hence it is possible to make a reasonable prediction of S_f for a uniform channel, a canal for example. The difficulty arises in predicting the effects of channel bends and non-uniformity in the slope and cross-section. Bathurst (1980a, 1980b) has discussed the problem for gravel bed rivers and mountain streams with extremely rough beds and derives equations for the total flow resistance from a detailed consideration of the flow processes. It is possible to determine the average effective resistances of a length of channel in the field by dilution gauging (Beven, Gilman and Newson, 1979).

7.1.2.2.2. For Hillslopes

The effective flow resistance of a hillslope depends on

(a) the resistance of the soil surface and vegetation

(b) the geometry of the hillslope

(c) the number of small streams which are included as part of the overland flow

The values of surface resistance for bare soil or vegetated surfaces can be established reasonably well by laboratory or small-scale field experiments. However very few field experiments have been made on an appropriate scale to establish the effects of geometry and small streams. Values of the effective roughness of grassland have been determined by Woolhiser *et al.* (1970) ($C = 1.8$ m$^{1/2}$ sec^{-1}) and Morris (1980) ($C = 0.46 \pm 0.04$ m$^{1/2}$ sec^{-1}) and are of the same order as values determined on experimental plots at a much smaller scale, ($C = 1.8 - 6.5$ m$^{1/2}$ sec^{-1}) but there is no reason to suppose that the effective roughness of other types of vegetation will be independent of scale. For example, the large scale roughness of a forested slope may be much lower than the roughness of a small plot of trees if there is a network of drainage ditches between the stands.

7.1.2.3. Solution of the Equations

7.1.2.3.1. Analytical

There are no analytical solutions for the Saint-Venant equations except in the special case where q is a power function of t. Series solutions then exist for parts of the surface unaffected by the boundary conditions (Brutsaert, 1968). For constant q these reduce to

$$h = qt$$

$$u = q^{\frac{1}{2}}C\sin^{\frac{1}{2}}\theta\, t^{\frac{1}{2}}\frac{I_2(\gamma t^{\frac{1}{2}})}{I_1(\gamma t^{\frac{1}{2}})} \tag{7.5}$$

where $\gamma = 2g\sin^{\frac{1}{2}}\theta/q^{\frac{1}{2}}C$ is a constant and I_υ are modified Bessel functions of the first kind. For $\gamma \gtrsim 10$ and t not very small

$$u \sim q^{\frac{1}{2}}C\sin^{\frac{1}{2}}\theta\, t^{\frac{1}{2}} \tag{7.6}$$

These solutions are often useful as a starting point for numerical solution of the Saint-Venant equations.

7.1.2.3.2. Numerical

Finite difference methods may be used to solve the Saint-Venant equations for u and h at a set of points $[i, j]$. These may lie on a rectangular $x - t$ grid or on a curvilinear characteristic grid defined by the equations

$$\frac{\mathrm{d}x}{\mathrm{d}t} = u \pm \{hg\cos\theta\}^{\frac{1}{2}} \tag{7.7}$$

The finite difference schemes used may be based on the original equations (7.1) and (7.2) or on transformed 'characteristic' equations. Explicit schemes have been used, but the size of time step is severely limited by the Courant condition for stability

$$Cr = \frac{\Delta t}{\Delta x}(u + \{ghc\cos\theta\}^{\frac{1}{2}}) \leqslant 1 \tag{7.8}$$

Implicit methods are stable at higher values of the Courant number Cr and are generally preferable. Liggett and Woolhiser (1967) have made a valuable survey of the various methods.

If the surface flow equations are to be solved in isolation it is most efficient to use the irregular characteristic grid. The density of grid points is highest where most detail is needed and is low where detailed solutions are not required. On the other hand if the surface flow model is to be linked to an infiltration model it is probably best to use a regular rectangular grid. Such a grid allows solutions of the two different sets of equations to be linked without interpolation, which can be expensive in computer time.

7.1.2.4. Simplified Forms of the Saint-Venant Equations

Many surface water modelling problems do not require use of the full Saint-Venant equations. The conditions under which the equations can be simplified become clear when they are written in the non-dimensional form (Woolhiser and Liggett, 1967)

$$\frac{\partial H}{\partial T} + U\frac{\partial H}{\partial X} + H\frac{\partial U}{\partial X} = 1 \tag{7.9}$$

$$\frac{\partial U}{\partial T} + U\frac{\partial U}{\partial X} + \frac{1}{F_0^2}\frac{\partial H}{\partial X} = k\left(1 - \frac{U^2}{H}\right) - \frac{U}{H} \tag{7.10}$$

The dimensionless variables are $H = h/h_0$, $U = u/u_0$, $X = x/L$ and $T = tu_0/L$. The normalizing parameters are related by

$$qL = h_0 u_0 ; \qquad \frac{u_0^2}{C^2 h_0} = \sin\theta \tag{7.11}$$

where h_0 and u_0 are the 'normal' depth and velocity which would be reached at the lower boundary $x = L$ given that q is a step function of t and there is no inflow across the upper boundary at $x = 0$. The Chézy expression for the friction slope has been used in the momentum equation.

It is apparent that there are only two independent parameters, the Froude number

$$F_0 = \frac{C\tan^{\frac{1}{2}}\theta}{g^{\frac{1}{2}}} \tag{7.12}$$

and a 'kinematic flow number' (cf. Woolhiser and Liggett, 1967)

$$k = \left\{\frac{g^3 L \tan\theta}{C^4 q^2}\right\}^{1/3} \tag{7.13}$$

when k is large equation (7.10) reduces to

$$U^2 = H \tag{7.14}$$

and when k is very small to

$$\frac{\partial U}{\partial T} + U\frac{\partial U}{\partial X} + \frac{1}{F_0^2}\frac{\partial H}{\partial X} = -\frac{U}{X} \tag{7.15}$$

These are respectively the *kinematic* and *gravity wave* approximations to the momentum equation. When F_0 is small but $F_0^2 k$ is not negligible the momentun equation reduces to

$$\frac{\partial H}{\partial X} = F_0^2 k\left(1 - \frac{U^2}{H}\right) \tag{7.16}$$

which is the *diffusion wave* approximation.

If there is no lateral inflow but there is an inflow I per unit width at the upper boundary one of the equations relating the normalizing parameters must be changed. Setting $h_0 u_0 = I$ the Saint-Venant equations become

$$\frac{\partial H}{\partial T} + U\frac{\partial H}{\partial X} + H\frac{\partial U}{\partial X} = 0 \tag{7.17}$$

$$\frac{\partial U}{\partial T} + U\frac{\partial U}{\partial X} + \frac{1}{F_0^2}\frac{\partial H}{\partial X} = k\left(1 - \frac{U^2}{H}\right) \tag{7.18}$$

where the kinematic flow number is now

$$k = \left(\frac{g^3 L^3 \tan \theta}{C^4 l^2} \right)^{1/3} \tag{7.19}$$

For overland flow in natural catchments, the values of k are usually very large and the kinematic approximation may be made. Even the extreme case of a small, flat smooth natural slope ($L = 10$ m, $\tan \theta = 0.01$, $C = 30$ m$^{\frac{1}{2}}$ sec^{-1}) with an intense rainfall of $q = 3 \times 10^{-5}$ m sec^{-1} gives a value of $k = 50$, with $F_0 = 0.96$. For more realistic slopes, k will be much larger. In urban catchments smaller values of k may be encountered but for most real problems the kinematic approximation may still be used.

A fairly wide range of values of F_0 and k occurs for flow in natural channels so in different cases the full momentum equation, the diffusion approximation or the kinematic approximation may be appropriate.

If for example the diffusion approximation is valid, equations (7.16) and (7.1) or (7.17) may be combined to give

$$\frac{\partial H}{\partial T} + c \frac{\partial H}{\partial X} - D \frac{\partial^2 H}{\partial X^2} = 1 \text{ or } 0 \tag{7.20}$$

where

$$c = \frac{3}{2} H^{\frac{1}{2}} \left(1 - \frac{1}{F_0^2 k} \frac{\partial H}{\partial X} \right)^{1/2} \tag{7.21}$$

and

$$D = \frac{H^{3/2}}{2 F_0^2 k} \left(1 - \frac{1}{F_0^2 k} \frac{\partial H}{\partial X} \right)^{-1/2} \tag{7.22}$$

Various empirical values for the convective, c, and diffusive, D, parameters in equation (7.20) have been used ranging from simple constant values for a representative depth (Hayami, 1951) to the full expressions (7.21) and (7.22) (Morris and Wool-hiser, 1980). If the kinematic equation is valid, equations (7.14) and (7.1) or (7.17) may be combined to give the kinematic wave equation

$$\frac{\partial H}{\partial T} + c \frac{\partial H}{\partial X} = 1 \text{ or } 0 \tag{7.23}$$

where

$$c = \frac{3}{2} H^{\frac{1}{2}} \tag{7.24}$$

Both constant and variable values for c, the kinematic wave speed, have been used.

7.1.2.5. Numerical Solution of Approximate Forms of the Saint-Venant Equations

Various numerical methods may be used to solve these approximate forms of the Saint-Venant equations each leading to a slightly different model. An explicit finite difference form of the kinematic wave equation for no lateral inflow is (Cunge, 1969),

$$\theta\left(\frac{h_i^{t+1} - h_i^t}{\Delta t}\right) + (1 - \theta)\left(\frac{h_{i+1}^{t+1} - h_{i+1}^t}{\Delta t}\right)$$

$$+ c\left\{\frac{h_{i+1}^{t+1} - h_i^{t+1} + h_{i+1}^t - h_i^t}{2\Delta x}\right\} = 0 \qquad (7.25)$$

Solving for h_{i+1}^{t+1} gives

$$h_{i+1}^{t+1} = \frac{1}{\left[\frac{\Delta x}{c}(1 - \theta) + \frac{\Delta t}{2}\right]}\left\{\left(\frac{\Delta t}{2} - \frac{\Delta x}{c}\theta\right)h_i^{t+1} + \left(\frac{\Delta x}{c}(1 - \theta) - \frac{\Delta t}{2}\right)h_{i+1}^t\right.$$

$$\left. + \left(\frac{\Delta x}{c}\theta + \frac{\Delta t}{2}\right)h_i^t\right\} \qquad (7.26)$$

This is similar in form to the Muskingum-Cunge channel flow routing equation:

$$O^{t+1} = C_1 I^{t+1} + C_2 I^t + C_3 O^t \qquad (7.27)$$

where I^t is the input and O^t is the output from a channel reach at time t. The Muskingum parameters are

$$C_1 = \frac{\left(-K\theta + \frac{\Delta t}{2}\right)}{\alpha}$$

$$C_2 = \frac{\left(K\theta + \frac{\Delta t}{2}\right)}{\alpha}$$

$$C_3 = \frac{\left(K(1 - \theta) - \frac{\Delta t}{2}\right)}{\alpha} \qquad (7.28)$$

$$\alpha = K(1 - \theta) + \frac{\Delta t}{2}$$

where K and θ are empirical coefficients to be determined for each reach. Although the Muskingum-Cunge routing model is based on an approximate form of the kinematic wave equation, it may in fact have a diffusive component. The behaviour of the model depends on the choice of the weighting parameter θ; with values of $\theta <$ 0.5 there is a numerical distortion which produces diffusion effects. Pure kinematic wave behaviour requires $\theta = 0.5$. The weighting parameter decreases with the length of the reach to be considered, so given a large number of small reaches θ can be set to zero. This produces a linked reservoir model of the Kalinin-Miljukov type. (Kalinin and Miljukov, 1958).

These routing equations for a whole reach are essentially lumped models and, although there is a formal analogy with the finite difference forms of the distributed partial differential equations, the parameters of the model are strongly dependent on 'scale' factors rather than purely on process. However, they have the great advantage of simplicity and are particularly useful as part of real-time forecasting models where flow forecasts must be quickly updated as new information is provided. The Muskingum–Cunge type of model may be called 'conceptual' as it is possible to link the parameters to some extent at least with flow processes.

A conceptual model may make no division between overland and channel flow and use only one equation to model all surface runoff within the catchment. For example, the instantaneous unit hydrograph for a catchment, $h(t)$, is used to relate effective input rainfall $x(t)$ to the surface runoff component $y(t)$ of the output hydrograph by the equation

$$y(t) = \int_0^\infty x(\tau)h(t-\tau)\mathrm{d}\tau \tag{7.29}$$

This method assumes that the system is *linear* (i.e. that the total response to a number of inputs can be determined by sinple summation of the individual responses to each input) and *time invariant* (i.e. that the form of the function $h(t)$ does not change with time). The form of the function $h(t)$ can be deduced from records of $x(t)$ and $y(t)$ for the catchment.

One method for determining $h(t)$ is given by Nash (1960) and is based on an expression relating the statistical moments of x, y, and h.

Nash showed that for a linear, time invariant system

$$U_R'(y) = \sum_{k=0}^{R} \binom{R}{k} U_k'(x)U_{R-k}'(h) \tag{7.30}$$

where $\binom{R}{k} = \dfrac{R!}{k!(R-k)!}$ and $U_R'(f)$, the statistical moment of a function f about the time origin is defined as

$$U_R'(f) = \int_0^\infty f(t)t^R \,\mathrm{d}t \tag{7.31}$$

Using equation (7.30) the moments of $h(t)$ can be calculated from those of $x(t)$ and $y(t)$.

Analytical forms for $h(t)$ have been suggested by many authors. One of these is the two-parameter gamma distribution, which may be thought of as the impulse response for a cascade of equal linear reservoirs, each with an individual response function

$$h' = \frac{1}{K} \exp\left(-\frac{t}{K}\right) \tag{7.32}$$

For a given catchment the appropriate number, n, of reservoirs to be used in the model and the best storage delay time, K, can be determined from the equations

$$U_1'(y) = U_1'(x) + nK$$

$$U_2(y) = U_2(x) + nK^2 .$$

(7.33)

where U_2 is the second moment about the centre of area

$$U_2(f) = \int_0^\infty f(t)\, (t - U_1'(f))^2\, dt \tag{7.34}$$

A large variety of linear lumped conceptual models have been built up using linear reservoirs, as in the Nash cascade model, and linear channel elements which produce a constant delay in the output.

The advantage of using combinations of linear elements is that the effect of a cascade of elements can be determined by mathematical convolution of the impulse responses of the individual elements. However, surface runoff can show strong non-linear features, especially in small catchments where overland flow is dominant, and non-linear conceptual models may often be preferable. For example, the Institute of Hydrology Conceptual Model (Douglas, 1974) uses the expression

$$RO(t + \text{RDEL}) = RK(\text{RSTORE}(t))^{RX} \tag{7.35}$$

to compute the output hydrograph RO. RSTORE (t) is the size of the surface run-off store at time t, RDEL is a delay time and RK and RX are parameters determined by optimization.

7.1.3. STOCHASTIC MODELS

The simpler flow routing models can be combined with models for predicting their errors to produce forecasting systems which can be updated using recent measured values of the surface flow. A description of the statistical methods of analysing the behaviour of the model errors is beyond the scope of this chapter but the general principles of the construction of such models can be illustrated by the following simple example. Jones and Moore (1980) have used a version of the Muskingum-Cunge equation for a real-time forecasting application on the River Dee. The calculated flow Q_t^n at time t and point n are related by

$$Q_t^n = \theta_t^n Q_{t-1}^{n-1} + (1 - \theta_t^n)Q_{t-1}^n + \theta_t^n q_t^n \tag{7.36}$$

where q_t^n are lateral inflows, and

$$\theta_t^n = a + bQ_{t-1}^k + c(Q_{t-1}^k)^2 + d(Q_{t-1}^k)^3 \tag{7.37}$$

where Q_{t-1}^k is the outflow at time $t - 1$. The parameter θ controls both the speed and attenuation of the hydrograph. Travel time thus varies with discharge.

Suppose the current time is τ and n observations of the outflow, Z_t, are available at times $t = \tau - n, \ldots, \tau - 1, \tau$. The errors in the calculated values Q_k^t are known up to time τ. Given these errors, ϵ_t, forecasts of the errors for $t > \tau$ can be made.

Writing subscript $t_1|t_2$ for a forecast for time t_1 based on information up to time t_2, an autoregressive moving average, (ARMA) expression

$$\epsilon_{T+\ell|\tau} = \sum_{i=1}^{p} \phi_i \epsilon_{T+\ell-i|\tau} + \sum_{j=\ell}^{q} \psi_j a_{T+\ell-j} \qquad (7.38)$$

where $a_t = \epsilon_t - \epsilon_{t|t-1}$; may be used to estimate ϵ_t, and hence outflow $Z_t = Q_t^k + \epsilon_t$, at time $t + 1$ from information at $t = \tau$

ϕ_i and ψ_j are parameters to be determined by optimization. Jones and Moore found that in the case of the Dee the orders of the ARMA components could be as low as $p = 2$ and $q = 0$, so that equation (7.38) could be simplified to a two parameter expression

$$\epsilon_{T+\ell|\tau} = \phi_1 \epsilon_{T+\ell-1|\tau} + \phi_2 \epsilon_{T+\ell-2|\tau} \qquad (7.39)$$

In the stochastic models the simplicity of the surface flow model is counteracted by full use of measured flow information up to the time of forecasting. These models can therefore produce highly successful short-term predictions for gauged catchments with characteristics which do not change over time.

7.1.4. APPLICATIONS

In some applications, for example the design of urban storm drains, the design of canals and flood routing in rivers with impervious beds, surface water flow models may be used alone. However, in most cases the models are used as parts of more general models which include several hydrological processes. These general hydrological models range from the physics-based, distributed models, such as the European Hydrological System (Jønch-Clausen, 1979), best used for forecasting the effects of catchment changes or forecasting flow in ungauged catchments; through the lumped, conceptual models, such as the Stanford Watershed Model (Crawford and Linsley, 1966), useful for a wide variety of problems provided that flow data is available for calibration; to the simple black-box models used for real-time forecasting. At the moment the choice of model for a particular study may be severely limited by lack of data. There may only be enough input information to run the simplest of black-box or lumped, conceptual models. However, developments in remote sensing and automatic ground instrumentation should lead to a considerable increase in the amount of data commonly available and we may expect the more complex distributed models to become increasingly important in the future.

REFERENCES

Bathurst, J. C., 1980a, Theoretical aspects of flow resistance, *Proceedings of the University of East Anglia/Institute of Hydrology (U.K.)/Colorado State University International Workshop on Engineering Problems in the Management of Gravel-Bed Rivers,* held at Newtown, Wales, June 23–27, 1980.

Bathurst, J. C., 1980b, Flow resistance in mountain streams, *Proceedings of the University International Workshop on Engineering Problems in the Management of Gravel-Bed Rivers*, held at Newtown, Wales, June 23-27, 1980.

Beven, K. J., Gilman, K., and Newson, M., 1979, Flow and flow routing in upland channel networks, *Bull. Hydrol. Sci.*, **24**, 3, 9, 303-325.

Brutsaert, W., 1968, The initial phase of the rising hydrograph of turbulent free surface flow with unsteady lateral inflow, *Water Resources Res.*, **4(6)**, 1189-1192.

Chow, V. T., and Ben-Zvi, A., 1973, Hydrodynamic Modelling of Two-dimensional Watershed flow, *J. Hydraulics Div.*, ASCE, 2023-2040.

Crawford, N. H., and Linsley, R. K., 1966, Digital simulation in hydrology, Stanford Watershed Model IV, *Technical Report No. 39*, Department of Civil Engineering, Stanford University, Stanford, California.

Cunge, J. A., 1969, On the subject of a flood propagation method (Muskingum method). *J. Hydraulic Res.*, International Association for Hydraulic Research, **7(2)**, 205-230.

Douglas, J. R., 1974, Conceptual modelling in hydrology, *Institute of Hydrology Report No. 24*.

Hayami, S., 1951, On the propagation of flood waves, *Bulletin No. 1*, Disaster Prevention Research Institute, Kyoto University, Japan, 16 pp.

Jønch-Clausen, T., 1979, A short description of SHE—Système Hydrologique Européen, *SHE Report no. 1*, Danish Hydraulic Institute, Horsholm, Denmark.

Jones, D. A., and Moore, R. J. A., 1980, Simple flow routing model, *Oxford Symposium on Hydrological Forecasting*, in press.

Kalinin, G. P., and Miljukov, P. I., 1958, Approximate computation of unsteady flow, *Trudy CIP*, **66**, Leningrad.

Liggett, J. A., and Woolhiser, D. A., 1967, Difference solutions of the shallow water equation, *J. Eng. Mech. Div. ASCE*, **93**, EM2, (Proc. Paper 5189), 39-71.

Morris, E. M., 1980, Forecasting flood flows in the Plynlimon catchments using a deterministic distributed mathematical model, *Oxford Symposium on Hydrological Forecasting*, in press.

Morris, E. M., and Woolhiser, D. A., 1980, Unsteady, one-dimensional flow over a plane—partial equilibrium and recession hydrographs, *Water Resources Res.*, in press.

Nash, J. E., 1960, A unit hydrograph study with particular reference to British catchments, *Proc. Inst. Civ. Engrs.*, **17**, November.

Woolhiser, D. A., and Liggett, J. A., 1967, Unsteady one-dimensional flow over a plane—the rising hydrograph, *Water Resources Res.*, **3(3)**, 753-771.

Woolhiser, D. A., Hanson, C. C., and Kuhlman, A. R., 1970, Overland flow on rangeland watersheds, *J. Hydrol. (New Zealand)*, **9(2)**, 336.

Tropical Agricultural Hydrology
Edited by R. Lal and E. W. Russell
© 1981, John Wiley & Sons Ltd.

7.2

Modelling Infiltration: The Key Process in Water Management, Runoff, and Erosion

D. C. SLACK AND C. L. LARSON

7.2.1. ABSTRACT

Infiltration from rainfall normally occurs in two stages, before and with surface ponding. With the latter there is a 'rainfall excess', some of which may be held in micro-relief storage. The remainder becomes runoff and causes soil erosion. Thus, modelling infiltration and surface storage are necessary steps in modelling runoff and erosion, also in modelling soil moisture levels for a variety of purposes.

This paper presents and illustrates the use of a physically based, two-stage infiltration model. The first stage of the model is the Mein–Larson equation, which predicts the time when surface ponding begins. The second stage utilizes the well-known Green–Ampt equation (corrected for delayed ponding) to determine infiltration rates after ponding begins. Both equations utilize soil characteristics that can be measured in the laboratory from soil samples, namely the saturated conductivity and the moisture release curve. The initial soil moisture content is utilized also. Since the model has no fitted parameters, infiltration tests are not required. Thus, the model is especially appealing where field data are limited or difficult to obtain.

The model has been verified analytically and experimentally, and has been shown to perform satisfactorily under field conditions. It can be applied to layered as well as uniform soils and with constant, variable, or intermittent rainfall. The model can account for special conditions such as entrapped air or surface sealing, and has been applied also to recently tilled soils with rough surfaces. In the latter case, the infiltration model was linked to a recently developed surface storage model.

Performance of the model under a variety of rainfall conditions is described in the paper.

7.2.2. INTRODUCTION

Infiltration of rainfall or irrigation water into the soil determines the amount of runoff and water available for crop growth and groundwater recharge. Thus, it is probably the most important hydrologic process within a watershed and has received considerable attention from research workers. The theory and process of infiltration has been reviewed by Philip (1957; 1969) Hillel (1971), Morel-Seytoux (1973), and Baver *et al.* (1972).

Although infiltration may involve water movement in two or three dimensions, for most purposes it can be treated as one-dimensional vertical flow and this is the case which will be discussed here.

Infiltration rate, f, is the rate at which water enters the soil at a given time. Infiltration is limited by soil characteristics and by the availability of water for infiltration. Infiltration capacity, f_p, is defined as the maximum rate at which water *can* infiltrate at a given time; i.e. if an ample supply of water is available. Infiltration capacity always decreases with time, mainly due to decreasing hydraulic gradient as infiltration continues. In some cases this decrease may also be a result of surface sealing and crusting. The water application rate, R, may be constant or variable. All three rates, f, f_p, and R, are normally expressed in the same units, e.g. cm hr^{-1}.

If the rainfall rate, R, at a given time is less than the infiltration capacity, all water being supplied is infiltration ($f = R$). In this case, $f < f_p$. If R is greater than the infiltrating capacity, infiltration occurs at capacity ($f = f_p$). In this case, $R > f$ and the remainder ($R - f_p$) accumulates on the surface or runs off or both. It is possible also that $R = f_p$, at least momentarily. At a given moment, therefore, there are three possibilities. The application rate can be less than, equal to, or more than the infiltration capacity.

If infiltration continues for a sufficiently long time under soil controlling conditions, the infiltration rate will approach a constant rate, f_c. The constant f_c has often been assumed to be equal to the saturated hydraulic conductivity of the soil, K_s, but is actually more closely approximated by the hydraulic conductivity at 'field saturation', K_{fs}. Field saturation is defined as the maximum moisture content behind the wetting front during infiltration and, due to entrapped air is less than saturation.

For rainfall or irrigation events, there are four possible cases for a constant application rate R (Figure 7.1), as follows:

A. Application rate, R_1, greater than the initial infiltration capacity, f_0. Immediate surface ponding.
B. Application rate, R_2, less than f_0 but application ends before f_p reduces to R_2. No surface ponding.
C. Two-stage infiltration event, beginning with application rate R_3, less than f_p, followed by period with f_p, less than R_3. Delayed surface ponding.
D. Application rate, R_4, less than K_{fs}. No ponding, regardless of duration.

Type A infiltration is what is measured with a flooded surface infiltrometer, but occurs rather seldom in nature. Most runoff events occur with two-stage (Type C) infiltration (Mein and Larson, 1973). The first stage (no ponding) can vary from a few minutes to more than an hour. Type B infiltration (no runoff) is also very common and is the design objective with sprinkler irrigation.

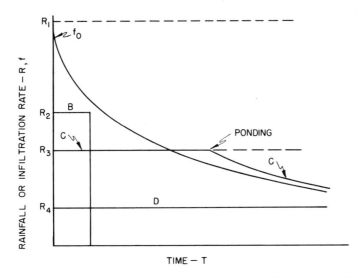

Figure 7.1 Types of infiltration events (adapted from Mein and Larson, 1971)

More commonly, the rainfall rate varies with time and, with moving sprinkler irrigation systems, the same is true. This complicates the picture (as compared to Figure 7.1) but the same general types of infiltration occur, i.e. with immediate, delayed or no ponding (Types A, C, and B or D, respectively). With intermittent application or widely varying application, numerous combinations are possible.

It should be noted that, in two-stage (or intermittent) infiltration the infiltration capacity curve after ponding begins does not coincide with the Type A curve. This is sometimes ignored, causing significant errors in predicting infiltration and time that ponding begins.

Infiltration rate is normally expressed in units of length (depth of water) per unit time (or volume per unit area per unit time, L^3/L^2t), e.g. in, cm, or mm hr^{-1}. Total infiltration volume or cumulative infiltration at any time t may be expressed as:

$$F(t) = \int_0^t f(t)dt \qquad (7.40)$$

where f is the infiltration rate.

7.2.3. INFILTRATION EQUATIONS AND MODELS

A number of infiltration equations may be found in the literature. They may be generally classed in two broad categories, those which are empirical in nature or require fitted parameters or both, and those which are derived from theory of flow in porous media and utilize measured parameters.

Equations in the first category have often involved simplified concepts which permit the infiltration rate or cumulative infiltration volume to be expressed algebraically as a function of time (t) and empirical constants or soil parameters.

One of the simplest equations in this category is that proposed by Kostiakov (1932):

$$f_p = Ct^{-\alpha}(t > 0) \tag{7.41}$$

where C and α are empirical constants depending on the soil and initial conditions. Another equation in this category is that of Horton (1939, 1940) which may be written as:

$$f_p = f_c + (f_0 - f_c)e^{-\beta t} \tag{7.42}$$

where β is a soil parameter (units t^{-1}) that controls the rate at which the infiltration rate decreases. These two equations are applicable only to Type A infiltration, i.e. with immediate ponding ($R > f_0$).

An equation developed by Holtan (1961) is also an empirical, fitted parameter, equation. This equation may be written as (Holtan and Lopez, 1971):

$$f_p = GI.a.S_a^{1.4} + f_c \tag{7.43}$$

where S_a is the available soil moisture storage at a given time for a specified soil depth, GI is the growth index of the crop in percent of the maturity and a is an index of surface connected porosity. The Holtan equation can be considered an infiltration model, since it can be applied to any type of infiltration event by simply adding the constraint $f \leq R$.

Perhaps the most theoretically basic equation is the general equation of flow for unsaturated porous media derived by Richards (1931). This is a second order, non-linear, partial differential equation which may be written as:

$$\frac{\partial \Theta}{\partial t} = \frac{\partial}{\partial z} K(\Theta)\frac{\partial \Psi}{\partial z} - \frac{\partial K(\Theta)}{\partial z} \tag{7.44}$$

where Θ is the volumetric moisture content, z distance below the soil surface, Ψ soil-water matric potential, and $K(\Theta)$ the unsaturated hydraulic conductivity. The equation may be solved by numerical methods using small increments of depth and time (Mein and Larson, 1971, 1973). The infiltration rate is given by $\partial \Theta / \partial t$ for the uppermost soil layer.

Philip (1957) solved equation (7.44) for a ponded surface and a deep homogeneous soil and proposed that the first two terms of the resulting series solution could be used as an infiltration equation, giving:

$$F = St^{1/2} + At \tag{7.45}$$

where F is the volume of infiltration at time t, and S (sorptivity) and A (transmissivity) are constants depending on soil properties and initial moisture content.

Green and Ampt (1911) derived an infiltration equation for ponded surfaces based on Darcy's Law. This equation has received considerable attention in recent years and has been shown to have a theoretically sound basis as well as measureable parameters. For the case of negligible ponding depth the equation may be written as (Mein and Larson, 1971):

$$f_p = K_{fs} \left(1 + \frac{D_i S_{av}}{F}\right)$$

(7.46)

or in integrated form

$$K_{fs}t = F - S_{av}D_i \log_e \left(1 + \frac{F}{D_i S_{av}}\right)$$

(7.46a)

where D_i is the initial soil moisture deficit $(\Theta_{fs} - \Theta_i)$ and S_{av} is the average suction at the wetting front (K_s was used by Mein and Larson instead of K_{fs}).

Morel-Seytoux and Khanji (1974) found that the form of equation (7.46) remained the same when simultaneous movement of both water and air is considered. They substituted H_c, the effective matric drive, for S_{av} and accounted for the resistance to air movement by the introduction of a viscous resistance correction factor, β, which was defined as a function of the soil and fluid properties. Wilson *et al.* (1979) have shown, however, that β is very sensitive to moisture content behind the wetting front and suggest that larger errors in calculation of infiltration are likely to result from variability of hydraulic conductivity than from neglecting the air viscous effects.

An infiltration model which would be applicable in all situations encountered in watershed modelling should utilize equations or algorithms which handle the four general cases of infiltration noted earlier, plus variable and intermittent application. In addition, applicability of the model is greatly enhanced if measureable parameters are used. Of the equations discussed in the preceding sections only the Richard's equation satisfies these criteria. However, Mein and Larson (1971; 1973) developed a two-stage model which utilizes a modified form of the Green and Ampt (1911) equation. This model fulfills the criteria noted above, although developed originally for constant application rate. It has the added advantage that the equations are relatively simple and, therefore, large amounts of computer time are not required as may be the case with the Richard's equation.

For brevity this model is termed the GAML model and will be discussed in considerably more detail.

7.2.3.1. The GAML Model

Mein and Larson (1971) utilized concepts of flow similar to those of Green and Ampt (1911) to develop a relationship for predicting infiltration volume prior to surface ponding for the case of an initial uniform moisture profile and constant rainfall rate. The relationship is:

$$F_p = \frac{S_{av}D_i}{R/K_{fs} - 1}$$

(7.47)

where F_p is the cumulative volume of infiltration at the instant of surface ponding and other parameters are as previously defined. For a constant rainfall rate the result of equation (7.47) may be used to calculate the time to surface ponding or

$$t_p = \frac{F_p}{R} \tag{7.48}$$

Mein and Larson (1971) then modified the Green and Ampt equation to make it applicable after surface ponding while accounting for the volume of water infiltrated prior to surface ponding. The modified equation applicable to this second stage of infiltration is (Mein and Larson, 1971):

$$K_{fs}(t - t_p + t_p') = F - D_i S_{av} \log_e \left(1 + \frac{F}{D_i S_{av}} \right) \tag{7.49}$$

where t_p' is the equivalent time to infiltrate volume F_p under ponded surface conditions.

Chu (1977) has shown how this method can be used for modelling infiltration under variable rain when rainfall can be assumed constant within any given discrete time interval. Slack (1977, 1978) has shown that the first stage of the model applies for continually varying application rates as might be the case under moving sprinkler irrigation systems. Although the Green and Ampt model was originally derived for a uniform soil, Bouwer (1969) has extended the approach to a soil profile in which the conductivity or moisture content varies with depth.

Results from application of equations (7.47) through (7.49) compare favourably with results from the Richard's equation (Mein and Larson, 1971) and with field data (Idike et al., 1977, Slack, 1977). Additional applications are presented later in this paper.

To apply equations (7.47)–(7.49) for a particular soil and event, it is necessary to evaluate the various terms in the equations. The term R is rainfall intensity and would be available either as data from a rainfall recorder or as output from a simulation model. The parameters, S_{av}, and K_{fs} are properties of the soil. D_i is simply the difference in water content (volume/volume) before and after wetting. The content after wetting will generally be less than saturated water content due to air entrapment. It is now generally believed that the hydraulic conductivity used should not be that at saturation (K_s) but rather the actual hydraulic conductivity in the wetted zone (K_{fs}) which is often less than K_s owing to air entrapped in the soil pores during the infiltration process (Bouwer, 1966, 1969; Morel-Seytoux and Khanji, 1974).

Mein and Larson (1971) have suggested that S_{av} can be obtained by taking the area under the capillary suction-relation conductivity curve (S-K_r) for the range of k_r between 0.01 and 1.0 where k is defined as:

$$k_r = \frac{K(\Theta)}{K_s} \tag{7.50}$$

The limits on k_r correspond to an extremely low moisture content and saturation and were selected to represent limits which might often be expected under field conditions. They do not account for reduced conductivity due to air entrapment. Moore (1979) has suggested that a more general expression for S_{av} would be:

$$S_{av} = \int\limits_{k_r(\Theta_i)}^{k_r(\Theta_{fs})} \frac{S(k_r)dk_r}{k_r(\Theta_{fs}) - k_r(\Theta_i)} \tag{7.51}$$

where $S(k_r)$ is matric suction (absolute value of matric potential) expressed as a function of relative conductivity, $k_r(\Theta_i)$ is the relative conductivity corresponding to initial moisture content and $k_r(\Theta_{fs})$ is the relative conductivity corresponding to the maximum volumetric moisture content attained in the wetted zone or 'field saturation'.

Campbell (1974) proposed a method for determining $k_r(\Theta)$ values from the moisture characteristic curve (S vs. Θ) for a given soil. The method is applicable if the S vs. Θ curve can be represented by a straight line on log-log paper. The general equation for such a line is:

$$S = A\Theta^{-b} \tag{7.52}$$

where $-b$ is the absolute slope of the line and A is a constant. One can then use the empirical equation developed by Campbell:

$$k_r(\Theta) = \left(\frac{\Theta}{\Theta_s}\right)^{2b+3} \tag{7.53}$$

where $k_r(\Theta)$ and b are defined by equations (7.50) and (7.52), respectively.

In equation (7.52), S or $S(\Theta)$ is the matric suction for sorption. These values can be approximated by dividing desorption values by 1.6 (Mein and Larson, 1971).

If experimental values of relative conductivity, $k_r(\Theta)$, vs. suction $S(\Theta)$ are available then equation (7.51) may be solved for S_{av} by graphical or numerical methods. If equation (7.53) is used to obtain $k_r(\Theta)$ then S may be expressed as a function of relative conductivity $S(k_r)$ using equations (7.52) and (7.53). The result may be integrated directly as shown in equation (7.51) which yields (Moore, 1979):

$$S_{av} = \frac{S_e(k_r(\Theta_{fs})^a - k_r(\Theta_i)^a)}{a(k_r(\Theta_{fs}) - k_r(\Theta_i))} \tag{7.54}$$

where:

$$a = \frac{b+3}{2b+3}$$

and S_e is the air entry suction.

The literature indicates that field saturation of agricultural soils varies between $0.8\,\Theta_s$ and $0.9\,\Theta_s$, (e.g. Wells and Skaggs, 1976; Poulovassilis and El-Ghamry, 1977; Topp and Miller, 1966). Based on these studies, it appears that coarse single grained structured soils exhibit a value of Θ_{fs} of about $0.8\,\Theta_s$ while for finer grained soils which may form aggregate structure Θ_{fs} will be in the neighbourhood of $0.9\,\Theta_s$. Field saturation probably varies with initial moisture content and rainfall intensity as well as soil texture. When possible Θ_{fs} should be evaluated experimentally. How-

Table 7.1. Moisture release data for Estherville sandy loam soil

Moisture content, θ (volume/volume)	Sorping suction, S (cm H_2O)
0.3296	63.7
0.2573	210.3
0.2198	637.3
0.1796	1912.0
0.1621	3824.0
0.1313	9559.9

ever, the range of values given above have given good results when used as estimates (Slack, 1977).

7.2.4. EXAMPLES OF MODEL APPLICATION

To illustrate evaluation of parameters for the GAML model and use of the model, three examples will be illustrated. All of these examples consider the soil surface to be protected so that surface sealing does not occur. This factor will be discussed in more detail in a following section as well as surface storage.

7.2.4.1. Example No. 1—Steady Rainfall

Rainfall at a constant rate of 5.0 cm hr^{-1} occurs on a deep Estherville sandy loam which has an initial moisture content of 0.31. The saturated moisture content of this soil is 0.436 and the saturated hydraulic conductivity K_s = 4.52 cm hr^{-1}. (Saturated hydraulic conductivity may be obtained for undisturbed cores using the method of Uhland and O'Neal, 1951). The water characteristic (moisture release) data for this soil is shown in Table 7.1. These data have been converted from desorption data by dividing the values of water potential (suction) by 1.6. A regression of the $\log_e$ of Θ on the $\log_e$ of the adjusted suction values yields coefficients for equation (7.52) of A = 0.121 cm H_2O and b = 5.613 with a coefficient of determination (R^2) of 0.996.

If field saturation is assumed to be Θ_{fs} = 0.9 Θ_s, then equation (7.53) can be utilized to calculate K_{fs} = 1.01 cm hr^{-1}. S_{av} may be calculated by equation (7.54) yielding S_{av} = 34.8 cm H_2O.

7.2.4.1.1. Stage I: Prediction of Time to Surface Ponding t_p

The initial moisture deficit D_i = Θ_{fs} − Θ_i = 0.39 − 0.31 = 0.08. Infiltration volume at the time of surface ponding, F_p is obtained from equation (7.47) as:

$$F_p = \frac{S_{av}D_i}{R/K_{fs} - 1} = \frac{34.8 \times 0.08}{5.0/1.01 - 1} = 0.70 \text{ cm}$$

at $t = t_p, f = R$ and $F_p = Rt_p$

thus:

$$t_p = \frac{F_p}{R} = \frac{0.70}{5.0} = 0.14 \text{ hr.}$$

7.2.4.1.2. Stage II: Prediction of Infiltration after Ponding

Equation (7.49) is used to determine $F(t)$ relationship after surface ponding. If the surface had been ponded at $t = 0$ the time, t'_p, required to infiltrate $F_p = 0.70$ cm is (from equation (7.46a)),

$$t'_p = \frac{0.70}{1.01} - \frac{(34.8)(0.08)}{1.01} \log_e \left(1 + \frac{0.70}{(0.8)(34.8)} \right)$$

$$t'_p = 0.075 \text{ hr.}$$

Now equation (7.49) may be solved for t for increasing values of F (beyond F_p):

$$t = t_p - t'_p + F/K_{fs} - \frac{D_i S_{av}}{K_{fs}} \log_e(1 + F/D_i S_{av})$$

with

$$t'_p = 0.075 \text{ hr}, \ t_p = 0.14 \text{ hr}, \ t_p - t'_p = 0.065 \text{ hr, and}$$

$$t = 0.065 + \frac{F}{1.01} - 2.76 \log_e \left(1 + \frac{F}{2.78} \right)$$

then for $F = 1.0$ cm

$$t = 0.065 + \frac{1}{1.01} - 2.76 \log_e \left(1 + \frac{1}{2.78} \right)$$

$$t = 0.21 \text{ hr}$$

Equation (7.46) is used to calculate the infiltration rate at this time as:

$$f = 1.01 \left(1 + \frac{0.8 \times 34.8}{1} \right) = 3.82 \text{ cm hr}^{-1}.$$

The above procedure may be repeated for successive values of infiltration volume F and rate f as a function of time thus easily obtained as shown in Table 7.2. This is a simple procedure for a small computer or a programmable calculator.

7.2.4.2. Example No. 2—Continually Varying Application Rate

Water is applied to the Estherville soil of the previous example by a moving sprinkler irrigation system. The application rate pattern is triangular in shape with a maxi-

Table 7.2. Calculated values of infiltration volume and rate into Estherville sandy loam for a steady rainfall rate of 5.0 cm hr^{-1}

	t(hr)	F(cm)	f(cm hr^{-1})
	0	0	5.00
Surface Ponding	0.14	0.70	5.00
	0.21	1.00	3.82
	0.36	1.50	2.88
	0.55	2.00	2.42
	0.77	2.50	2.13

mum rate of 5.08 cm hr^{-1} and a duration of 1.8 hr. The peak rate occurs at 0.9 hr; thus the pattern is symmetrical about the peak (Figure 7.2). The rate for the ascending portion of the pattern is αt, in which α is the slope and is found to be 5.64. For the descending portion, the application rate is found to be $(10.15 - 5.64\, t)$. If ponding occurs on the ascending leg then:

$$F_p = \int_0^{t_p} \alpha t \, dt = \frac{\alpha t_p^2}{2} = 2.82 \, t_p^2 \tag{7.55}$$

7.2.4.2.1. Stage I: Prediction of Time to Ponding t_p

Given the same soil parameters as in the previous example time to ponding can be determined by first equating equation (7.55) to equation (7.47). Rearranging terms one obtains

$$\alpha t_p^3 - K_{fs} t_p^2 = 2 \frac{S_{av} D_i \cdot K_{fs}}{\alpha} \tag{7.56}$$

Equation (7.56) may now be solved for t_p by iterative techniques. Substituting values previously calculated:

$$5.64 t_p^3 - 1.01 t_p^2 = 0.997$$

Solving for t_p yields $t_p = 0.628$ hr and from equation (7.55) $F_p = 1.11$ cm.

7.2.4.2.2. Stage II: Prediction of Infiltration After Ponding

The solution of this portion of the model proceeds as in the previous example except that at some point on the descending leg of the pattern the application rate will fall below the infiltration capacity. The time at which this occurs is denoted t_x and may be determined by taking small increments of infiltrated volume as R approaches f. In this case (by equation (7.46c)):

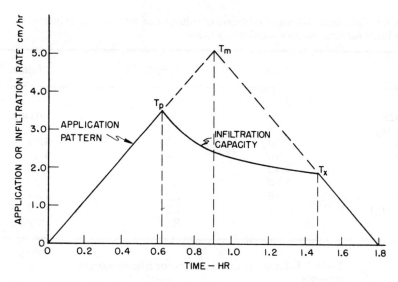

Figure 7.2 Application and infiltration rate curves for a moving sprinkler irrigation system (Estherville sandy loam soil)

$$t'_p = \frac{1.11}{1.01} - \frac{(34.8)(0.08)}{1.01} \log_e \left(1 + \frac{1.11}{0.08(34.8)}\right)$$

$$t'_p = 0.174 \text{ hr}$$

with $t'_p = 0.174$ hr and $t_p = 0.628$ hr

$$t = 0.454 + \frac{F}{1.01} - 2.76 \log_e \left(1 + \frac{F}{2.78}\right)$$

Values of infiltration volume and rate as a function of time are shown in Table 7.3 together with the corresponding application rate. It can be seen from this table that as t approaches t_x, f approaches R. For times greater than t_x, $f = R$. Results of calculation are shown in Figure 7.2 with the application rate pattern.

7.2.4.3. Example No. 3–Unsteady Intermittent Rainfall

Another useful application of the GAML model is to an unsteady rain with constant rainfall rates for short discrete time intervals. For this example, the same soil properties used in the previous example are used. Rainfall rate is given for fifteen minute intervals in Table 7.4.

7.2.4.3.1. Stage I: Prediction of Time to Surface Ponding

For this type of rainfall pattern, time to surface ponding is calculated on a trial and error basis starting with the initial rate. Thus, if ponding does not occur in the first

Table 7.3. Calculated values of infiltration volume and rate into an Estherville
sandy loam for time varying application rates

t(hr)	F(cm)	F(cm hr^{-1})	R(cm hr^{-1})
0	0	0	0
0–0.628		R	varying
0.228 (t_p)	1.11	3.54	3.54
0.748	1.50	2.88	4.22
0.938	2.00	2.42	4.87
1.159	2.50	2.13	3.62
1.404	3.00	1.94	2.23
1.456	3.10	1.92	1.94
1.461 (t_x)	3.11	1.91	1.92
$> t_x$		R	varying

Table 7.4. Example rainfall rates for fifteen minute
intervals

Time interval (hr)	R (cm hr^{-1})
0–0.25	1.0
0.25–0.50	0.8
0.50–0.75	5.0
0.75–1.00	5.0
1.00–1.25	0
1.25–1.50	1.5
1.50–1.75	3.0
1.75–2.00	1.0

time interval then the rate for the second interval is used in equations (7.47) and
(7.48) and this process repeated until time calculated falls within the interval used.
For the first and second intervals, it can be seen that ponding will not occur since
$R < K_{fs}$ for these two intervals. For the third interval with an R of 5.0 cm hr^{-1}:

$$F_p = \frac{34.8(0.08)}{5.0/1.01 - 1} = 0.70 \text{ cm}$$

the corresponding time is not 0.70/5.0 but $0.5 + \dfrac{(0.70 - 0.25 - 0.2)}{5.0} = 0.55$ hr.
Since this does fall within the time interval, ponding does occur and $t_p = 0.55$ hr.

7.2.4.3.2. Stage II: Prediction of Infiltration After Ponding

t'_p is calculated as in previous examples, thus:

Table 7.5. Calculated values of infiltration volume and
rate into an Estherville sandy loam for unsteady rainfall

t(hr)	F(cm)	f(cm hr^{-1})	R(cm hr^{-1})
0	0	0.00/1.00	0.00/1.00
0.25	0.25	0.80	0.80
0.50	0.45	5.00	5.00
0.55 (t_p)	0.70	5.00	5.00
0.62	1.00	3.81	5.00
0.96	2.00	2.42	5.00
1.00	2.10	2.32/0.00	0.00
1.25	2.10	1.50	1.50
1.50	2.47	2.15	3.00
1.75 (t_x)	2.96	1.96	1.00
2.00	3.21	1.00/0.00	1.00/0.00

$$t_p' = \frac{0.70}{1.01} - \frac{(34.8)(0.08)}{1.01} \log_e \left(1 + \frac{0.70}{(0.08)(34.8)}\right)$$

$$t_p' = 0.075$$

and:

$$t = 0.475 + \frac{F}{1.01} - 2.76 \log_e \left(1 + \frac{F}{2.78}\right)$$

Values of infiltration volume and rate as a function of time are calculated as before and results shown in Table 7.5. Note that as in the previous example if R becomes less than F the infiltration is controlled by rainfall rate. These results are also shown in Figure 7.3. In this example at the end of one hour, the rainfall rate drops to zero for the next 0.25 hr interval and resumes again at 1.25 hr at a rate of 1.5 cm hr^{-1}. Assume that surface storage is zero. Thus, none of the rainfall excess from the previous time infiltrates during this no rainfall period. It is also assumed that redistribution is negligible. It appears from Figure 7.3 that during the interval 1.25–1.50 hr $f_p > R$ thus $f = R$ or 1.5 cm hr^{-1}. This may be verified by adding the volume of rainfall during this period to the volume infiltrated at the end of the fourth interval or 2.10 + 0.37 = 2.47 cm. The corresponding rate for this volume calculated from equation (7.46) is 2.15 cm hr^{-1} which verifies the above assumption. During the period 1.50–1.75 hr $R > f_p$ resulting in a rainfall excess again.

From Table 7.5, it can be seen that rainfall became the controlling factor for the final time interval so the total amount of water infiltrated will be 2.96 + 0.25 = 3.21 cm. Figure 7.3 illustrates the results of the foregoing calculations superimposed on the rainfall intensity curve.

The three examples presented illustrate some of the possible applications and results of the GAML model when surface storage and surface sealing are negligible.

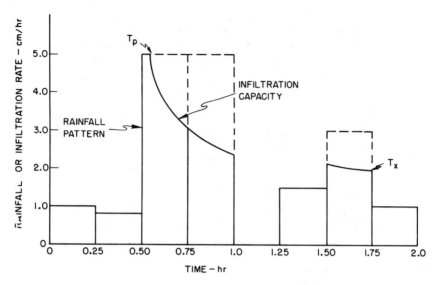

Figure 7.3 Infiltration rate curve for intermittent rainfall (Estherville sandy loam soil)

These conditions cannot be handled by direct modification of the model rather they may be treated by using the GAML model in conjunction with the models discussed in the following section.

7.2.4.4. Application of the GAML Model to Soil and Water Management

A goal of watershed management is to obtain maximum benefit from the water and soil resources of the watershed. In agricultural watersheds these benefits are generally realized in the form of agricultural production. This production is dependent upon proper management of the soil and water resources within the watershed. From a production standpoint, it is beneficial to maintain adequate moisture in the soil and to minimize erosion. High erosion rates are generally associated with high runoff, therefore, maintaining or increasing infiltration serves the dual goals of erosion reduction and maintenance of soil moisture levels. As has been illustrated, the GAML model may be used to predict infiltration when surface effects such as surface storage or sealing are not significant. However, generally an increase in the surface storage within a watershed will tend to increase infiltration thereby serving the goals mentioned earlier. This is often best accomplished by tillage practices which increase the micro-relief within a given field. At the same time, tillage exposes bare soil and until crop or mulch recovers, the susceptability to surface sealing is increased.

   Moore and Larson (1979a) developed a surface storage and routing model based on micro-relief as quantified by random roughness. A simplified version of the model was incorporated into a broader model for predicting runoff from small plots (Moore and Larson, 1979b). In this broader model, the GAML model as previously described was utilized as the infiltration component except in this case the infiltration was impeded by formation of a seal, the hydraulic conductivity of which was described as:

$$K_t = K_c + (K_i - K_c)e^{-bt} \tag{7.57}$$

where K_t is the saturated hydraulic conductivity of the seal, K_i the initial saturated hydraulic conductivity of the seal, K_c is the final or steady state hydraulic conductivity of a stable well established seal, t is time of exposure to rainfall and b is a constant. The equation was developed from data for 5 mm thick crusts (Mannering, 1967; Edwards, 1967). Moore and Larson indicated that it should be applicable to a wide variety of soils subject to surface sealing. Moore and Larson (1979b) found that, on bare soils, a seal developed very rapidly. According to various studies the conductivity of the seal is typically one-tenth that of the loose soil, or less. Thus, the seal conductivity becomes the controlling factor governing infiltration and quite early in the event. Darcy's Law was used to describe flow through the seal utilizing the conductivity of Equation (7.57). The GAML model was then used to describe infiltration into clods within the ploughed layer and the subsoil mass using flow through the seal as the application rate. The method of modelling flow through surface seals presented by Moore (1979) and Moore and Larson (1979b) shows considerable promise for modelling infiltration on soils where surface seals develop, but requires data for evaluating the parameters.

The surface storage and routing model developed by Moore and Larson (1979a) indicates that even for soils which have considerable surface storage some runoff will begin as soon as surface ponding occurs. Thus, it is not necessary to satisfy all or some of the surface storage before runoff begins. Of course, some storage is 'dead' storage and water retained in such storage will be available for infiltration. The model developed by Moore and Larson is a micro-model, being quite detailed in terms of input data and the model itself. On a field or watershed scale, it would have to be applied to plot size areas representing larger areas. Mitchell and Jones (1978) have developed a storage model with similar characteristics, but less detailed, which may be adequate for this purpose.

Ridging, furrowing, or listing, or all three can be used to create macrosurface storage and hold several cm of water which would otherwise run off and erode the soil. Knowing the surface configuration, one can write a submodel to represent this as a dynamic storage and combine it with an infiltration model, such as the GAML model. As illustrated in Figure 7.3, the model has the flexibility needed to handle this. With substantial surface storage, all or part of the excess water from the third and fourth time steps of the example would be stored and all or part of this would infiltrate during the next two time steps.

Application of the GAML model to the evaluation or design of sprinkler irrigation systems has been briefly illustrated in the second example presented earlier. A goal of such analysis is to determine an application rate pattern which will allow the maximum application rate without producing runoff. Such an analysis is especially useful for determining the suitability of high intensity (low energy) irrigation systems for heavy textured soils. For example, Slack (1977) has shown that moving sprinkler lines (e.g. center pivots) can be designed for peak application rates considerably higher than the steady state infiltration rate, f_c.

7.2.5. SUMMARY

A physically-based infiltration model which utilizes measureable parameters is described. The model is a two-stage model utilizing the Mein and Larson (1971) equation to predict time to surface ponding and a modified form of the Green and Ampt (1911) equation to calculate infiltration after ponding. The combination is termed the GAML model.

Methods of evaluating parameters for the equations of the model are presented and illustrated. The parameters can be evaluated from soil moisture retention (water characteristic) data and saturated hydraulic conductivity. The effect of air entrapment can be handled by methods described in the paper.

Several examples are presented illustrating how the GAML model may be applied to various rainfall patterns. Utilization of the model in combination with models for surface storage and surface sealing is also discussed.

The basic simplicity of the GAML model together with its sound theoretical basis and flexibility and the fact that the parameters can be evaluated from measurable soil characteristics should make it a very useful tool for modelling a wide variety of infiltration events.

7.2.6. ACKNOWLEDGEMENT

This paper has been approved for publication by the Minnesota Agricultural Experiment Station as *Scientific Journal Series Paper No. 10,948.*

REFERENCES

Baver, L. D., Gardner, W. H., and Gardner, W. R., 1972, *Soil Physics,* 4th ed. Wiley and Sons, New York, NY.

Bouwer, H., 1966, Rapid field measurement of air entry value and hydraulic conductivity of soil as significant parameters in flow system analysis, *Water Resources Research,* 2(4), 729–738.

Bouwer, H., 1969, Infiltration of water into nonuniform soil, *J. Irrigation and Drainage Division, ASCE,* 95(IR4), 451–462.

Campbell, G. S., 1974, A simple method for determining unsaturated conductivity from moisture retention data, *Soil Science,* 117(6), 313–331.

Chu, S. T., 1977, Modeling infiltration during a variable rain, *ASAE Paper No. 77-2063,* Presented at the 1977 Summer Meeting of ASAE, Raleigh, NC, June 1977.

Edwards, W. M., 1967, Infiltration of water into soils as influenced by surface conditions, *Unpub. Ph.D. Thesis,* Iowa State University, Ames, IA.

Green, W. H., and Ampt, G., 1911, Studies of soil physics, part I.–the flow of air and water through soils, *J. Agricultural Science,* **4,** 1-24.

Hillel, D., 1971, *Soil and Water–Physical Principles and Processes,* Academic Press, NY.

Holtan, H. N., 1961, A concept for infiltration estimates in watershed engineering, *USDA ARS Bull.,* **41-51,** 25.

Holtan, H. N., and Lopez, N. C., 1971, USDAHL-70 Model of watershed hydrology, *Tech. Bull. No. 1435,* USDA-ARS.

Horton, R. E., 1939, Analysis of runoff plot experiments with varying infiltration capacity, *Trans. Am. Geophys. Union,* **Part IV,** 693-694.

Horton, R. E., 1940, An approach towards a physical interpretation of infiltration capacity, *Soil Sci. Am. Proc.,* **5,** 399-417.

Idike, F., Larson, C. L., Slack, D. C., and Young, R. A., 1977, Experimental Evaluation of Two Infiltration Equations, *ASAE Paper No. 77-2558,* presented at 1977 Winter ASAE Meeting Chicago, IL. 21 pp.

Kostiakov, A. N., 1932, On the dynamics of the coefficient of water-percolation in soils and on the necessity for studying it from a dynamic point of view for purposes of amelioration, *Trans. 6th Comm. Intern. Soil Sci. Soc.,* Russian Part A, 17-21.

Mannering, J. V., 1967, The relationship of some physical and chemical properties of soils to surface sealing, *Unpub. Ph.D. Thesis,* Purdue University, Lafayette, IN.

Mein, R. G., and Larson, C. L., 1971, Modeling infiltration component of the rainfall-runoff process, Bulletin 43, *Water Resour. Research Center,* University of Minnesota, Minneapolis, MN, 72 pp.

Mein, R. G., and Larson, C. L., 1973, Modeling infiltration during a steady rain, *Water Resour. Res.,* **9(2),** 384-394.

Mitchell, J. K., and Jones, B. A., 1978, Micro-relief surface depression storage: Changes during rainfall events and their application to rainfall-runoff models, *Water Resources Bulletin,* **14(4),** 777-802.

Moore, I. D., 1979, Infiltration into tillage affected soils, *Unpub. Ph.D. Thesis,* University of Minnesota, St. Paul, MN.

Moore, I. D., and Larson, C. L., 1979a, Estimating micro-relief surface storage from point data, *Trans. ASAE* (in press).

Moore, I. D., and Larson, C. L., 1979b, An infiltration model for soils with disturbed surfaces, *ASAE Paper No. 79-2044,* Presented at the 1979 Joint Meeting of ASAE-CSAE, Winnipeg, Manitoba, June 1979.

Morel-Seytoux, H. J., 1973, Two phase flows in porous media, *Advances in Hydroscience,* **9,** 119-202.

Morel-Seytoux, H. J., and Khanji, J., 1974, Derivation of an equation of infiltration, Water Resources Research **10(4),** 795-800.

Philip, J. R., 1957, The theory of infiltration: 4. Sorptivity and algebraic infiltration equations, *Soil Science,* **84,** 257-264.

Philip, J. R., 1969, Theory of infiltration, *Advances in Hydroscience,* **5,** 215-296.

Poulovassilis, A., and El-Ghamry, W. M., 1977, Hysteretic steady state soil water profiles, *Water Resources Research,* **13,** (3) 5-49-557.

Richards, L. A., 1931, Capillary conduction through porous mediums, *Physics,* **1,** 318-313.

Slack, D. C., 1977, Modeling infiltration under moving sprinkler irrigation systems, *ASAE Paper No. 77-2557,* Presented at the 1977 Winter Meeting of ASAE, Chicago, Il., December 1977.

Slack, D. C., 1978, Predicting ponding under moving irrigation systems, *ASCE J. Irrig. and Drain. Div.,* **104(IR4),** 446-451.

Topp, G. C., and Miller, E. E., 1966, Hysteretic moisture characteristics and hydraulic conductivities of glass-bead media, *Soil Sci. Soc. Amer Proc.,* **30**, 156–162.

Uhland, R. E., and O'Neal, A. M., 1951, Soil permeability determination for use in soil and water conservation, *USDA,* Soil Conservation Service, SCS-TP-101.

Wells, L. G., and Skaggs, R. W., 1976, Upward water movement in field cores, *Trans. ASAE,* **19(2),** 275–283.

Wilson, B. N., Slack, D. C., and Larson, C. L., 1979, Development and evaluation of infiltration model parameters, *ASAE Paper No. 79-2045,* Presented at the 1979 Joint Meeting of ASAE-CSAE, Winnipeg, Manitoba, June 1979.

Tropical Agricultural Hydrology
Edited by R. Lal and E. W. Russell
© 1981, John Wiley & Sons Ltd.

7.3

Applicability of Different Models to Nigerian Watersheds

N. Egbuniwe

7.3.1. INTRODUCTION

Recognition of water resources management as one of the short-cuts to economic and social advance has helped hasten water resources exploitation in Nigeria. Studies of watershed hydrology are necessary for an optimal utilization of the water resources.

The availability of high speed digital computers in Nigeria has made possible the use of new techniques to estimate the hydrologic regime. Simulation models for synthesizing the hydrologic cycle are some of the new techniques.

7.3.2. HYDROLOGIC SIMULATION

Simulation is the indirect investigation of the response of behaviour of a system.

Simulation modelling is classified into two principal categories—parametric and stochastic. Parametric modelling is the development of relationships among physical parameters involved in the hydrologic events and the use of these relationships to generate, or synthesize, non-recorded hydrologic sequences. Such models are often called deterministic (Dawdy, 1969) because the system response at any time due to a given input is uniquely determined (Chow, 1969). Stochastic modelling is the use of statistical characteristics of hydrologic variables to solve hydrologic problems (Amorocho and Hart, 1964). This often involves the generation of non-historic sequences to which certain levels of probability can be attached. Parametric modelling usually requires input data with considerable detail in time and it models transient responses. Stochastic simulation is widely used for predictions for time periods which average out the transient responses. Statistical parameters used to describe the responses make it difficult to model transients.

Although the various methods of hydrologic system investigations have certain basic differences, they share two characteristics of prime importance: (1) their dependence on historical records for the values of the parameters and (2) the assumption of stationarity or time invariance of hydrologic systems (Amorocho and Hart, 1964). Both of these characteristics place definite limitations on the generality of the solutions.

A distinction is made in parametric modelling of the black box approach and conceptual approach. In the black box approach, the internal workings of the hydrologic system is not used in the analysis; the identification of the system is based only on the information given by the data in the input and output series (Zand and Harder, 1973). A black-box model of hydrologic system is not a practical means of elucidating the components of the hydrologic process. In the conceptual approach the hydrologic system is modelled by a series of equations which is equivalent to the actual internal physics of the systems (Crawford and Linsley, 1966). Conceptual modelling is limited by the knowledge of the physics of the hydrologic system.

Two principal current methods of parametric hydrology are the linear and non-linear systems analyses. The memory period is the duration of time during which input elements significantly influence the output. A linear system is one in which each input over each distinguishable interval in the memory period, produces an output that is independent of every other input. There is no interaction between parts of the input. Thus, linearity implies superposition. A non-linear system admits not only the possibility that given input over a given time interval may produce an output that is not linear with respect to itself but all the elements of the input may interact with each other; thus, each describable combination of input elements may produce its own output (Zand and Harder, 1973). Linear system analysis is the basis of the classical unit hydrograph method. The benefits of linear analysis lie not in the precision of the representation of the phenomena, but rather in the increased understanding of the basic processes and in the computational economy with which wide ranges of behaviour can be explored while retaining a high degree of physical realism (Eagleson, 1969). Many investigations have recognized the importance of non-linear system analysis as a tool of solving hydrologic problems (Bidwell, 1971; Amorocho and Brandstetter, 1971; Cawood *et al.*, 1971; Zand and Harder, 1973). The non-linear systems analysis is more complex than that of linear systems analysis by several orders of magnitude. The non-linear analysis is essentially independent of any qualitative or quantitative judgement based on incomplete knowledge of the physical hydrology of the watersheds under study. Since the method is a black-box type of analysis, it does not give any, not even qualitative information on the role played by any of the components of the hydrologic cycle.

Parametric models have been developed to help in the solution of very practical operational problems. They have demonstrated their utility in many different types of situations. Examples of their use are for the design of the capacity of a small reservoir (Linsley and Crawford, 1966), to determine the effect of channel improvements upon the flood frequency characteristics of a catchment (Nash, 1959), the effect of urbanization of flood peaks (James, 1965), and for extending stream-flow records for small basins on the basis of rainfall records (Lichty, Dawdy, and Berg-mann, 1968), as well as for problems concerning aquifer response.

Two typical approaches used in stochastic hydrology are described briefly. The first utilizes the Monte Carlo methods and the second utilizes the concepts of tran-sition probability developed under the general theory of Markov processes.

Stochastic models have been widely used for the simulation of inputs to com-plex systems, particularly for reservoir planning studies (Hufschmidt and Fiering, 1966; Beard, 1967).

The following models have been studied at Nsukka (Sections 7.3.3-7.3.5).

7.3.3. STANFORD WATERSHED MODEL

Stanford Watershed Model was developed at Stanford University (Linsley and Craw-ford, 1966).

The Stanford Watershed Model represents the processes in the hydrologic cycle from precipitation to outflow from a watershed as a series of mathematical expres-sions.

The model is essentially a soil water accounting process that has been expressed by: (Shanholtz *et al.*, 1972)

$$\frac{\delta SM}{\delta t} = \frac{\delta (P + ML\text{g})}{\delta t} - \frac{\delta (O + PC + ET + MLL)}{\delta t} \qquad (7.58)$$

where SM is the soil water status, P is all applied water, ML is minor gains or losses, O is discharge, PC is deep percolation and ET is evapotranspiration.

A solution of Equation (7.58) can be obtained by solving for individual compon-ents over a preselected time increment. Crawford and Linsley utilized two hypo-thetical dimensionless storage upper zone and lower zone—together with ground water storage to represent soil moisture and ground water conditions. The upper zone was used to simulate the dynamic hydrologic processes that occur on and very near the soil surface, while the lower zone was used to model conditions between the soil surface and the water table.

The Stanford watershed model adapted to the Nigeria's conditions has been tested on Nigerian watersheds (Egbuniwe and Todd, 1976). Malendo Watershed is one of the watersheds studied.

For the simulation of Malendo watershed characteristics the recorded flows are those for the gauge at Malendo Bridge. The hourly precipitation and Class A evapo-ration data for Yelwa were used. The watershed hydrology was simulated for the water years (1968-1973).

The annual values obtained by simulation are plotted against those observed. The agreement is good (coefficient of correlation = 0.97). The first year of simula-tion is not included in the analysis of results, since it is observed that for the water-shed tested in this study, the soil moisture conditions reach equilibrium in the second year. Simulated monthly streamflow versus recorded flow is plotted. The coefficient of correlation, is 0.91. The agreement is fairly good, although there are some months with large discrepancies.

The main merit of the Stanford Watershed Model is that when the parameters have been adjusted and the model reproduces closely the recorded streamflow, the data input or individual model parameters can be varied and the effect of these variations on the output can be examined. This characteristic can make the model useful; in extension of a short streamflow recorded by inputting available rainfall data to simulate flows when there was no streamflow record: in the evaluation of watershed changes—e.g. urbanization by changing the impervious area factor; in planning irrigation projects by making use of soil moisture and evapotranspiration outputs; and in estimating streamflow for ungauged watersheds.

7.3.4. RAINFALL MODEL

Most of the rainfall data in Nigeria are daily totals. Hourly rainfall inputs are necessary for some models.

The parameter estimate of storm duration, time between storms and storm depth can be used to construct a model that can generate synthetic hourly rainfalls which represent the distribution of an observed daily rain according to the storm interior. The storm depth is the total rainfall for a rainfall duration. The storm interior is the pattern of the rainfall intensities. The procedure as applied to rainfall data for Yelwa is as follows:

(a) The starting hour of any rainday is found from the piecewise linear distribution by the Monte Carlo technique (Figure 7.4).

(b) The Storm Duration: The values of storm duration are selected from the exponential distribution (Figure 7.5)

$$f(t_d) = 0.49 \exp(-0.49 t_d) \tag{7.59}$$

(c) The Storm Depth: Once the storm duration is selected, the storm depth is found as follows: the transformed percent residual is selected from the Weibull distribution (Figure 7.6).

$$f(t) = 0.0094 \frac{(t-6)^{-0.18}}{86.48} \exp\frac{(-(t-6)^{0.82})}{86.48} \tag{7.60}$$

by Monte Carlo techniques.

If R is the number selected from this Weibull distribution and the expected depth $(\bar{D})$ from the regression analysis is

$$D = -1.57 + 16.29 \text{ (DURATION)} \tag{7.61}$$

and

$$\text{Storm Depth } (D) = \left\{ 1 + \frac{P_1 - 100}{100} \right\} \bar{D} \tag{7.62}$$

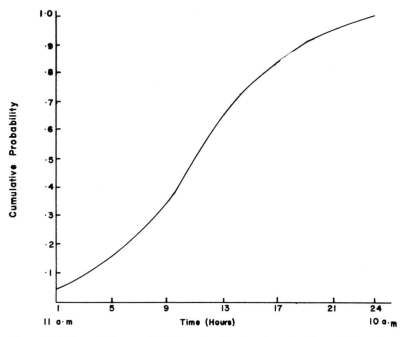

Figure 7.4 Cumulative probability distribution for the historical starting hour of rainfall (Yelwa, Nigeria)

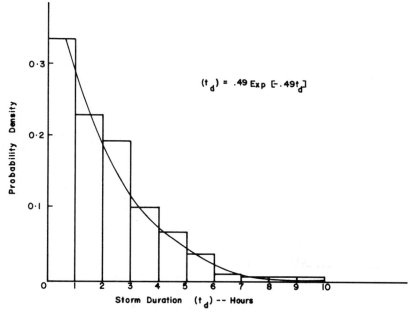

$(t_d) = .49 \, Exp \, [-.49t_d]$

Figure 7.5 Distribution of storm duration (Yelwa, Nigeria)

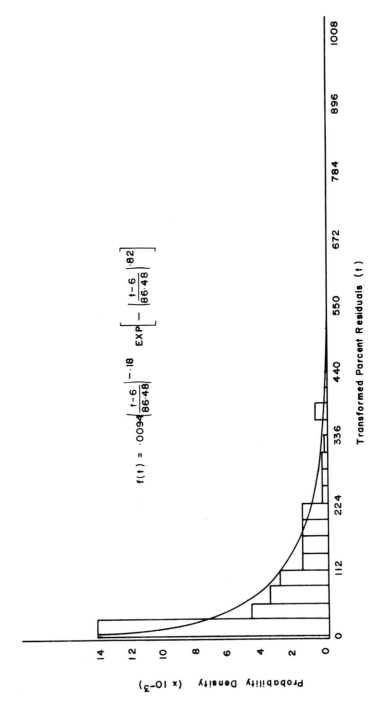

$$f(t) = \cdot 0094 \left| \frac{t-6}{86 \cdot 48} \right|^{-\cdot 18} \, EXP \left[- \left| \frac{t-6}{86 \cdot 48} \right|^{\cdot 82} \right]$$

Figure 7.6 Fit of Weibull distribution to transformed percentage residuals

where P_1 is the transformed percent residual $(P_1 = P + 100)$

$$P = \frac{D - \overline{D}}{\overline{D}} \times 100 \qquad (7.63)$$

then

$$D = \frac{(1 + R - 100)}{\overline{D}} \times -1.57 + 16.29 \text{ (DURATION)} \qquad (7.64)$$

(d) Storm Interior: The storm depth is distributed over the storm duration according to the triangular storm pattern. This is in accord with various studies (Eagleson, 1970).

(e) Time between storms: The storm duration is added to the starting hour of rainfall and if the sum is less than 24 hours and storm depth found is less than the daily rainfall, then the time between storms is found. This is done by Monte Carlo techniques by sampling from Weibull distribution (Figure 7.7).

$$f(t_b) = 0.266 \frac{t_b^{0.02}}{3.831} \exp\left(-\frac{t_b^{1.02}}{3.831}\right) \qquad (7.65)$$

The whole procedure is repeated until all the daily rainfall has been distributed. The rainfall model was applied to daily rainfall data for Yelwa (1968-1972). The data were used to simulate streamflow for Malendo Watershed. The synthetic data was input into the Nigerian version of the Stanford Watershed Model. Table 7.6 lists the percentage of times during which mean daily streamflows generated by the watershed model from the historic and synthetic rainfall were equalled or exceeded. These flow duration characteristics show good agreement.

7.3.5. SOIL CONSERVATION SERVICE MODEL

As most of the rainfall data in Nigeria are daily totals a model based on the Soil Conservation Service (1972) runoff curve number techniques that take inputs of daily rainfall to predict runoff is being studied. The study is along the line of that done by Williams and La Seur (1976).

Runoff is predicted for daily rainfall using Soil Conservation Service equation

$$Q = \frac{(P - 0.25)^2}{P + 0.8S} \qquad (7.66)$$

Where Q is the daily runoff; P is the daily rainfall; and S is a retention parameter. S must be estimated for each storm. The Soil Conservation Service uses antecedent rainfall index to estimate three antecedent moisture conditions (1-dry, 2-normal, 3-wet). The relation between rainfall and runoff for these three conditions is expressed as curve numbers CN. To predict runoff CN is related to S by

$$S = \frac{1000}{CN} - 10 \qquad (7.67)$$

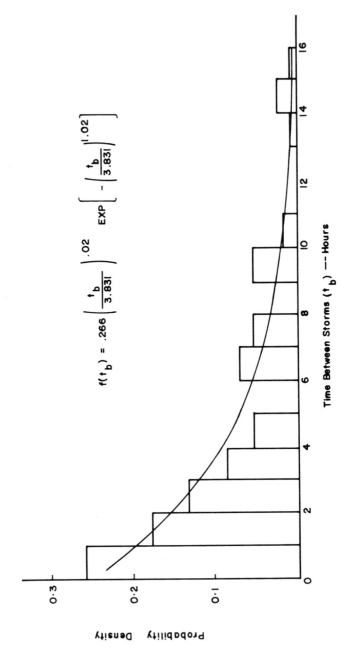

$$f(t_b) = .266 \left(\frac{t_b}{3.831} \right)^{.02} \text{EXP} \left[-\left(\frac{t_b}{3.831} \right)^{1.02} \right]$$

Time Between Storms (t_b) — Hours

Probability Density

Figure 7.7 Distribution of time between storms (Yelwa, Nigeria)

Table 7.6. Flow duration characteristics of synthesized flow for Malendo Watershed. Percentage of time indicated flows were equalled or exceeded

Flow interval (CFSD)	From historical rain	From generated rain
1	100	100
10	55.7	55.7
100	40.4	41.3
200	30.3	32.2
300	27.2	29.1
400	25.2	25.9
500	23.6	24.7
600	22.5	23.7
700	21.7	22.9
800	21.3	22.4
900	20.7	21.8
1000	20.1	21.4
2000	19.7	20.8
3000	16.1	16.8
4000	14.2	14.6
6000	12.6	12.9
8000	10.1	10.6
10000	8.2	9.1
12000	6.6	7.8
14000	4.3	5.6
16000	2.4	4.2
20000	1.2	2.0
30000	0.4	0.8

1 CFSD = 1 cusec day^{-1} = 2450 m^3 day^{-1}

The 2-condition curve number is determined by considering soil types and agricultural practices. The 1- and 3-condition curve numbers are a function of antecedent rainfall and 2-condition curve number.

The model is being applied to some Nigerian Watersheds.

7.3.6. PROBLEMS OF APPLYING WATERSHED MODELS

Models need inputs of historical records. Hydrometeorological data are being collected by the Nigerian Meteorological Department through synoptic, climatological, agromet, and rainfall stations which are maintained by the Department or by other Federal and State agencies. Few stations have recording gauges and a few record evaporation, temperature, humidity, radiation, and wind. The hydrological data like river stages and discharge measurements are collected by the Inland Waterways Department, State agencies and National Electric Power Authority. There is no network for assessing the ground water potential of the country. Sediment and water quality measurements are done only when required by consulting firms. The inadequacy of reporting stations (one station in 5000 km^2) leaves a great deal of uncertainty as to the real distribution of the recorded data.

A general hydrological data collection network should be designed for the country so that data collection is not done in a haphazard manner by different Federal and State agencies. There is a great need for standardization of methods of observation and recording of data and of instruments. There is also the need for making arrangements for systematic scrutiny, processing, and regular publication of such data.

7.3.7. CONCLUSION

Some models have been successfully applied to some Nigerian Watersheds within the constraints of the serviceability of the computers, the availability of trained personnel and the reliability of the hydrometeorological data. Representative and experimental watersheds should be set up in many parts of the country to facilitate the accumulation of data for testing different models.

REFERENCES

Amorocho, J., and Brandstetter, A., 1971, *Determination of Non-linear Functional Response Functions in Rainfall–Runoff Processes,* Water Resources Research, Vol. 7, No. 5.

Amorocho, J., and Hart, W. E., 1964, *A Critique of Current Methods in Hydrologic Systems Investigation,* Trans-American Geophysical Union, Vol. 45, No. 2.

Beard, L. R., 1967, Hydrologic simulation in water yield analysis, *Proc. Paper 5134, ASCE, IRI,* 33–42.

Bidwell, J., 1971, *Regression Analysis of Non-linear Systems,* Water Resources Research, Vol. 7, No. 5.

Cawood, P. B., Thunvik, R., and Nilsson, L.Y., 1971, *Hydrologic Modelling,* Department of Land Improvement and Drainage, Royal Institute of Technology, Stockholm, Sweden.

Chow, Ven Te, 1969, System approaches in hydrology and water resources, *Proc. First Internat. Seminar Hydrol. Profs.,* University of Illinois, Urbana.

Crawford, N. H., and Linsley, R. K., 1966, Digital simulation in hydrology, Stanford Watershed Model IV, *Technical Report No. 39,* Department of Civil Engineering, Stanford University., Stanford, California.

Dawdy, D. R., 1969, Mathematical modelling in hydrology, *Proc. First Internat. Seminar Hydrol. Profs.,* University of Illinois, Urbana.

Eagleson, P. S., 1969, Determination—linear hydrologic systems, *Proc. First Internat. Seminar Hydrol. Profs.,* University of Illinois, Urbana.

Eagleson, P. S., 1970, *Dynamic Hydrology,* New York, McGraw Hill.

Egbuniwe, N., and Todd, D. K., 1976, *Application of the Stanford Watershed Model to Nigerian Watersheds,* Water Resources Bulletin, Vol. 12, No. 3.

Hufschmidt, M. M., and Fiering, M. B., 1966, *Simulation Techniques for Design of Water-resource Systems,* Harvard University Press p. 212.

James, L. D., 1965, *Using a Digital Computer to Estimate the Effects of Urban Development on Flood Peaks,* Water Resources Research, Vol. 1. No. 2.

Lichty, R. W., Dawdy, D. R., and Bergmann, J. M., 1968, Rainfall runoff model for small basin flood hydrograph simulation, *IASH Pub. No. 81,* pp. 365–367.

Linsley, R. K., and Crawford, N. H., 1966, Computation of a synthetic streamflow record on a digital computer, *IASH Pub. No. 51,* pp. 526–538.

Nash, J. E., 1959, The effects of Flood-elimination Works on the Flood Frequency of the River Wandle, *Proc. Inst. Civ. Engrs.,* 13, 317–338.

Soil Conservation Service, 1972, National Engineering Handbook, *US Department of Agriculture,* Section 4, Chapters 4–10.

Shanholtz, V. O., Burford, J. B., and Lillard, J. H., 1972, *Evaluation of Deterministic Model for Predicting Water yields from small Agricultural Watersheds in Virginia,* UPI and State University, Blacksburg, Virginia.

Williams, J. R., and La Seur, V. W., 1976, Water Yield Model Using SCS Curve Numbers, *J. Hydraul. Div.,* ASCE, 102, No. HY9.

Zand, S. M., and Harder, J. A., 1973, Application of non-linear systems to the Lower Mekong River, *Water Resources Res.,* 9, No. 2.

PART 8

Research and Development Needs

Tropical Agricultural Hydrology
Edited by R. Lal and E. W. Russell
© 1981, John Wiley & Sons Ltd.

8.1

Future Trends in Watershed Management and Land Development Research

CHARLES PEREIRA

The soil and water management is an urgent top priority problem for research and development. Fortunately, that urgency is becoming recognized by many countries, though this problem is not being given enough priority in many regions of the tropics.

Forty years ago, in 1939, Jacks and White startled the world by their dramatically titled, but accurate book *The Rape of the Earth*. A vast amount of research and development, of successful experience and some hard economic necessities, have led to sound policies of watershed management under conditions of high technology where good management skills are available. Scientifically we have more understanding of the physical facts and processes governing the environments of agriculture and forestry. We have better equipment for recording rainfall, stream-flow, and suspended sediment (but we still lack methods of measuring bedload and soil-laden torrent-flows). Above all, we now have the computer and can make better use of large amounts of data, both by analysis and by modelling.

There are many examples of successful schemes concerning appropriate land use and watershed management in many tropical climates. But there is an alarming unanimity in the observations of experienced professional workers in many developing countries that there is still a considerable destruction of the basic soil and water resources from which the expanding populations must be fed.

8.1.1. IS THIS DUE TO LACK OF RESEARCH?

It may be true that there is some lack of data. There is indeed a lack of measurements in the fields of climate and hydrology and there is a scarcity of adequate equipment for recording rainfall and stream-flow in tropical regions. But can we honestly describe the massive soil erosion damage visibly occurring in watersheds in Africa, Asia, and South America to lack of data? Is it true that there are no technical solutions available? The results presented in this volume provide ample demonstrations that good technical solutions are indeed available. There are encouraging accounts of success over a wide range of environments from steep semi-arid revines to water-logged plains.

The single-factor variables are already measured well enough for practical use and a few watershed studies in the tropics have succeeded in integrating these variables over long enough periods to be sure that we can calculate combined energy and water budgets under tropical conditions and can model these results on computers. There are plenty of problems remaining, such as the interesting examples of unusual situations in cloud-forests, but we are now, I believe, able to give sound quantitative advice for land-use policies on watersheds which are farmed under high levels of technology or are managed as uninhabited forest reserves.

But the massive soil erosion damage to watersheds in Africa, Asia, and South America are not occurring under high technology. The damage is occurring in watersheds heavily populated by subsistence farmers and their livestock. Here we have very meagre evidence of proven successes to offer to the administrator. Certainly our evidence is not yet good enough to encourage governments to take unpopular decisions in land use regulation.

This, I believe, is the present focus of the world problems of watershed management and this has emerged clearly from the results presented in this book. The problems which we should now address in the next generation of watershed studies are vital to the lives of hundreds of millions and will become much more acute in the next two decades. I do not believe that we can solve these problems by more advanced studies of single factors, although these studies must be continued. We are not even using competently the research information already available in the tropics. In both Asia and Africa I have heard frequently statements that 'There is no data of the effect of X and Y' when the lack could be remedied by a week or two in a good library.

The clear requirement now is for field studies of the treatment of watersheds now eroding under subsistnce agriculture. We need to study quantitatively the combined effects of agricultural improvements and soil conservation safeguards by making hydrological measurements on the outflows of both water and transported soil. This requires strong multi-disciplinary teams, for experience has already shown that in over-grazed and over-cultivated watersheds, mechanical structures alone cannot win the battle against erosion. Only visible improvement in the success and profitability of the farming system throughout a complete watershed can be expected to reverse the present destructive trends.

Such improvements are most easily organized when introducing new cash crops, but with increasing food shortages in the developing world and a potential for increases of 200 to 300 per cent in yield of the main food crops these need only the organization of inputs and of markets to provide a basis for prosperity. Some small element of subsidy is needed for soil conservation structures; these have been found necessary in the USA.

Watershed improvement projects need active participation by water-engineers for the design and operation of the gauging structures, since only quantitative evidence of the pattern and quality of water outflow and sediment transport can give definitive resources. The agronomists and livestock managers and extention specialists may be departmentals, the marketing economists and the sociological field teams are more likely to come from the Universities; together with part-time contributions from specialists in other fields. Cooperation and participation by Forest Departments, Meteorological Departments, and River Basin Authorities will be needed. An active full-time executive team, containing soil scientists and hydrologists, together with some essential administrative support, can probably be based most effectively on the resources of a research institute. All of the technologies discussed herein will be needed; gulley stopping, tree planting, terracing, contours cultivation, fodder production, livestock husbandry, and the full packet of improved practices for crops are all interdependent for success.

In response to the question 'where do we go from here?' I believe that priority should be given to the following:

I   To organize multi-disciplinary practical studies of improvement both of the agriculture and of the hydrology on small eroding catchments inhabited by a few hundred small farmers.

II   To obtain quantitative hydrological evidence of the rate and the extent by which watershed stability may be improved under these conditions, and thus to provide evidence to justify government investment lower in the catchments, such as irrigation and power developments downstream.

III To combine the well-known concepts of the engineers' 'representative basins' with the watershed application of 'intensive conservation areas'.

Author Index

469

Subject Index

DATE DUE

ÉCHÉANCE